Student Solutions Manual

Prealgebra

FIFTH EDITION

Alan S. Tussy
Citrus College

Diane R. Koenig
Rock Valley College

Prepared by

Rhoda Oden

CENGAGE Learning

Australia • Brazil • Mexico • Singapore • United Kingdom • United States

CENGAGE Learning

ISBN-13: 978-1-285-73807-9
ISBN-10: 1-285-73807-1

Cengage Learning
200 First Stamford Place, 4th Floor
Stamford, CT 06902
USA

Cengage Learning is a leading provider of customized learning solutions with office locations around the globe, including Singapore, the United Kingdom, Australia, Mexico, Brazil, and Japan. Locate your local office at: **www.cengage.com/global**.

Cengage Learning products are represented in Canada by Nelson Education, Ltd.

To learn more about Cengage Learning Solutions, visit **www.cengage.com**.

Purchase any of our products at your local college store or at our preferred online store **www.cengagebrain.com**.

Printed in the United States of America
1 2 3 4 5 6 7 17 16 15 14 13

Table of Contents

ACKNOWLEDGEMENTS

Special Thanks to Ellen Sawyer, College of DuPage

CHAPTER 1 Whole Numbers

Section 1.1: An Introduction to the Whole Numbers

VOCABULARY

1. The numbers 0, 1, 2, 3, 4, 5, 6, 7, 8, and 9 are the digits.

3. When we write five thousand eighty-nine as 5,089, we are writing the number in standard form.

5. When 297 is written as 200 + 90 + 7, we are writing 297 in expanded form.

7. The symbols < and > are inequality symbols.

CONCEPTS

9.

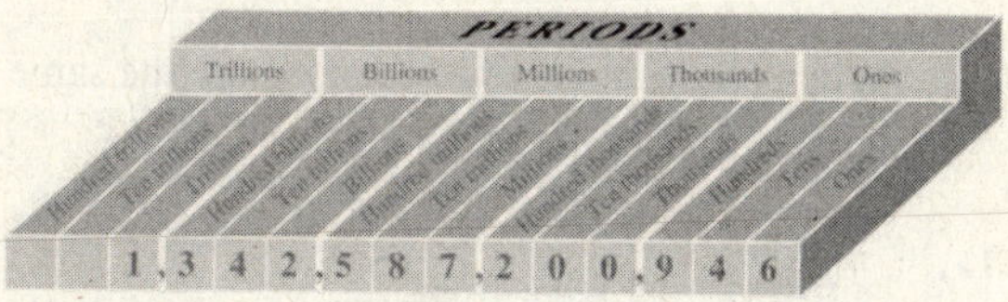

11. a. forty
b. ninety
c. sixty-eight
d. fifteen

13.

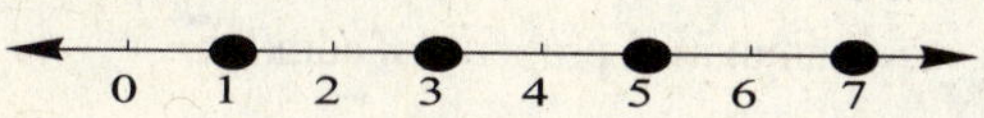

15.

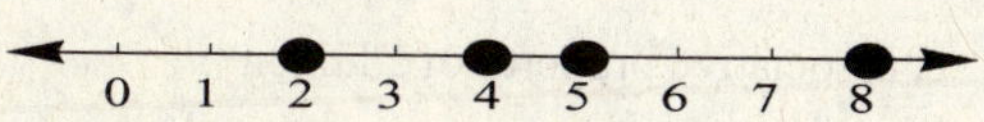

17.

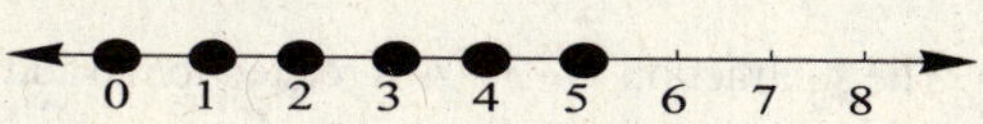

19.

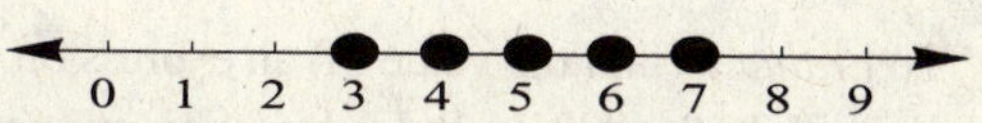

NOTATION

21. The symbols { }, called braces, are used when writing a set.

GUIDED PRACTICE

23. a. 3 tens
b. 7
c. 6 hundreds
d. 5

25. a. 6 millions
b. 7
c. 3 ten millions
d. 9

27. 93 = ninety-three

29. 732 = seven hundred thirty-two

31. 154,302 = one hundred fifty-four thousand, three hundred two

33. 14,432,500 = fourteen million, four hundred thirty-two thousand, five hundred

35. 970,031,500,104 = nine hundred seventy billion, thirty-one million, five hundred thousand, one hundred four

37. 82,000,415 = eighty-two million, four hundred fifteen

39. 3,737

41. 930

43. 7,021

45. 26,000,432

47. 245 = 200 + 40 + 5

49. 3,609 = 3,000 + 600 + 9

51. 72,533 = 70,000 + 2,000 + 500 + 30 +3

53 104,401 = 100,000 + 4,000 + 400 + 1

55. 8,403,613 = 8,000,000 + 400,000 + 3,000 + 600 + 10 + 3

57. 26,000,156 = 20,000,000 + 6,000,000 + 100 + 50 + 6

59. a. 11 > 8 b. 29 < 54

61. a. 12,321 > 12,209 b. 23,223 < 23,231

63. 98,150, since $4 < 5$

65. 512,970, since $7 \geq 5$

67. 8,400, since $5 \geq 5$

69. 32,400, since $3 < 5$

71. 66,000, (since $8 \geq 5$, 981 rounds to 1,000)

73. 2,581,000, (since $5 \geq 5$, 952 rounds to 1,000)

75. 53,000 ; 50,000

77. 77,000 ; 80,000

79. 816,000 ; 820,000

81. 297,000 ; 300,000

83. a. 79,590 b. 79,600
c. 80,000 d. 80,000

85. a. $419,160 b. $419,200
c. $419,000 d. $420,000

87. 40,025

89. 202,036

91. 27,598

93. 10,700,506

APPLICATIONS

95. Aisha is the closest to $4,745 without being over.

97. a. The 1970s, with 7 successful missions
b. The 1960s, with 9 unsuccessful missions
c. The 1960s, with 12 missions
d. The 1980s

99.

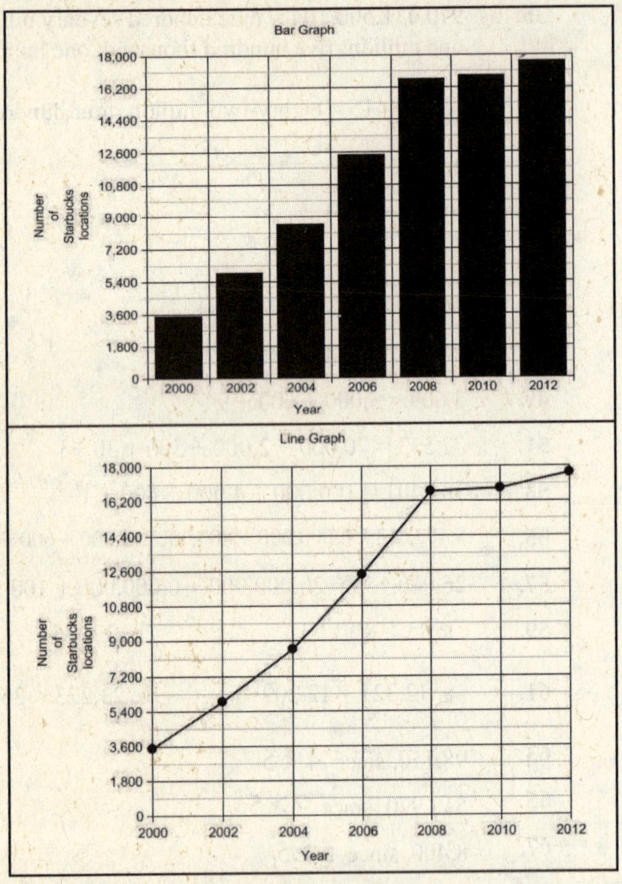

101. Fifteen thousand, six hundred one
Three thousand, four hundred thirty-three

103. 1,865,593 ; 482,880; 1,503; 269; 43,449

105. a. hundred thousands
b. 980,000,000; 900,000,000+80,000,000
c. 1,000,000,000; one billion

WRITING

107. To round 687 to the nearest ten, look to the right of the tens place. Since this digit is 7, increase the 8 in the tens place to 9 and make the ones place a zero. So, to the nearest ten, 687 is approximately 690.

109. Because **1**,000 (3 zeros) is a thousand 1s, so **1,000**,000 is a thousand thousands.

111. 2 , 10 , 0 , 1,000 , 80
12 , 3 , 100 , 2 , 0

113. Two hours is too long to wait!

Section 1.2: Adding and Subtracting Whole Numbers

VOCABULARY

1. $\underset{\textit{addend}}{10} + \underset{\textit{addend}}{15} = \underset{\textit{sum}}{25}$

3. The commutative property of addition states that the order in which whole numbers are added does not change their sum.

5. To see whether the result of an addition is reasonable, we can round the addends and estimate the sum.

7. The figure on the left is an example of a rectangle. The figure on the right is an example of a square.

9. When all the sides of a rectangle are the same length, we call the rectangle a square.

11. $\underset{\text{minuend}}{25} - \underset{\text{subtrahend}}{10} = \underset{\text{difference}}{15}$

13. Every subtraction has a related addition statement. For example $7 - 2 = 5$ because $5 + 2 = 7$.

CONCEPTS

15. a. commutative property of addition

b. associative property of addition

c. associative property of addition

d. commutative property of addition

17. The subtraction $7 - 3 = 4$ is related to the addition statement $4 + 3 = 7$.

19. To *evaluate* (find the value of) an expression that contains both addition and subtraction, we perform the operations as they occur from left to right.

NOTATION

21. The symbols ( ) are called parentheses. It is standard practice to perform the operations within them first.

23. $$\begin{aligned} 12+(15+2) &= 12+17 \\ &= 29 \end{aligned}$$

GUIDED PRACTICE

25. $$\begin{array}{r} 25 \\ +13 \\ \hline 38 \end{array}$$

27.
$$\begin{array}{r} 406 \\ +283 \\ \hline 689 \end{array}$$

29.
$$\begin{array}{r} {\scriptstyle 1} \\ 156 \\ +305 \\ \hline 461 \end{array}$$

31.
$$\begin{array}{r} {\scriptstyle 1\ \ 1\,1} \\ 4,301 \\ 789 \\ +3,847 \\ \hline 8,937 \end{array}$$

33. $(13+8)+12=13+(8+12)=13+20=33$

35. $94+(6+37)=(94+6)+37$
$=100+37=137$

37.
$$\begin{array}{r} {\scriptstyle 1} \\ 700 \\ 800 \\ 10,000 \\ 20,000 \\ +\ \ 6,000 \\ \hline 37,500 \end{array}$$

39.
$$\begin{array}{r} 600,000 \\ 20,000 \\ 300,000 \\ +\ \ 100,000 \\ \hline 1,020,000 \end{array}$$

41. $32+12+32+12=88\,ft$

43. $17+17+17+17=68in$

45. $94+94+94+94=376mi$

47. $87+6+87+6=186cm$

49.
$$\begin{array}{r} 7,989 \\ -347 \\ \hline 7,642 \end{array}$$

51.
$$\begin{array}{r} 2,967 \\ -405 \\ \hline 2,562 \end{array}$$

53.
$$\begin{array}{r} {\scriptstyle 6\ 13\ 16} \\ 8,746 \\ -\ \ 289 \\ \hline 8,457 \end{array}$$

55.
$$\begin{array}{r} {\scriptstyle 8\ 15\ 11} \\ 6,961 \\ -\ \ 478 \\ \hline 6,483 \end{array}$$

57.
$$\begin{array}{r} {\scriptstyle 3\ 14\ 9\ 16} \\ 54,506 \\ -\ \ 2,829 \\ \hline 51,677 \end{array}$$

59.
$$\begin{array}{r} {\scriptstyle 7\ 13\ 9\ 12} \\ 48,402 \\ -\ \ 3,958 \\ \hline 44,444 \end{array}$$

61. $123+175=298$ correct

63. 1,364+3,275=4,629 incorrect

65.
$$\begin{array}{r} 70,000 \\ -\ \ 4,000 \\ \hline 66,000 \end{array}$$

67.
$$\begin{array}{r} 80,000 \\ -30,000 \\ \hline 50,000 \end{array}$$

69. $35-12+6=23+6=29$

71. $574+47-13=621-13=608$

TRY IT YOURSELF

73. $\begin{array}{r} 8,539 \\ +\ 7,368 \\ \hline 15,907 \end{array}$

75. $\begin{array}{r} 3,430 \\ -\ 529 \\ \hline 2,901 \end{array}$

77. $\begin{array}{r} 51,246 \\ 578 \\ 37 \\ +\ 4,599 \\ \hline 56,460 \end{array}$

79. $633-598+30=35+30=65$

81. $(45+16)+4=45+20=65$

83. $20,007-78=19,929$

85. $852-695+40=157+40=197$

87. $\begin{array}{r} 632 \\ +347 \\ \hline 979 \end{array}$

89. $\begin{array}{r} 15,700 \\ -15,397 \\ \hline 303 \end{array}$

91. $16,427+13,573=30,000$

92. $18,788+13,567=32,355$

93. $\begin{array}{r} 50,009 \\ -\ 1,249 \\ \hline 48,760 \end{array}$

APPLICATIONS

95. 24 + 35 + 16 + 16 = 91 ft.

97. 241,288,000 + 57,253,000 = 298,541,000 visitors

99. 1,621 + 4,052 + 2,162 + 540 + 4,055 + 2,972 + 11,619 = $27,021

101. 64 + 34 + 64 + 34 = 196 inches of fringe

103. 48 – 2(11) = 48 – 22 = 26 inches each side

4(26) = 104 inches total

105. 2,009 – 291 = 1,718 lbs.

107. 71,649 – 70,154 = 1,495 miles

109. 5,577,717 – 2,010,879 = 3,566,838 magazines

111. $1,947-183=1,764°F$

113. a. $49,565

b. 50,887 – 49,565 = $1,322

WRITING

115. Benefit : faster ; Tradeoff : less accurate

117. Because taking 2 things from 3 things is not equivalent to taking 3 things from 2 things.

REVIEW

119. a. 3,000 + 100 + 20 + 5

b. 60,000 + 30 + 7

121. a. 5,370,650

b. 5,370,000

c. 5,400,000

Section 1.3: Multiplying Whole Numbers

VOCABULARY

1. $\underset{\textit{factor}}{5} \cdot \underset{\textit{factor}}{10} = \underset{\textit{product}}{50}$

3. The <u>commutative</u> property of multiplication states that the order in which whole numbers are multiplied does not change their product. The <u>associative</u> property of multiplication states that the way in which whole numbers are grouped does not change their product.

5. If a square measures 1 inch on a side, its area is 1 square inch.

CONCEPTS

7. a. $4\cdot 8$

b. $15+15+15+15+15+15+15$

9. a. 3 b. 5

11. a. area b. perimeter

c. area d. perimeter

NOTATION

13. $\times$, $\cdot$, ( )

15. $A=l\cdot w$ or $A=lw$

GUIDED PRACTICE

17.
$$\begin{array}{r} \overset{3}{1}5 \\ \times\ \ 7 \\ \hline 105 \end{array}$$

19.
$$\begin{array}{r} \overset{3}{3}4 \\ \times\ \ 8 \\ \hline 272 \end{array}$$

21. 100 has 2 zeros : attach 2 zeros : 3,700

23. 10 has 1 zero : attach 1 zero : 750

25. 10,000 has 4 zeros : attach 4 zeros : 1,070,000

27. 1,000 has 3 zeros : attach 3 zeros : 512,000

29. $68\cdot 4=272$

$68\cdot 40=2,720$

31. $56\cdot 2=112$

$56\cdot 200=11,200$

33. $13\cdot 3=39$

$130(3,000)=390,000$

35. $27\cdot 4=108$

$2,700(40,000)=108,000,000$

37.
$$\begin{array}{r} 128 \\ \times\ \ 73 \\ \hline 384 \\ 8,960 \\ \hline 9,344 \end{array}$$

39.
$$\begin{array}{r} 287 \\ \times\ \ 64 \\ \hline 1,148 \\ 17,220 \\ \hline 18,368 \end{array}$$

41. $602\cdot 679=600\cdot 679+2\cdot 679$
$=407,400+1,358=408,758$

43. $3,002(5,619)=3,000(5,619)+2(5,619)$
$=16,857,000+11,238=16,868,238$

45. $(18\cdot 20)\cdot 5=18\cdot(20\cdot 5)=18\cdot 100=1,800$

47. $250\cdot(4\cdot 135)=(250\cdot 4)\cdot 135$
$=1000\cdot 135=135,000$

49. $90\cdot 200=18,000$

51. $200\cdot 2,000=400,000$

53. $6\cdot 14=84in^2$

55. $12\cdot 12=144in^2$

TRY IT YOURSELF

57.
$$\begin{array}{r} 2\overset{2}{1}3 \\ \times\ \ 7 \\ \hline 1,491 \end{array}$$

59.
$$\begin{array}{r} 34,\overset{1}{4}74 \\ \times\ \ 2 \\ \hline 68,948 \end{array}$$

61.
$$\begin{array}{r} 99 \\ \times \quad 77 \\ \hline 693 \\ 6{,}930 \\ \hline 7{,}623 \end{array}$$

63. $44(55)(0)=0$

65. $53\cdot 3=159$ so $53\cdot 30=1{,}590$

67.
$$\begin{array}{r} 754 \\ \times \quad 59 \\ \hline 6786 \\ +37700 \\ \hline 44{,}486 \end{array}$$

69.
$$\begin{array}{r} 2978 \\ \times 3004 \\ \hline 11\,912 \\ 00\,000 \\ 000\,000 \\ +8\,934\,000 \\ \hline 8{,}945{,}912 \end{array}$$

71.
$$\begin{array}{r} 916 \\ \times \quad 409 \\ \hline 8\,244 \\ 00\,000 \\ +366\,400 \\ \hline 374{,}644 \end{array}$$

73. $25\cdot(4\cdot 99)=(25\cdot 4)\cdot 99=100\cdot 99=9{,}900$

75. $48\cdot 5=240$ so $4{,}800\cdot 500=2{,}400{,}000$

77.
$$\begin{array}{r} 2{,}779 \\ \times \quad 128 \\ \hline 22232 \\ 55\,580 \\ +277\,900 \\ \hline 355{,}712 \end{array}$$

79.
$$\begin{array}{r} 370 \\ \times 450 \\ \hline 000 \\ 18\,500 \\ +148\,000 \\ \hline 166{,}500 \end{array}$$

APPLICATIONS

81. $2\cdot 36=72$ cups of raisins

83. $12\cdot 17=204$ grams of fat

85. $60\cdot 65=3{,}900$ times per minute

87. $12\cdot 5{,}280=63{,}360\,in$ in a mile

89. $250\cdot 308=77{,}000$ words

91. $36\cdot 174{,}900=\$6{,}296{,}400$ per year

93. $8\cdot 9=72$ entries

95. $17\cdot 33=561$: There are 561 students and 1 instructor, so since $562<570$ they are O.K.

97. $3\cdot 6=18$ hours asleep

99. $28{,}000\cdot 6=168{,}000$ attacks

101. $2\cdot 3\cdot 14=6\cdot 14=84$ pills

103. $3\cdot 18=54\,ft^2$

105. Perimeter: $360+270+360+270=1{,}260mi$
Area: $360\cdot 270=97{,}200mi^2$

WRITING

107. 1 foot is a unit of length, while 1 square foot is a unit of area.

REVIEW

109. $10{,}357+9{,}809+476=20{,}642$

Section 1.4: Dividing Whole Numbers

VOCABULARY

1. $\underset{dividend}{12} \div \underset{divisor}{4} = \underset{quotient}{3}$

$$\text{divisor} \rightarrow 4\overline{)12} \leftarrow \text{dividend}, \quad 3 \leftarrow \text{quotient}$$

$$\frac{\text{dividend} \rightarrow 12}{\text{divisor} \rightarrow 4} = 3 \leftarrow \text{quotient}$$

3. The problem $6\overline{)246}$ is written in long division form.

5. One number is divisible by another number if, when we divide them, the remainder is 0.

CONCEPTS

7. a. 7 groups of 3

b. 5 groups of 4, 2 left over

9. a. $\frac{25}{25} = 1$ b. $\frac{6}{1} = 6$

c. $\frac{100}{0}$ is undefined d. $\frac{0}{12} = 0$

11. a. $5\overline{)1147}$, 2 b. $9\overline{)587}$, 6

c. $23\overline{)7501}$, 3 d. $16\overline{)892}$, 5

13.
$$\begin{array}{r} 37 \\ \times\ 9 \\ \hline 333 \end{array}$$

15. a. 0 or 5

b. 2 and 3

c. sum

d. 10

NOTATION

17. $\div$, $\overline{)\ }$, $-$

GUIDED PRACTICE

19. $9\overline{)45}$ = 5 because $5 \cdot 9 = 45$.

21. $44 \div 11 = 4$ because $4 \cdot 11 = 44$.

23. $7 \cdot 3 = 21$

25. $6 \cdot 12 = 72$

27.
$$\begin{array}{r} 16 \\ 6\overline{)96} \\ -\underline{6} \\ 36 \\ -\underline{36} \\ 0 \end{array}$$

Check: $6(16) = 96$

29.
$$\begin{array}{r} 29 \\ 3\overline{)87} \\ -\underline{6} \\ 27 \\ -\underline{27} \\ 0 \end{array}$$

Check: $3(29) = 87$

31.
$$\begin{array}{r} 325 \\ 7\overline{)2275} \\ -\underline{21} \\ 17 \\ -\underline{14} \\ 35 \\ -\underline{35} \\ 0 \end{array}$$

Check: $7(325) = 2275$

33.
$$9\overline{)1962}$$ quotient 218
-18
16
-9
72
-72
0
Check: $9(218)=1962$

35.
$$62\overline{)31248}$$ quotient 504
-310
24
-0
248
-248
0
Check: $62(504)=31,248$

37.
$$37\overline{)22274}$$ quotient 602
-222
07
-0
74
-74
0
Check: $37(602)=22,274$

39.
$$24\overline{)951}$$ quotient 39
-72
231
-216
15
39 R 15
Check : $39\cdot 24+15=951$

41.
$$46\overline{)999}$$ quotient 21
-92
79
-46
33
21 R 33
Check : $21\cdot 46+33=999$

43.
$$524\overline{)24714}$$ quotient 47
-2096
3754
-3668
86
Check : $47\cdot 524+86=24,714$

45.
$$178\overline{)3514}$$ quotient 19
-178
1734
-1602
132
Check : $19\cdot 178+132=3,514$

	Divisible by	2	3	4	5	6	9	10
47.	2,940	Y	Y	Y	Y	Y		Y
49.	43,785		Y		Y		Y	
51.	181,223							
53.	9,499,200	Y	Y	Y	Y	Y		Y

55. 10 has 1 zero; take away 1 zero: 70

57. Begin by cancelling a zero from each.

$$\begin{array}{r} 22 \\ 45\overline{)990} \\ \underline{-90} \\ 90 \\ \underline{-90} \\ 0 \end{array}$$

59. $360,000 \div 40 = 9,000$

61. $50,000 \div 1,000 = 50$

TRY IT YOURSELF

63.

$$\begin{array}{r} 4325 \\ 6\overline{)25950} \\ \underline{-24} \\ 19 \\ \underline{-18} \\ 15 \\ \underline{-12} \\ 30 \\ \underline{-30} \\ 0 \end{array}$$

65.

$$\begin{array}{r} 6 \\ 9\overline{)54} \\ \underline{-54} \\ 0 \end{array}$$

67.

$$\begin{array}{r} 8 \\ 31\overline{)273} \\ \underline{-248} \\ 25 \end{array}$$

8 R 25

69. Begin by cancelling 2 zeros from each.

$$\begin{array}{r} 160 \\ 4\overline{)640} \\ \underline{-4} \\ 24 \\ \underline{-24} \\ 0 \end{array}$$

71.

$$\begin{array}{r} 106 \\ 7\overline{)745} \\ \underline{-7} \\ 04 \\ \underline{-0} \\ 45 \\ \underline{-42} \\ 3 \end{array}$$

106 R 3

73.

$$\begin{array}{r} 509 \\ 29\overline{)14761} \\ \underline{-145} \\ 26 \\ \underline{-0} \\ 261 \\ \underline{-261} \\ 0 \end{array}$$

75.

$$\begin{array}{r} 3080 \\ 175\overline{)539000} \\ \underline{-525} \\ 140 \\ \underline{-0} \\ 1400 \\ \underline{-1400} \\ 0 \end{array}$$

77.

$$\begin{array}{r} 5 \\ 15\overline{)75} \\ \underline{-75} \\ 0 \end{array}$$

79.

$$\begin{array}{r} 23 \\ 212\overline{)5087} \\ \underline{-424} \\ 847 \\ \underline{-636} \\ 211 \end{array}$$

23 R 211

81.
$$\begin{array}{r} 30 \\ 42\overline{)1273} \\ -126 \\ \hline 13 \\ -0 \\ \hline 13 \end{array}$$
30 R 13

83. 1,000 has 3 zeros : remove 3 zeros : 89

85.
$$\begin{array}{r} 7 \\ 8\overline{)57} \\ -56 \\ \hline 1 \end{array}$$
7 R 1

APPLICATIONS

87. $2500 \div 4 = 625$ tickets

89. $405 \div 15 = 27$ trips

91. $50 \div 23 = 2R4$ Each student got 2 cartons, with 4 left over.

93. $640 \div 68 = 9R28$ It can be filled 9 times with 28 oz. left.

95. $58{,}000 \div 4 = 14{,}500$ There are 14,500 lbs. on each jack.

97. $25{,}200 \div 240 = \$105$ per book

99. $700 \div 140 = 5$ miles per gallon

101. $156 \div 12 = 13$ They should order 13 dozen donuts.

103. $216 \div 7 \approx 30.86$ - teams won't have the same number

$216 \div 8 = 27$ - not an even number of teams

$216 \div 9 = 24$ GOOD CHOICE

$216 \div 10 = 21.6$ - teams won't have the same number

There are 24 teams with 9 girls each.

105. Divide each by 12: Health Sciences: \$3,880; Business: \$4,295; Social Sciences: \$3,069

WRITING

107. Find out how many times you must subtract 6 from 24 to get 0.

109. $30 - 2 \cdot 8 = 30 - 16 = 14$

Since 14 is divisible by 7, 308 is also.

REVIEW

111. $2{,}903 + 378 = 3{,}281$

113. $2{,}903 \times 378 = 1{,}097{,}334$

Section 1.5: Prime Factors and Exponents

VOCABULARY

1. Numbers that are multiplied together are called factors.

3. A prime number is a whole number greater than 1 that has only 1 and itself as factors.

5. To prime factor a number means to write it as a product of only prime numbers.

7. In the exponential expression 6^4, the number 6 is the base and 4 is the exponent.

CONCEPTS

9. $1 \cdot 45 = 45 \quad 3 \cdot 15 = 45 \quad 5 \cdot 9 = 45$

The factors of 45, in order from least to greatest, are 1, 3, 5, 9, 15, 45.

11. yes

13. a. even, odd

b. 0, 2, 4, 6, 8, 10, 12, 14, 16, 18

c. 1, 3 , 5, 7, 9, 11, 13, 15, 17, 19

15. The blank should be a 6.

The prime factorization of 150 is $2 \cdot 3 \cdot 5 \cdot 5$.

17.
$$\begin{array}{l} 2\underline{|150} \\ \quad 3\underline{|75} \\ \quad\quad 5\underline{|25} \\ \quad\quad\quad 5 \end{array}$$

The prime factorization of 150 is $2 \cdot 3 \cdot 5 \cdot 5$.

NOTATION

19. a. base 7; exponent 6

b. base 15; exponent 1

GUIDED PRACTICE

21. 1, 2, 5, 10

23. 1, 2, 4, 5, 8, 10, 20, 40

25. 1, 2, 3, 6, 9, 18

27. 1, 2, 4, 11, 22, 44

29. 1, 7, 11, 77

31. 1, 2, 4, 5, 10, 20, 25, 50, 100

33. $2 \cdot 4$

35. $3 \cdot 9$

37. $7 \cdot 7$

39. $2 \cdot 10$ or $4 \cdot 5$

41. $30 = 2 \cdot 15 = 2 \cdot 3 \cdot 5$

43. $63 = 3 \cdot 21 = 3 \cdot 3 \cdot 7$

45. $54 = 6 \cdot 9 = 2 \cdot 3 \cdot 9$ or $3 \cdot 3 \cdot 6$

47. $60 = 2 \cdot 3 \cdot 10 = 2 \cdot 5 \cdot 6 = 2 \cdot 2 \cdot 15 = 3 \cdot 4 \cdot 5$

49. 11 : 1 and 11

51. 37 : 1 and 37

53. Yes

55. No $(3 \cdot 3 \cdot 11)$

57. No $(3 \cdot 17)$

59. Yes

61. $30 = 6 \cdot 5 = 2 \cdot 3 \cdot 5$

63. $39 = 3 \cdot 13$

65. $99 = 9 \cdot 11 = 3 \cdot 3 \cdot 11 = 3^2 \cdot 11$

67. $162 = 2 \cdot 81 = 2 \cdot 9 \cdot 9$
$= 2 \cdot 3 \cdot 3 \cdot 3 \cdot 3 = 2 \cdot 3^4$

69. $64 = 8 \cdot 8 = 2 \cdot 4 \cdot 2 \cdot 4$
$= 2 \cdot 2 \cdot 2 \cdot 2 \cdot 2 \cdot 2 = 2^6$

71. $147 = 3 \cdot 49 = 3 \cdot 7 \cdot 7 = 3 \cdot 7^2$

73. $220 = 22 \cdot 10 = 2 \cdot 11 \cdot 2 \cdot 5$
$= 2^2 \cdot 5 \cdot 11$

75. $102 = 2 \cdot 51 = 2 \cdot 3 \cdot 17$

77. $2 \cdot 2 \cdot 2 \cdot 2 \cdot 2 = 2^5$

79. $5 \cdot 5 \cdot 5 \cdot 5 = 5^4$

81. $4(4)(8)(8)(8) = 4^2(8^3)$

83. $7 \cdot 7 \cdot 7 \cdot 9 \cdot 9 \cdot 7 \cdot 7 \cdot 7 \cdot 7 = 7^7 \cdot 9^2$

85. a. $3^4 = 3 \cdot 3 \cdot 3 \cdot 3 = 81$

b. $4^3 = 4 \cdot 4 \cdot 4 = 64$

87. a. $2^5 = 2 \cdot 2 \cdot 2 \cdot 2 \cdot 2 = 32$

b. $5^2 = 5 \cdot 5 = 25$

89. a. $7^3 = 7 \cdot 7 \cdot 7 = 343$

b. $3^7 = 3 \cdot 3 \cdot 3 \cdot 3 \cdot 3 \cdot 3 \cdot 3 = 2{,}187$

91. a. $9^1 = 9$

b. $1^9 = 1$

93. $2 \cdot 3 \cdot 3 \cdot 5 = 90$

95. $7 \cdot 11^2 = 7 \cdot 121 = 847$

97. $3^2 \cdot 5^2 = 9 \cdot 25 = 225$

99. $2^3 \cdot 3^3 \cdot 13 = 8 \cdot 27 \cdot 13 = 2{,}808$

APPLICATIONS

101. Factors of 28: 1, 2, 4, 7, 14, 28

$1 + 2 + 4 + 7 + 14 = 28$

103. 2^2 square units, 3^2 square units, 4^2 square units

WRITING

105. Multiply the factors together to verify you get the original number.

107. $1^2 = 1^3 = 1^4 = 1$. Any power of 1 is 1.

REVIEW

109. $8 \cdot 15 + 5 = 120 + 5 = 125$ band members

Section 1.6: The Least Common Multiple and the Greatest Common Factor

VOCABULARY

1. The multiples of a number are the products of that number and 1, 2, 3, 4, 5, and so on.

3. One number is divisible by another number if, when dividing them, we get a remainder of 0.

CONCEPTS

5. a. 12

b. In general, the LCM of two whole numbers is the smallest whole number that is divisible by both numbers.

7. a. 20

b. 20

9. a. 2 appears twice with 36.

b. 3 appears twice with 90 and with 36.

c. 5 appears once with 90.

d. LCM = $2 \cdot 2 \cdot 3 \cdot 3 \cdot 5 = 180$

11. a. 2 appears twice with 12.

b. 3 appears three times with 54.

c. LCM = $2^2 \cdot 3^3 = 108$

13. a. 2, 3, and 5 are common to both.

b. GCF = $2 \cdot 3 \cdot 5 = 30$

NOTATION

15. a. The abbreviation for the greatest common factor is GCF.

b. The abbreviation for the least common multiple is LCM.

GUIDED PRACTICE

17. 4, 8, 12, 16, 20, 24, 28, 32

19. 11, 22, 33, 44, 55, 66, 77, 88

21. 8, 16, 24, 32, 40, 48, 56, 64

23. 20, 40, 60, 80, 100, 120, 140, 160

25. 5 is not divisible by 3.

10 is not divisible by 3.

15 is divisible by 3.

LCM(3,5) = 15

27. 12 is not divisible by 8.

24 is divisible by 8.

LCM(8,12)=24

29. 11 is not divisible by 5.

22 is not divisible by 5.

33 is not divisible by 5.

44 is not divisible by 5.

55 is divisible by 5.

LCM(5,11) = 55

31. 7 is not divisible by 4.

14 is not divisible by 4.

21 is not divisible by 4.

28 is divisible by 4.

LCM(4,7) = 28

33. 6 is not divisible by 3 and 4.

12 is divisible by 3 and 4.

LCM(3,4,6) = 12

35. 10 is not divisible by 2 and 3.

20 is not divisible by 2 and 3.

30 is divisible by 2 and 3.

LCM(2,3,10) = 30

37. $16 = 2^4$

$20 = 2^2 \cdot 5$

$LCM = 2^4 \cdot 5 = 16 \cdot 5 = 80$

39. $30 = 2 \cdot 3 \cdot 5$

$50 = 2 \cdot 5^2$

$LCM = 2 \cdot 3 \cdot 5^2 = 6 \cdot 25 = 150$

41. $35 = 5 \cdot 7$

$45 = 3^2 \cdot 5$

$LCM = 3^2 \cdot 5 \cdot 7 = 9 \cdot 35 = 315$

43. $100 = 2^2 \cdot 5^2$

$120 = 2^3 \cdot 3 \cdot 5$

$LCM = 2^3 \cdot 3 \cdot 5^2 = 600$

45. $6 = 2 \cdot 3$

$24 = 2^3 \cdot 3$

$36 = 2^2 \cdot 3^2$

$LCM = 2^3 \cdot 3^2 = 72$

47. $5 = 5$

$12 = 2^2 \cdot 3$

$15 = 3 \cdot 5$

$LCM = 2^2 \cdot 3 \cdot 5 = 60$

49. $4 = \underline{2} \cdot 2$

$6 = \underline{2} \cdot 3$

$GCF = 2$

51. $9 = 3 \cdot \underline{3}$

$12 = 2 \cdot 2 \cdot \underline{3}$

$GCF = 3$

53. $22 = 2 \cdot \underline{11}$

$33 = 3 \cdot \underline{11}$

$GCF = 11$

55. $15 = \underline{3} \cdot \underline{5}$

$30 = 2 \cdot \underline{3} \cdot \underline{5}$

$GCF = 3 \cdot 5 = 15$

57. $18 = \underline{2} \cdot \underline{3} \cdot 3$

$96 = \underline{2} \cdot 2 \cdot 2 \cdot 2 \cdot 2 \cdot \underline{3}$

$GCF = 2 \cdot 3 = 6$

59. $28 = \underline{2} \cdot 2 \cdot \underline{7}$

$42 = \underline{2} \cdot 3 \cdot \underline{7}$

$GCF = 2 \cdot 7 = 14$

61. $16 = 2 \cdot 2 \cdot 2 \cdot 2$

$51 = 3 \cdot 17$

$GCF = 1$

63. $81 = 3 \cdot 3 \cdot 3 \cdot 3$

$125 = 5 \cdot 5 \cdot 5$

$GCF = 1$

65. $12 = 2 \cdot 2 \cdot 3$

$68 = 2 \cdot 2 \cdot 17$

$92 = 2 \cdot 2 \cdot 23$

$GCF = 2 \cdot 2 = 4$

67. $72 = 2 \cdot 2 \cdot 2 \cdot 3 \cdot 3$

$108 = 2 \cdot 2 \cdot 3 \cdot 3 \cdot 3$

$144 = 2 \cdot 2 \cdot 2 \cdot 2 \cdot 3 \cdot 3$

$GCF = 2 \cdot 2 \cdot 3 \cdot 3 = 36$

TRY IT YOURSELF

69. $100 = 2 \cdot 2 \cdot 5 \cdot 5$

$120 = 2 \cdot 2 \cdot 2 \cdot 3 \cdot 5$

$LCM = 2 \cdot 2 \cdot 2 \cdot 3 \cdot 5 \cdot 5 = 600$

$GCF = 2 \cdot 2 \cdot 5 = 20$

71. $14 = 2 \cdot 7$

$140 = 2 \cdot 2 \cdot 5 \cdot 7$

$LCM = 2 \cdot 2 \cdot 5 \cdot 7 = 140$

$GCF = 2 \cdot 7 = 14$

73. $66 = 2 \cdot 3 \cdot 11$
$198 = 2 \cdot 3 \cdot 3 \cdot 11$
$242 = 2 \cdot 11 \cdot 11$
$LCM = 2 \cdot 3 \cdot 3 \cdot 11 \cdot 11 = 2{,}178$
$GCF = 2 \cdot 11 = 22$

75. $8 = 2 \cdot 2 \cdot 2$
$9 = 3 \cdot 3$
$49 = 7 \cdot 7$
$LCM = 2 \cdot 2 \cdot 2 \cdot 3 \cdot 3 \cdot 7 \cdot 7 = 3{,}538$
$GCF = 1$

77. $120 = 2 \cdot 2 \cdot 2 \cdot 3 \cdot 5$
$125 = 5 \cdot 5 \cdot 5$
$LCM = 2 \cdot 2 \cdot 2 \cdot 3 \cdot 5 \cdot 5 \cdot 5 = 3{,}000$
$GCF = 5$

79. $34 = 2 \cdot 17$
$68 = 2 \cdot 2 \cdot 17$
$102 = 2 \cdot 3 \cdot 17$
$LCM = 2 \cdot 2 \cdot 3 \cdot 17 = 204$
$GCF = 2 \cdot 17 = 34$

81. $46 = 2 \cdot 23$
$69 = 3 \cdot 23$
$LCM = 2 \cdot 3 \cdot 23 = 138$
$GCF = 23$

83. $50 = 2 \cdot 5 \cdot 5$
$81 = 3 \cdot 3 \cdot 3 \cdot 3$
$LCM = 2 \cdot 3 \cdot 3 \cdot 3 \cdot 3 \cdot 5 \cdot 5 = 4{,}050$
$GCF = 1$

APPLICATIONS

85.

1st	2nd	3rd
7,500mi	15,000mi	22,500mi
4th	5th	6th
30,000mi	37,500mi	45,000mi

87. LCM(45,60) = 180 minutes, or 3 hours

89. LCM of 10 and 12 = 60
5 packs of buns, 6 packs of hot dogs

91. LCM(6,8) = 24
4 sheets wide by 3 sheets tall = 12 sheets

93. a. GCF(28,21,63) = 7

The most that the art supplies cost a student is $7.

b. 4 students, 3 students, 9 students

WRITING

95. Find the prime factorization of both 8 and 28, then multiply each factor present the largest number of times it appears.

97. Since each factor has only one 3, the LCM should only have one 3.

REVIEW

99. $9{,}999 + 1{,}111 = 11{,}110$

101. $305 \cdot 50 = 15{,}250$

Section 1.7: Order of Operations

VOCABULARY

1. Numbers are combined with the operations of addition, subtraction, multiplication, and division to create expressions.

3. The grouping symbols ( ) are called parentheses, and the symbols [ ] are called brackets.

5. In the expression $9+6\left[8+6(4-1)\right]$, the parentheses are the inner most grouping symbols and the brackets are the outer most grouping symbols.

CONCEPTS

7. a. $5(2)^2 - 1$: square, multiply, subtract

b. $15 + 90 - (2 \cdot 2)^3$: multiply, cube, add, subtract

c. $7 \cdot 4^2$: square, multiply

d. $(7 \cdot 4)^2$: multiply, square

9. multiply; square

NOTATION

11. The fraction bar groups the numerator and denominator.

13. We read the expression $16-(4+9)$ as "16 minus the quantity of 4 plus 9."

15. $7\cdot4-5(2)^2=7\cdot4-5(4)$
$=28-20$
$=8$

17. $[4(2+7)]-4^2=[4(9)]-4^2$
$=36-4^2$
$=36-16$
$=20$

GUIDED PRACTICE

19. $3\cdot5^2-28=3\cdot25-28$
$=75-28=47$

21. $6\cdot3^2-41=6\cdot9-41$
$=54-41=13$

23. $52-6\cdot3+4=52-18+4$
$=34+4=38$

25. $32-9\cdot3+31=32-27+31$
$=5+31=36$

27. $192\div4-4(2)3=48-24=24$

29. $252\div3-6(2)6=84-72=12$

31. a. $26-2+9=24+9=33$

b. $26-(2+9)=26-11=15$

33. a. $51-16+8=35+8=43$

b. $51-(16+8)=51-24=27$

35. $(4+6)^2=10^2=100$

37. $(3+5)^3=8^3=512$

39. $8+4(29-5\cdot3)=8+4(29-15)$
$=8+4(14)=8+56=64$

41. $77+9(38-4\cdot6)=77+9(38-24)$
$=77+9(14)=77+126=203$

43. $46+3[5^2-4(9-5)]=46+3[25-4(4)]$
$=46+3[25-16]=46+3[9]=46+27=73$

45. $81+9[7^2-7(11-4)]=81+9[49-7(7)]$
$=81+9[49-49]=81+9[0]=81$

47. $\dfrac{2(50)-4}{2(4^2)}=\dfrac{100-4}{2\cdot16}=\dfrac{96}{32}=3$

49. $\dfrac{25(8)-8}{6(2^3)}=\dfrac{200-8}{6\cdot8}=\dfrac{192}{48}=4$

51. $\dfrac{6+9+4+3+8}{5}=\dfrac{30}{5}=6$

53. $\dfrac{3+5+9+1+7+5}{6}=\dfrac{30}{6}=5$

55. $\dfrac{19+15+17+13}{4}=\dfrac{64}{4}=16$

57. $\dfrac{5+8+7+0+3+1}{6}=\dfrac{24}{6}=4$

TRY IT YOURSELF

59. $(8-6)^2+(4-3)^2=2^2+1^2=4+1=5$

61. $2\cdot3^4=2\cdot81=162$

63. $7+4\cdot5=7+20=27$

65. $(7-4)^2+1=3^2+1=9+1=10$

67. $\dfrac{10+5}{52-47}=\dfrac{15}{5}=3$

69. $5\cdot10^3+2\cdot10^2+3\cdot10^1+9$
$=5\cdot1000+2\cdot100+3\cdot10+9$
$=5{,}000+200+30+9=5{,}239$

71. $20-10+5=10+5=15$

73. $25 \div 5 \cdot 5 = 5 \cdot 5 = 25$

75. $150 - 2(2 \cdot 6 - 4)^2 = 150 - 2(12 - 4)^2$
$= 150 - 2 \cdot 8^2 = 150 - 2 \cdot 64 = 150 - 128 = 22$

77. $190 - 2[10^2 - (5 + 2^2)] + 45$
$= 190 - 2[100 - (5 + 4)] + 45$
$= 190 - 2[100 - 9] + 45 = 190 - 2[91] + 45$
$= 190 - 182 + 45 = 53$

79. $2 + 3(0) = 2$

81. $\frac{(5-3)^2 + 2}{4^2 - (8+2)} = \frac{2^2 + 2}{4^2 - 10} = \frac{4+2}{16-10} = \frac{6}{6} = 1$

83. $4^2 + 3^2 = 16 + 9 = 25$

85. $3 + 2 \cdot 3^4 \cdot 5 = 3 + 2 \cdot 81 \cdot 5$
$= 3 + 810 = 813$

87. $60 - \left(6 + \frac{40}{2^3}\right) = 60 - \left(6 + \frac{40}{8}\right)$
$= 60 - (6 + 5) = 60 - 11 = 49$

89. $\frac{(3+5)^2 + 2}{2(8-5)} = \frac{8^2 + 2}{2(3)} = \frac{64+2}{6}$
$= \frac{66}{6} = 11$

91. $(18 - 12)^3 - 5^2 = 6^3 - 5^2 = 216 - 25 = 191$

93. $30(1)^2 - 4(2) + 12 = 30 - 8 + 12 = 34$

95. $16^2 - \frac{25}{5} + 6(3)4 = 256 - 5 + 72 = 323$

97. $\frac{3^2 - 2^2}{(3-3)^2} = \frac{9-4}{0^2} = \frac{5}{0}$: undefined

101. $4[50 - (3^3 - 5^2)] = 4[50 - (27 - 25)]$
$= 4[50 - 2] = 4[48] = 192$

103. $80 - 2[12 - (5 + 4)] = 80 - 2[12 - 9]$
$= 80 - 2[3] = 80 - 6 = 74$

APPLICATIONS

105. $3 \cdot 7 + 4 \cdot 4 + 2 \cdot 3 = 21 + 16 + 6 = \$43$

107. $3(8 + 7 + 8 + 8 + 7) = 3(38) = 114$

109. brick: $3 \cdot 3 + 1 + 1 + 3 + 3 \cdot 5 = 29$

aphid: $3[1 + 2(3) + 4 + 1 + 2] = 42$

111. $2^2 + 3^2 + 5^2 + 7^2 = 4 + 9 + 25 + 49 = 87$

113. $\frac{75 + 80 + 83 + 80 + 77 + 72 + 86}{7}$
$= \frac{553}{7} = 79°$

115. $\frac{39 + 40 + \cdots + 42}{12} = \frac{372}{12} = 31$ therms

117. $\frac{230 + 280 + \cdots 375}{8} = \frac{2400}{8}$
$= 300$ calories

119. a. $1 + 4 + 35 + 85 = 125$

b. $1 \cdot 2{,}500 + 4 \cdot 500 + 35 \cdot 150 + 85 \cdot 25$
$= 2{,}500 + 2{,}000 + 5{,}250 + 2{,}125$
$= 11{,}875$

c. $\frac{11{,}875}{125} = \$95$

WRITING

121. Order of operations is necessary so that different people don't come up with different answers to the same problem.

123. The multiplication of 2 and 3 takes precedence over the addition.

REVIEW

125. Two hundred fifty-four thousand, three hundred nine

Section 1.8: Solving Equations Using Addition and Subtraction

VOCABULARY

1. An equation is a statement indicating that two expressions are equal. All equations contain an = symbol.

3. To solve an equation means to find all values of the variable that make the equation true.

5. Equations with the same solutions are called equivalent equations.

CONCEPTS

7. a. $x+6$

b. neither

c. $5+6\neq 12$, so no.

d. $6+6=12$, so yes.

9. a. The addition property of equality: Adding the same number to both sides of an equation does not change its solution.

b. If $a=b$, then $a+c=b+c$.

11. a. To solve $x-8=24$, we add 8 to both sides of the equation.

b. To solve $x+4=11$, we subtract 4 from both sides of the equation.

NOTATION

13. $x-5=45$

$x-5+5=45+5$

$x=50$

Check: $x-5=45$

$50-5\stackrel{?}{=}45$

$45=45$ True

50 is the solution.

15. is possibly equal to

GUIDED PRACTICE

17. $x+2=3$

$1+2\stackrel{?}{=}3$

$3=3$

1 is a solution.

19. $a-7=0$

$7-7\stackrel{?}{=}0$

$0=0$

7 is a solution.

21. $50=y-8$

$50\stackrel{?}{=}40-8$

$50=32$

40 is not a solution.

23. $1=x+2$

$1\stackrel{?}{=}2+2$

$1=4$

2 is not a solution.

25. $x-7=3$

$x-7+7=3+7$

$x=10$

Check: $x-7=3$

$10-7\stackrel{?}{=}3$

$3=3$

The solution checks.

27. $a-20=50$

$a-20+20=50+20$

$a=70$

Check: $a-20=50$

$70-20\stackrel{?}{=}50$

$50=50$

The solution checks.

29. $1 = b - 2$

$1 + 2 = b - 2 + 2$

$3 = b$

Check: $1 = b - 2$

$1 \stackrel{?}{=} 3 - 2$

$1 = 1$

The solution checks.

31. $19 = n - 42$

$19 + 42 = n - 42 + 42$

$61 = n$

Check: $19 = n - 42$

$19 \stackrel{?}{=} 61 - 42$

$19 = 19$

The solution checks.

33. $x + 9 = 12$

$x + 9 - 9 = 12 - 9$

$x = 3$

Check: $x + 9 = 12$

$3 + 9 \stackrel{?}{=} 12$

$12 = 12$

The solution checks.

35. $y + 7 = 12$

$y + 7 - 7 = 12 - 7$

$y = 5$

Check: $y + 7 = 12$

$5 + 7 \stackrel{?}{=} 12$

$12 = 12$

The solution checks.

37. Analyze

* The scroll is 1,700 years old.

* The scroll is 425 years older than the jar.

* How old is the jar?

Assign Let x = the age of the jar.

Form Now we look for a key word or phrase in the problem.
Key phrase: older than

Translation: addition

Scroll's age is 425 years plus jar's age

1,700 = 425 + x

Solve

$1,700 = 425 + x$

$1,700 - 425 = 425 + x - 425$

$1,275 = x$

State The jar is 1,275 years old.

Check

$$\begin{array}{r} 1,275 \\ +\ \underline{425} \\ 1,700 \end{array}$$

The result checks.

TRY IT YOURSELF

39. $s + 55 = 100$

$s + 55 - 55 = 100 - 55$

$x = 45$

Check: $s + 55 = 100$

$45 + 55 \stackrel{?}{=} 100$

$100 = 100$

The solution checks.

41. $x-4=0$

$x-4+4=0+4$

$x=4$

Check: $x-4=0$

$4-4\stackrel{?}{=}0$

$0=0$

The solution checks.

43. $y-7=6$

$y-7+7=6+7$

$y=13$

Check: $y-7=6$

$13-7\stackrel{?}{=}6$

$6=6$

The solution checks.

45. $70=x-5$

$70+5=x-5+5$

$75=x$

Check: $70=x-5$

$70\stackrel{?}{=}75-5$

$75=75$

The solution checks.

47. $312=x-428$

$312+428=x-428+428$

$740=x$

Check: $312=x-428$

$312\stackrel{?}{=}740-428$

$312=312$

The solution checks.

49. $x-117=222$

$x-117+117=222+117$

$x=339$

Check: $x-117=222$

$339-117\stackrel{?}{=}222$

$222=222$

The solution checks.

51. $t+19=28$

$t+19-19=28-19$

$t=9$

Check: $t+19=28$

$9+19\stackrel{?}{=}28$

$28=28$

The solution checks.

53. $23+x=33$

$23+x-23=33-23$

$x=10$

Check $23+x=33$

$23+10\stackrel{?}{=}33$

$33=33$

The solution checks.

55. $5=4+c$

$5-4=4+c-4$

$1=c$

Check: $5=4+c$

$5\stackrel{?}{=}4+1$

$5=5$

The solution checks.

57. $99=r+43$

$99-43=r+43-43$

$56=r$

Check: $99=r+43$

$99\stackrel{?}{=}56+43$

$99=99$

The solution checks.

59. $512=428+x$

$512-428=428+x-428$

$84=x$

Check: $512=428+x$

$512\stackrel{?}{=}428+84$

$512=512$

The solution checks.

61. $x+117=222$

$x+117-117=222-117$

$x=105$

Check: $x+117=222$

$105+117\stackrel{?}{=}222$

$222=222$

The solution checks.

63. $3+x=7$

$3+x-3=7-3$

$x=4$

Check: $3+x=7$

$3+4\stackrel{?}{=}7$

$7=7$

The solution checks.

65. $y-5=7$

$y-5+5=7+5$

$y=12$

Check $y-5=7$

$12-5\stackrel{?}{=}7$

$7=7$

The solution checks.

67. $4+a=12$

$4+a-4=12-4$

$a=8$

Check: $4+a=12$

$4+8\stackrel{?}{=}12$

$12=12$

The solution checks.

69. $x-13=34$

$x-13+13=34+13$

$x=47$

Check: $x-13=34$

$47-13\stackrel{?}{=}34$

$34=34$

The solution checks.

APPLICATIONS

71. x = the amount of money she needs to borrow

$68,500+x=316,500$

$68,500+x-68,500=316,500-68,500$

$x=248,000$

She needs to borrow \$248,000.

73. x = the amount Dr. Dre earned in 2011

$x-72=38$

$x-72+72=38+72$

$x=110$

Dr. Dre earned \$110 million in 2011.

75. r = the amount of reduction

$81+r=110$

$81+r-81=110-81$

$r=29$

The earplugs reduce the noise by 29 decibels.

77. x = the amount of increase that will shut down the system

$60+x=85$

$60+x-60=85-60$

$x=25$

There must be an increase of 25 units to trigger a shutdown.

79. g = the amount she would have paid at the gas station

$g-29=190$

$g-29+29=190+29$

$g=219$

The gas station would have charged \$219.

81. j = Jimmy's age when he had the number one single

$j+55=67$

$j+55-55=67-55$

$j=12$

Jimmy was 12 years old when he had the hit single.

WRITING

83. It means that the number makes the equation true when substituted for the variable.

85. They are showing that doing the same thing on both sides of an equation preserves the equality.

87. Bob felt isolated from the rest of the company because of his body odor.

89. That is true because you must perform the opposite action of the operation given in the problem.

REVIEW

91. 325,780

93. $2\cdot3^2\cdot5=2\cdot9\cdot5=18\cdot5=90$

95. $8-2\left(2^2-1\right)+1^3=8-2\left(4-1\right)+1$

$=8-2\left(3\right)+1=8-6+1=2+1=3$

Section 1.9: Solving Equations Using Multiplication and Division

VOCABULARY

1. To <u>solve</u> an equation means to find all values of the variable that make the equation true.

3. To solve an equation, we <u>isolate</u> the variable on one side of the equal symbol.

CONCEPTS

5. a. The multiplication property of equality: Multiplying both sides of an equation by the <u>same</u> nonzero number does not change its solution.

b. If $a=b$, then $ca=cb$. (provided c is not 0)

7. a. If we multiply x by 6 and then divide that product by 6, the result is x.

b. If we divide x by 8 and then multiply that quotient by 8, the result is x.

9.
a. To solve $\frac{x}{5}=10$, we <u>multiply</u> both sides of the equation by 5.

b. To solve $5x=10$, we <u>divide</u> both sides of the equation by 5.

c. To solve $x-5=10$, we <u>add</u> 5 to both sides of the equation.

d. To solve $x+5=10$, we <u>subtract</u> 5 from both sides of the equation.

NOTATION

11.
$\frac{x}{5}=9$

$5\cdot\frac{x}{5}=5\cdot9$

$x=45$

Check: $\frac{x}{5}=9$

$\frac{45}{5}\overset{?}{=}9$

$9=9$ True

45 is the solution.

GUIDED PRACTICE

13.
$\frac{x}{7}=2$

$7\cdot\frac{x}{7}=7\cdot2$

$x=14$

Check: $\frac{x}{7}=2$

$\frac{14}{7}\overset{?}{=}2$

$2=2$

The solution checks.

15.
$\frac{y}{14}=3$

$14\cdot\frac{y}{14}=14\cdot3$

$y=42$

Check: $\frac{y}{14}=3$

$\frac{42}{14}\overset{?}{=}3$

$3=3$

The solution checks.

17. $16 = \frac{x}{24}$

$24 \cdot 16 = 24 \cdot \frac{x}{24}$

$384 = x$

Check: $16 = \frac{x}{24}$

$16 \stackrel{?}{=} \frac{384}{24}$

$16 = 16$

The solution checks.

19. $31 = \frac{t}{11}$

$11 \cdot 31 = 11 \cdot \frac{t}{11}$

$341 = t$

Check: $31 = \frac{x}{11}$

$31 \stackrel{?}{=} \frac{341}{11}$

$31 = 31$

The solution checks.

21. $3x = 3$

$\frac{3x}{3} = \frac{3}{3}$

$x = 1$

Check: $3x = 3$

$3 \cdot 1 \stackrel{?}{=} 3$

$3 = 3$

The solution checks.

23. $9z = 90$

$\frac{9z}{9} = \frac{90}{9}$

$z = 10$

Check: $9z = 90$

$9 \cdot 10 \stackrel{?}{=} 90$

$90 = 90$

The solution checks.

25. **Analyze**

* 3 people shared the cash award equally.

* Each person received $318,500.

* What was the amount of the Nobel Prize cash award?

Assign

Let x = the amount of the Nobel Prize award.

Form

We now look for a key phrase in the problem.

Key phrase: *shared the prize money equally*

Translation: division

The cash award divided by the number of winners is equal to each person's share.

x	$\div$	3	=	318,500

Solve

$\frac{x}{3} = 318{,}500$

$3 \cdot \frac{x}{3} = 3 \cdot 318{,}500$

$x = 955{,}500$

State The amount of the Nobel Prize cash award was $955,500.

Check: $\frac{\$955{,}500}{3} = \$318{,}500$

The result checks.

TRY IT YOURSELF

27. $100 = 100x$

$\frac{100}{100} = \frac{100x}{100}$

$1 = x$

Check: $100 = 100x$

$100 \stackrel{?}{=} 100 \cdot 1$

$100 = 100$

The solution checks.

29. $\dfrac{a}{15}=5$

$15\cdot\dfrac{a}{15}=15\cdot 5$

$a=75$

Check: $\dfrac{a}{15}=5$

$\dfrac{75}{15}\stackrel{?}{=}5$

$5=5$

The solution checks.

31. $16=8r$

$\dfrac{16}{8}=\dfrac{8r}{8}$

$2=r$

Check: $16=8r$

$16\stackrel{?}{=}8\cdot 2$

$16=16$

The solution checks.

33. $21s=210$

$\dfrac{21s}{21}=\dfrac{210}{21}$

$s=10$

Check: $21s=210$

$21\cdot 10\stackrel{?}{=}210$

$210=210$

The solution checks.

35. $\dfrac{c}{1{,}000}=3$

$1{,}000\cdot\dfrac{c}{1{,}000}=1{,}000\cdot 3$

$c=3{,}000$

Check: $\dfrac{c}{1{,}000}=3$

$\dfrac{3{,}000}{1{,}000}\stackrel{?}{=}3$

$3=3$

The solution checks.

37. $1=\dfrac{x}{50}$

$50\cdot 1=50\cdot\dfrac{x}{50}$

$50=x$

Check: $1=\dfrac{x}{50}$

$1\stackrel{?}{=}\dfrac{50}{50}$

$1=1$

The solution checks.

39. $7=\dfrac{t}{7}$

$7\cdot 7=7\cdot\dfrac{t}{7}$

$49=t$

Check: $7=\dfrac{t}{7}$

$7\stackrel{?}{=}\dfrac{49}{7}$

$7=7$

The solution checks.

41. $7x=21$

$\dfrac{7x}{7}=\dfrac{21}{7}$

$x=3$

Check: $7x=21$

$7\cdot 3\stackrel{?}{=}21$

$21=21$

The solution checks.

43. $172=43t$

$\dfrac{172}{43}=\dfrac{43t}{43}$

$4=t$

Check: $172=43t$

$172\stackrel{?}{=}43\cdot 4$

$172=172$

The solution checks.

45. $\frac{d}{20}=201$

$20\cdot\frac{d}{20}=20\cdot 201$

$d=4{,}020$

Check: $\frac{d}{20}=201$

$\frac{4{,}020}{20}\stackrel{?}{=}201$

$201=201$

The solution checks.

47. $417=\frac{t}{3}$

$3\cdot 417=3\cdot\frac{t}{3}$

$1{,}251=t$

Check:

$417=\frac{t}{3}$

$417=\frac{1{,}251}{3}$

$417=417$

The solution checks.

49. $170y=5{,}100$

$\frac{170y}{170}=\frac{5{,}100}{170}$

$y=30$

Check:

$170y=5{,}100$

$170\cdot 30\stackrel{?}{=}5{,}100$

$5{,}100=5{,}100$

The solution checks.

51. $\frac{t}{3}=47$

$3\cdot\frac{t}{3}=3\cdot 47$

$t=141$

Check: $\frac{t}{3}=47$

$\frac{141}{3}\stackrel{?}{=}47$

$47=47$

The solution checks.

53. $34y=204$

$\frac{34y}{34}=\frac{204}{34}$

$y=6$

Check: $34y=204$

$34\cdot 6\stackrel{?}{=}204$

$204=204$

The solution checks.

APPLICATIONS

55. x = the number of words per minute she could read prior to the course

$3x=399$

$\frac{3x}{3}=\frac{399}{3}$

$x=133$

Alicia could read 133 words per minute before the course.

57. x = the original estimate

$10x=540$

$\frac{10x}{10}=\frac{540}{10}$

$x=54$

The original estimate was $54 million.

59. r = the number of rows in the spreadsheet

$6r=294$

$\frac{6r}{6}=\frac{294}{6}$

$r=49$

There are 49 rows in the spreadsheet.

61. s = the number of the shelter's calls after being on the news

$4 \cdot 8 = s$

$32 = s$

The shelter received 32 calls per day after the news feature.

63. w = the weight of the object on the moon

$6w = 330$

$\frac{6w}{6} = \frac{330}{6}$

$w = 55$

The scale would register 55 pounds on the moon.

65. g = the life span of a guinea pig

$13g = 104$

$\frac{13g}{13} = \frac{104}{13}$

$g = 8$

The average life span of a guinea pig is 8 years.

WRITING

67. They are showing that multiplying both sides by the same thing preserves the balance, or equality.

69. It means to find all values of the variable which make the equation true.

REVIEW

71. $P = 2l + 2w = 2 \cdot 8 + 2 \cdot 16$

$= 16 + 32 = 48cm$

73. $120 = 4 \cdot 30 = 2 \cdot 2 \cdot 3 \cdot 10$

$= 2 \cdot 2 \cdot 3 \cdot 2 \cdot 5 = 2^3 \cdot 3 \cdot 5$

75. $3^2 \cdot 2^3 = 9 \cdot 8 = 72$

Chapter 1 Review

1. a. 6

b. 7

c. 1 billion

d. 8

2. a. ninety-seven thousand, two hundred eighty – three

b. five billion, four hundred forty-four million, sixty thousand, seventeen

3. a. 3,207

b. 23,253,412

4. 61,204

5. 500,000+70,000+300+2

6. 30,000,000+7,000,000+300,000

+9,000+100+50+4

7.

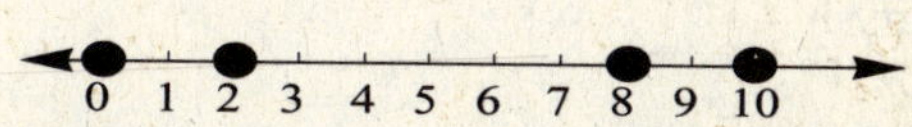

8.

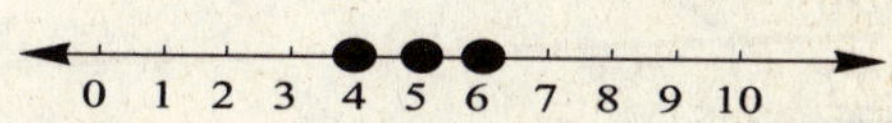

9. $9 > 7$

10. $301 < 310$

11. a. 2,507,300

b. 2,510,000

c. 2,507,350

d. 3,000,000

12. a. 970,000

b. 1,000,000

13.

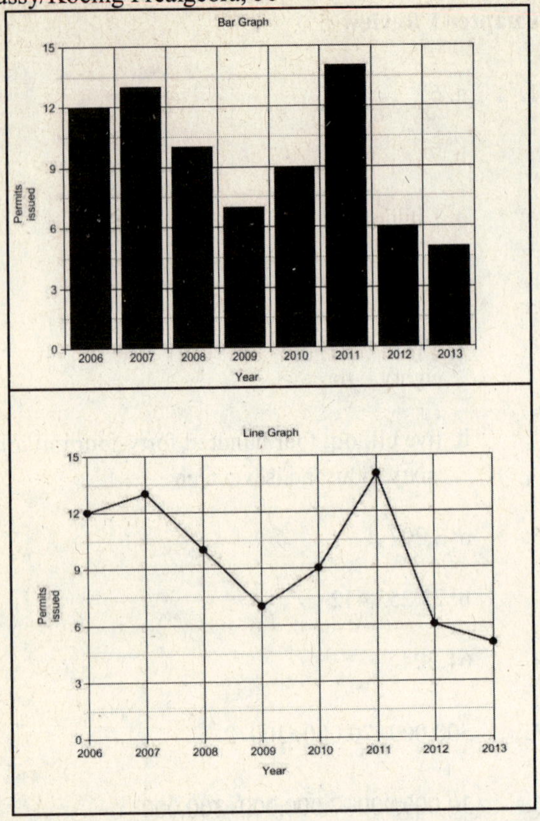

14. Nile, Amazon, Yangtze, Mississippi-Missouri, Ob-Irtysh

15.

$$\begin{array}{r} \scriptstyle 1 \\ 436 \\ +27 \\ \hline 463 \end{array}$$

16. $4+(36+19)=4+(55)=59$

17.

$$\begin{array}{r} \scriptstyle 1\;1\;1 \\ 5{,}345 \\ +\ 655 \\ \hline 6{,}000 \end{array}$$

18. $2+1+38+3+6=3+38+3+6$
$=41+3+6=44+6=50$

19.

$$\begin{array}{r} \scriptstyle 1\;\;2\;2 \\ 4{,}447 \\ 7{,}478 \\ +\ 676 \\ \hline 12{,}601 \end{array}$$

20.

$$\begin{array}{r} \scriptstyle 1\;1 \\ 32{,}812 \\ 65{,}034 \\ +54{,}323 \\ \hline 152{,}169 \end{array}$$

21. $600+800+10{,}000+40{,}000+8{,}000$
$=1{,}400+50{,}000+8{,}000$
$=51{,}400+8{,}000=59{,}400$

22. a. $24+61=61+24$

b. $9+(91+29)=(9+91)+29$

23.

$$\begin{array}{r} \scriptstyle 1\;1\;\;1\;1\;\;\;\;1 \\ 92{,}389{,}023 \\ 66{,}701{,}241 \\ +61{,}862{,}052 \\ \hline 220{,}952{,}316 \end{array}$$

24.

$$\begin{array}{r} \scriptstyle 2\;2 \\ 226 \\ 859 \\ 345 \\ +1{,}291 \\ \hline 2{,}721 \end{array}$$

It is not correct.

25.

$$\begin{array}{r} \scriptstyle 1\;\;\;1 \\ 10{,}955 \\ +3{,}706 \\ \hline 14{,}661 \end{array}$$

26.

$$\begin{array}{r} \scriptstyle 1\;1 \\ 327{,}891 \\ +451{,}775 \\ \hline 779{,}666 \end{array}$$

27.

$$\begin{array}{r} 1{,}695 \\ +785 \\ \hline 2{,}480\ sq.\ ft. \end{array}$$

28.
$$\begin{array}{r} \overset{1}{6}42 \\ 731 \\ 642 \\ +731 \\ \hline 2,746\,ft \end{array}$$

29.
$$\begin{array}{r} 148 \\ -87 \\ \hline 61 \end{array}$$

30.
$$\begin{array}{r} 10,4\overset{2}{\cancel{3}}\,\overset{15}{\cancel{5}} \\ -10,2\,1\,8 \\ \hline 2\,1\,7 \end{array}$$

31. $750-259+14=491+14=505$

32.
$$\begin{array}{r} 7,\overset{7}{\cancel{8}}\,\overset{9}{\cancel{10}}\,\overset{10}{\cancel{0}} \\ -5,7\,2\,5 \\ \hline 2,0\,7\,5 \end{array}$$

33.
$$\begin{array}{r} \overset{1\,1\,1}{1,168} \\ +6,949 \\ \hline 8,117 \end{array}$$

The subtraction is incorrect.

34. $20-8=12$ because $12+8=20.$

35. $200,000-40,000=160,000$

36.
$$\begin{array}{r} \overset{5}{\cancel{6}},\overset{12}{\cancel{3}}\,\overset{12}{\cancel{2}}3,4\overset{7}{\cancel{8}}\,\overset{12}{\cancel{2}} \\ -3,5\,5\,1,0\,2\,3 \\ \hline 2,7\,72,4\,5\,9mi^2 \end{array}$$

37. $12,975-3,800+4,270=9,175+4,270$
$=\$13,445$

38.
$$\begin{array}{r} \cancel{2}\,\overset{13}{\cancel{14}}\,\overset{12}{\cancel{2}} \\ -1\,8\,8 \\ \hline 5\,4 \text{ days} \end{array}$$

39.
$$\begin{array}{r} \overset{6}{4}7 \\ \times 9 \\ \hline 423 \end{array}$$

40. $5\cdot(7\cdot6)=5\cdot42=210$

41. $72\cdot10,000$: Since there are 4 zeros, move the decimal point 4 units to the right: 720,000

42.
$$\begin{array}{r} 157 \\ \times 59 \\ \hline 1,413 \\ +7,850 \\ \hline 9,263 \end{array}$$

43.
$$\begin{array}{r} 5,624 \\ \times 281 \\ \hline 5,624 \\ 449,920 \\ +1,124,800 \\ \hline 1,580,344 \end{array}$$

44.
$$\begin{array}{r} 502 \\ \times 459 \\ \hline 4,518 \\ 25,100 \\ +200,800 \\ \hline 230,418 \end{array}$$

45. $7,000\cdot400=2,800,000$

46. a. $7+7+7+7+7=5\cdot7$

b. $2\cdot t=2t$

c. $m\cdot n=mn$

47. a. $8\cdot0=0$ b. $7\cdot1=7$

48. a. Associative Property of Multiplication

b. Commutative Property of Multiplication

49. $A = l \cdot w$

$A = 8 \cdot 4$

$A = 32cm^2$

50. $A = l \cdot w$

$A = 78 \cdot 78$

$A = 6,084in^2$

51. a. $365 \cdot 7 = 2,555hr$

b. $365 \cdot 9 = 3,285hr$

52.

$$\begin{array}{r} 22 \\ \times 15 \\ \hline 110 \\ +220 \\ \hline 330 \end{array}$$

There were 330 people in the graduating class.

53. Sarah: $12 \cdot 9 = \$108$

Santiago: $14 \cdot 8 = \$112$

Santiago earned more money.

54. $12 \cdot 12 \cdot 100 = 144 \cdot 100$

$= 14,400\text{eggs}$

55. $\frac{72}{4} = \frac{\cancel{4} \cdot 18}{\cancel{4}} = 18$

56.

$$\begin{array}{r} 37 \\ 39\overline{)1443} \\ -117 \\ \hline 273 \\ -273 \\ \hline 0 \end{array}$$

57.

$$\begin{array}{r} 307 \\ 68\overline{)20876} \\ -204 \\ \hline 47 \\ -0 \\ \hline 476 \\ -476 \\ \hline 0 \end{array}$$

58.

$$\begin{array}{r} 19 \\ 21\overline{)405} \\ -21 \\ \hline 195 \\ -189 \\ \hline 6 \end{array}$$

$19R6$

59. $\frac{0}{10} = 0$

60. $\frac{165}{0}$ is undefined

61.

$$\begin{array}{r} 42 \\ 127\overline{)5347} \\ -508 \\ \hline 267 \\ -254 \\ \hline 13 \end{array}$$

$42R13$

62. Begin by dividing both by 100 to eliminate two zeros:

$$\begin{array}{r} 380 \\ 39\overline{)14820} \\ -117 \\ \hline 312 \\ -312 \\ \hline 0 \end{array}$$

63. $40 \cdot 4 = 160$

64. $45 \cdot 7 + 6 = 315 + 6 = 321$

It is not correct.

65. 364,545 is divisible by 3, 5, and 9.

66. $200{,}000 \div 50 = 4{,}000$

67.

$$\begin{array}{r} 16 \\ 45\overline{)745} \\ \underline{-45} \\ 295 \\ \underline{-270} \\ 25 \end{array}$$

Each child will get 16 candies, with 25 left over.

68. Begin by cancelling 3 zeros from each number.

$$\begin{array}{r} 34 \\ 25\overline{)850} \\ \underline{-75} \\ 100 \\ \underline{-100} \\ 0 \end{array}$$

They can purchase 34 cars.

69. $1, 2, 3, 6, 9, 18$

70. $1, 3, 5, 15, 25, 75$

71. $20 = 2 \cdot 10 = 4 \cdot 5$

72. $54 = 2 \cdot 3 \cdot 9 = 3 \cdot 3 \cdot 6$

73. a. prime

b. composite

c. neither

d. neither

e. composite

f. prime

74. a. odd

b. even

c. even

d. odd

75. $42 = 2 \cdot 21 = 2 \cdot 3 \cdot 7$

76. $75 = 3 \cdot 25 = 3 \cdot 5 \cdot 5 = 3 \cdot 5^2$

77. $220 = 10 \cdot 22 = 2 \cdot 5 \cdot 2 \cdot 11 = 2^2 \cdot 5 \cdot 11$

78. $140 = 10 \cdot 14 = 2 \cdot 5 \cdot 2 \cdot 7 = 2^2 \cdot 5 \cdot 7$

79. $6 \cdot 6 \cdot 6 \cdot 6 = 6^4$

80. $5(5)(5)(13)(13) = 5^3 \cdot 13^2$

81. $5^3 = 5 \cdot 5 \cdot 5 = 25 \cdot 5 = 125$

82. $11^2 = 11 \cdot 11 = 121$

83. $2^4 \cdot 7^2 = 2 \cdot 2 \cdot 2 \cdot 2 \cdot 7 \cdot 7$

$= 4 \cdot 4 \cdot 49 = 16 \cdot 49 = 784$

84. $2^2 \cdot 3^3 \cdot 5^2 = 4 \cdot 27 \cdot 25$

$= 108 \cdot 25 = 2{,}700$

85. $9, 18, 27, 36, 45, 54, 63, 72, 81, 90$

86. a. 24 and 48 are in both sets of multiples.

b. 1 and 2 are in both sets of factors.

87. $4 = 2 \cdot 2$

$6 = 2 \cdot 3$

$LCM(4, 6) = 2 \cdot 2 \cdot 3 = 12$

88. $3 = 3$

$4 = 2 \cdot 2$

$LCM(3, 4) = 2 \cdot 2 \cdot 3 = 12$

89. $9 = 3 \cdot 3$

$15 = 3 \cdot 5$

$LCM(9, 15) = 3 \cdot 3 \cdot 5 = 45$

90. $12 = 2 \cdot 2 \cdot 3$

$18 = 2 \cdot 3 \cdot 3$

$LCM(12, 18) = 2 \cdot 2 \cdot 3 \cdot 3 = 36$

91. $18 = 2 \cdot 3 \cdot 3$

$21 = 3 \cdot 7$

$LCM(18,21) = 2 \cdot 3 \cdot 3 \cdot 7 = 126$

92. $24 = 2 \cdot 2 \cdot 2 \cdot 3$

$45 = 3 \cdot 3 \cdot 5$

$LCM(24,45) = 2 \cdot 2 \cdot 2 \cdot 3 \cdot 3 \cdot 5 = 360$

93. $4 = 2 \cdot 2$

$14 = 2 \cdot 7$

$20 = 2 \cdot 2 \cdot 5$

$LCM(4,14,20) = 2 \cdot 2 \cdot 5 \cdot 7 = 140$

94. $21 = 3 \cdot 7$

$28 = 2 \cdot 2 \cdot 7$

$42 = 2 \cdot 3 \cdot 7$

$LCM(21,28,42) = 2 \cdot 2 \cdot 3 \cdot 7 = 84$

95. $8 = 2 \cdot 2 \cdot 2$

$12 = 2 \cdot 2 \cdot 3$

$GCF(8,12) = 2 \cdot 2 = 4$

96. $9 = 3 \cdot 3$

$12 = 2 \cdot 2 \cdot 3$

$GCF(9,12) = 3$

97. $30 = 2 \cdot 3 \cdot 5$

$40 = 2 \cdot 2 \cdot 2 \cdot 5$

$GCF(30,40) = 2 \cdot 5 = 10$

98. $30 = 2 \cdot 3 \cdot 5$

$45 = 3 \cdot 3 \cdot 5$

$GCF(30,45) = 3 \cdot 5 = 15$

99. $63 = 3 \cdot 3 \cdot 7$

$84 = 2 \cdot 2 \cdot 3 \cdot 7$

$GCF(63,84) = 3 \cdot 7 = 21$

100. $112 = 2 \cdot 2 \cdot 2 \cdot 2 \cdot 7$

$196 = 2 \cdot 2 \cdot 7 \cdot 7$

$GCF(112,196) = 2 \cdot 2 \cdot 7 = 28$

101. $48 = 2 \cdot 2 \cdot 2 \cdot 2 \cdot 3$

$72 = 2 \cdot 2 \cdot 2 \cdot 3 \cdot 3$

$120 = 2 \cdot 2 \cdot 2 \cdot 3 \cdot 5$

$GCF(48,72,120) = 2 \cdot 2 \cdot 2 \cdot 3 = 24$

102. $88 = 2 \cdot 2 \cdot 2 \cdot 11$

$132 = 2 \cdot 2 \cdot 3 \cdot 11$

$176 = 2 \cdot 2 \cdot 2 \cdot 2 \cdot 11$

$GCF(88,132,176) = 2 \cdot 2 \cdot 11 = 44$

103. $14 = 2 \cdot 7$

$21 = 3 \cdot 7$

$LCM(14,21) = 2 \cdot 3 \cdot 7 = 42$

They will meet on the same day 42 days later.

104. $32 = 2 \cdot 2 \cdot 2 \cdot 2 \cdot 2$

$24 = 2 \cdot 2 \cdot 2 \cdot 3$

$16 = 2 \cdot 2 \cdot 2 \cdot 2$

$GCF(32,24,16) = 2 \cdot 2 \cdot 2 = 8$

a. 8 arrangements

b. 4 red, 3 white, 2 blue (the 'remainder' when the GCF is taken out)

105. $3^2 + 12 \cdot 3 = 9 + 36 = 45$

106. $35 - 5 \cdot 3 + 3 = 35 - 15 + 3$

$= 20 + 3 = 23$

107. $(6 \div 2 \cdot 3)^2 \cdot 3 = (3 \cdot 3)^2 \cdot 3 = 9^2 \cdot 3$

$= 81 \cdot 3 = 243$

108. $(35 - 5 \cdot 3) \div 5 = (35 - 15) \div 5$

$= 20 \div 5 = 4$

109. $2^3 \cdot 5 - 4 \div 2 \cdot 4 = 8 \cdot 5 - 2 \cdot 4$

$= 40 - 8 = 32$

110. $8 \cdot (5 - 4 \div 2)^2 = 8 \cdot (5 - 2)^2$

$= 8 \cdot 3^2 = 8 \cdot 9 = 72$

111. $2 + 3\left(\frac{100}{10} - 2^2 \cdot 2\right) = 2 + 3(10 - 4 \cdot 2)$

$= 2 + 3(10 - 8) = 2 + 3(2) = 2 + 6 = 8$

112. $4(4^2-5\cdot3+2)-4=4(16-15+2)-4$
$=4(1+2)-4=4(3)-4=12-4=8$

113. $\dfrac{4(6)-6}{2(3^2)}=\dfrac{24-6}{2(9)}=\dfrac{18}{18}=1$

114. $\dfrac{6\cdot2+3\cdot7}{5^2-2(7)}=\dfrac{12+21}{25-14}=\dfrac{33}{11}=3$

115. $7+3[3^3-10(4-2)]=7+3[27-10(2)]$
$=7+3[27-20]=7+3[7]=7+21=28$

116. $5+2\left[\left(2^4-3\cdot\dfrac{8}{2}\right)-2\right]=5+2[(16-3\cdot4)-2]$
$=5+2[(16-12)-2]=5+2[4-2]$
$=5+2[2]=5+4=9$

117. $\dfrac{80+74+66+88}{4}=\dfrac{308}{4}=77$

118. $\dfrac{73+77+81+0+69}{5}=\dfrac{300}{5}=60$

119. $x+2=13$
$5+2\stackrel{?}{=}13$
$7=13$
5 is not a solution.

120. $x-3=1$
$4-3\stackrel{?}{=}1$
$1=1$
4 is a solution.

121. $x-7=2$
$x-7+7=2+7$
$x=9$
Check: $x-7=2$
$9-7\stackrel{?}{=}2$
$2=2$
The solution checks.

122. $x-11=20$
$x-11+11=20+11$
$x=31$
Check: $x-11=20$
$31-11\stackrel{?}{=}20$
$20=20$
The solution checks.

123. $225=y-115$
$225+115=y-115+115$
$340=y$
Check: $225=y-115$
$225\stackrel{?}{=}340-115$
$225=225$
The solution checks.

124. $101=p-32$
$101+32=p-32+32$
$133=p$
Check: $101=p-32$
$101\stackrel{?}{=}133-32$
$101=101$
The solution checks.

125. $x+9=18$
$x+9-9=18-9$
$x=9$
Check: $x+9=18$
$9+9\stackrel{?}{=}18$
$18=18$
The solution checks.

126. $b+12=26$
$b+12-12=26-12$
$b=14$
Check: $b+12=26$
$14+12\stackrel{?}{=}26$
$26=26$
The solution checks.

127. $175 = p + 55$

$175 - 55 = p + 55 - 55$

$120 = p$

Check: $175 = p + 55$

$175 \stackrel{?}{=} 120 + 55$

$175 = 175$

The solution checks.

128. $212 = m + 207$

$212 - 207 = m + 207 - 207$

$5 = m$

Check: $212 = m + 207$

$212 \stackrel{?}{=} 5 + 207$

$212 = 212$

The solution checks.

129. x = how much they needed to borrow

$x + 25{,}500 = 122{,}750$

$x + 25{,}500 - 25{,}500 = 122{,}750 - 25{,}500$

$x = 97{,}250$

They had to borrow $97,250.

130. x = the original number of patients

$x - 13 = 172$

$x - 13 + 13 = 172 + 13$

$x = 185$

He had 185 patients originally.

131. $3x = 12$

$\frac{3x}{3} = \frac{12}{3}$

$x = 4$

Check: $3x = 12$

$3 \cdot 4 \stackrel{?}{=} 12$

$12 = 12$

The solution checks.

132. $15y = 45$

$\frac{15y}{15} = \frac{45}{15}$

$y = 3$

Check: $15y = 45$

$15 \cdot 3 \stackrel{?}{=} 45$

$45 = 45$

The solution checks.

133. $105 = 5r$

$\frac{105}{5} = \frac{5r}{5}$

$21 = r$

Check: $105 = 5r$

$105 \stackrel{?}{=} 5 \cdot 21$

$105 = 105$

The solution checks.

134. $224 = 16q$

$\frac{224}{16} = \frac{16q}{16}$

$14 = q$

Check: $224 = 16q$

$224 \stackrel{?}{=} 16 \cdot 14$

$224 = 224$

The solution checks.

135. $\frac{x}{7} = 3$

$7 \cdot \frac{x}{7} = 7 \cdot 3$

$x = 21$

Check: $\frac{x}{7} = 3$

$\frac{21}{7} \stackrel{?}{=} 3$

$3 = 3$

The solution checks.

136. $\frac{a}{3}=12$

$3\cdot\frac{a}{3}=3\cdot12$

$a=36$

Check: $\frac{a}{3}=12$

$\frac{36}{3}\stackrel{?}{=}12$

$12=12$

The solution checks.

137. $15=\frac{s}{21}$

$15\cdot21=\frac{s}{21}\cdot21$

$315=s$

Check: $15=\frac{s}{21}$

$15\stackrel{?}{=}\frac{315}{21}$

$15=15$

The solution checks.

138. $25=\frac{d}{17}$

$25\cdot17=\frac{d}{17}\cdot17$

$425=d$

Check: $25=\frac{d}{17}$

$25\stackrel{?}{=}\frac{425}{17}$

$25=25$

The solution checks.

139. x = the number of orders the previous week

$2x=364$

$\frac{2x}{2}=\frac{364}{2}$

$x=182$

They received 182 orders the previous week.

140. x = the total amount cost

$\frac{x}{4}=32$

$\frac{x}{4}\cdot4=32\cdot4$

$x=128$

The chain cost $128.

Chapter 1 Test

1. a. The set of <u>whole</u> numbers is $\{0,1,2,3,4,5,...\}$.

b. The symbols > and < are <u>inequality</u> symbols.

c. The <u>area</u> of a rectangle is a measure of the amount of surface it encloses.

d. The grouping symbols ( ) are called <u>parentheses</u>, and the symbols [ ] are called <u>brackets</u>.

e. A <u>prime</u> number is a whole number greater than 1 that has only 1 and itself as factors.

f. An <u>equation</u> is a statement indicating that two expressions are equal.

g. A number that makes an equation true when substituted for a variable is called a <u>solution</u> of the equation.

h. In this chapter, we used the addition, subtraction, multiplication, and division properties of <u>equality</u> to solve equations.

2.

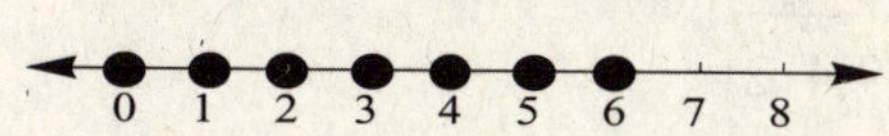

3. a. 1 hundred

b. 0

4. a. seven million, eighteen thousand, six hundred forty-one

b. 1,385,266

c. 90,000 + 2,000 + 500 + 60 + 1

5. a. $15>10$ b. $1,247<1,427$

6. a. 35,000,000

b. 34,800,000

c. 34,760,000

7.

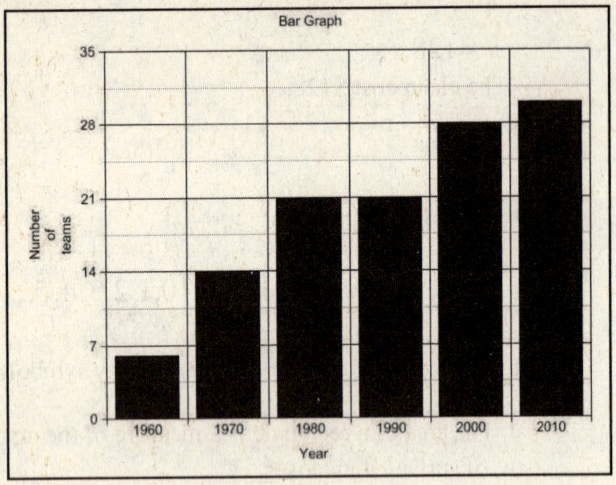

8.

$$\begin{array}{r} \overset{4}{\cancel{5}}\ \overset{\overset{12}{\cancel{13}}}{\cancel{3}}\ \overset{15}{\cancel{5}} \\ -2\ 8\ 7 \\ \hline 2\ 4\ 8 \end{array}$$

Check: $248+287=535$

9.

$$\begin{array}{r} \scriptstyle 1\,1\,1\ \ 1\,1 \\ 136{,}231 \\ 82{,}574 \\ +\ \ 6{,}359 \\ \hline 225{,}164 \end{array}$$

10.

$$\begin{array}{r} \overset{3}{\cancel{4}},\overset{\overset{14}{\cancel{15}}}{\cancel{5}}\ \overset{\overset{11}{\cancel{12}}}{\cancel{2}}\ \overset{11}{\cancel{1}} \\ -3{,}5\ 7\ 9 \\ \hline 9\ 4\ 2 \end{array}$$

11.

$$\begin{array}{r} \scriptstyle 2\ \ \\ 53 \\ \times\ \ 8 \\ \hline 424 \end{array}$$

12.

$$\begin{array}{r} 562 \\ \times\quad 74 \\ \hline 2{,}248 \\ +39{,}340 \\ \hline 41{,}588 \end{array}$$

13.

$$\begin{array}{r} 72 \\ 6\overline{)432} \\ \underline{-42} \\ 12 \\ \underline{-12} \\ 0 \end{array}$$

14.

$$\begin{array}{r} 114 \\ 73\overline{)8379} \\ \underline{-73} \\ 107 \\ \underline{-73} \\ 349 \\ \underline{-292} \\ 57 \end{array}$$

$114R57$

Check: $(73\cdot 114)+57$
$=8{,}322+57$
$=8{,}379$

15. $23\cdot 6=138$, now attach 5 zeros: $13{,}800{,}000$

16. Begin by dividing out 2 zeros from each:

$$\begin{array}{r} 250 \\ 5\overline{)1250} \\ \underline{-10} \\ 25 \\ \underline{-25} \\ 0 \end{array}$$

17. $50{,}000-7{,}000=43{,}000$

18. $327+757+327+757=1{,}084+1{,}084$
$=2{,}168in$

19. $23\cdot 23=529in^2$

20. a. 1, 2, 3, 4, 6, 12

b. 4, 8, 12, 16, 20, 24

c. $8\cdot 5$

21. $1260 = 10 \cdot 126 = 2 \cdot 5 \cdot 9 \cdot 14$
$= 2 \cdot 5 \cdot 3 \cdot 3 \cdot 2 \cdot 7 = 2^2 \cdot 3^2 \cdot 5 \cdot 7$

22. $10,000 - 5,067 = 4,933$ tails

23. $12 \cdot 8 = 96$ students

24. $12,255 \div 3 = 4,085\, ft^2$

25. $23 \cdot 18 = 414 mi$

26. a. Associative Property of Multiplication

b. Commutative Property of Addition

27. a. 0

b. 0

c. 1

d. undefined

28. $15 = 3 \cdot 5$
$18 = 2 \cdot 3 \cdot 3$
$LCM(15,18) = 2 \cdot 3 \cdot 3 \cdot 5 = 90$

29. $8 = 2 \cdot 2 \cdot 2$
$9 = 3 \cdot 3$
$12 = 2 \cdot 2 \cdot 3$
$LCM(8,9,12) = 2 \cdot 2 \cdot 2 \cdot 3 \cdot 3 = 72$

30. $30 = 2 \cdot 3 \cdot 5$
$54 = 2 \cdot 3 \cdot 3 \cdot 3$
$GCF(30,54) = 2 \cdot 3 = 6$

31. $24 = 2 \cdot 2 \cdot 2 \cdot 3$
$28 = 2 \cdot 2 \cdot 7$
$36 = 2 \cdot 2 \cdot 3 \cdot 3$
$GCF(24,28,36) = 2 \cdot 2 = 4$

32. $8 = 2 \cdot 2 \cdot 2$
$10 = 2 \cdot 5$
$LCM(8,10) = 2 \cdot 2 \cdot 2 \cdot 5 = 40$
a. 40 inches

b. 5 boxes of rice, 4 boxes of potatoes

33. It is divisible by 2, 3, 4, 5, 6, and 10.

34. $\frac{73 + 52 + 95 + 70 + 0}{5} = \frac{290}{5} = 58$

35. $9 + 4 \cdot 5 = 9 + 20 = 29$

36. $3^4 \cdot 10 - 2(6)(4) = 81 \cdot 10 - 12(4)$
$= 810 - 48 = 762$

37. $20 + 2\left[4^2 - 2\left(6 - 2^2\right)\right] = 20 + 2\left[16 - 2(6 - 4)\right]$
$= 20 + 2\left[16 - 2(2)\right] = 20 + 2[16 - 4]$
$= 20 + 2[12] = 20 + 24 = 44$

38. $\frac{3^3 - 2(15 - 14)^2}{33 - 9 + 1} = \frac{27 - 2(1)^2}{24 + 1}$
$= \frac{27 - 2}{24 + 1} = \frac{25}{25} = 1$

39. $x + 13 = 16$
$3 + 13 \stackrel{?}{=} 16$
$16 = 16$
3 is a solution.

40. To solve an equation means to find all the values of the variable that, when substituted into the equation, make a true statement.

41. $100 = x + 1$
$100 - 1 = x + 1 - 1$
$99 = x$
Check: $100 = x + 1$
$100 \stackrel{?}{=} 99 + 1$
$100 = 100$
The solution checks.

42. $y - 12 = 18$
$y - 12 + 12 = 18 + 12$
$y = 30$
Check: $y - 12 = 18$
$30 - 12 \stackrel{?}{=} 18$
$18 = 18$
The solution checks.

43. $5m = 55$

$$\frac{5m}{5} = \frac{55}{5}$$

$$m = 11$$

Check: $5m = 55$

$$5 \cdot 11 \stackrel{?}{=} 55$$

$$55 = 55$$

The solution checks.

44. $\frac{q}{3} = 27$

$$3 \cdot \frac{q}{3} = 3 \cdot 27$$

$$q = 81$$

Check: $\frac{q}{3} = 27$

$$\frac{81}{3} \stackrel{?}{=} 27$$

$$27 = 27$$

The solution checks.

45. x = the original number of spots

$$3x = 6,240$$

$$\frac{3x}{3} = \frac{6,240}{3}$$

$$x = 2,080$$

At the time, the college had 2,080 spaces.

46. x = the unfiltered intensity

$$x - 41 = 73$$

$$x - 41 + 41 = 73 + 41$$

$$x = 114$$

The sound intensity is 114 decibels.

47. x = the size of the class

$$\frac{x}{6} = 12$$

$$6 \cdot \frac{x}{6} = 6 \cdot 12$$

$$x = 72$$

There are 72 students in the class.

48. x = the amount she needs to borrow

$$x + 12,500 = 27,250$$

$$x + 12,500 - 12,500 = 27,250 - 12,500$$

$$x = 14,750$$

She needs to borrow $14,750.

CHAPTER 2 The Integers

Section 2.1: An Introduction to the Integers

CABULARY

1. Positive numbers are greater than 0 and negative numbers are less than 0.

3. To graph an integer means to locate it on the number line and highlight it with a dot.

5. The absolute value of a number is the distance between the number and 0 on the number line.

CONCEPTS

7. a. -$225

b. -10 sec.

c. -3°

d. -$12,000

e. -1 mi.

9. a. The spacing is not uniform.

b. The numbering is not uniform.

c. Zero is missing.

d. The arrowheads aren't drawn.

11. a. -4 is 3 units right of -7.

b. -2 is 4 units left of 2.

13. a. -7 is closer to -3 than 2.

b. 8 is further from 1 than -5.

15. a. $15 > -12$

b. $-5 < -4$

17.

Number	Opposite	Absolute value
-25	25	25
39	-39	39
0	0	0

NOTATION

19. a. $-(-8)$

b. $|-8|$

c. $8-8$

d. $-|-8|$

21. a. We read $\geq$ as "is greater than or equal to."

b. We read $\leq$ as "is less than or equal to."

GUIDED PRACTICE

23.

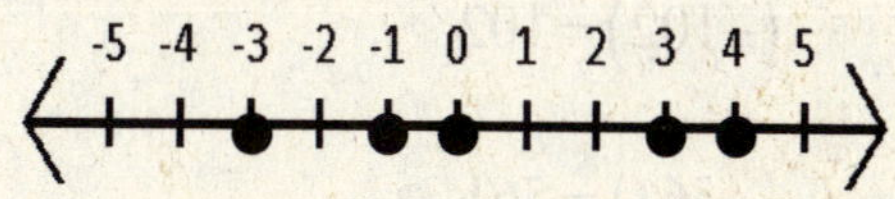

25.

27.

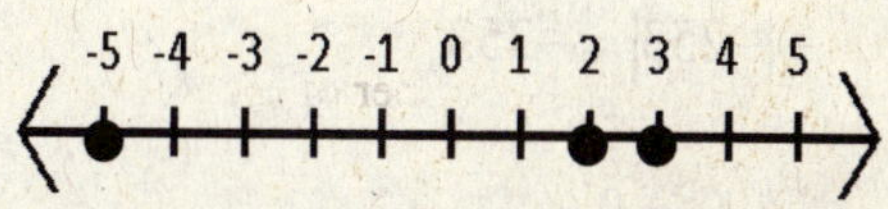

29.

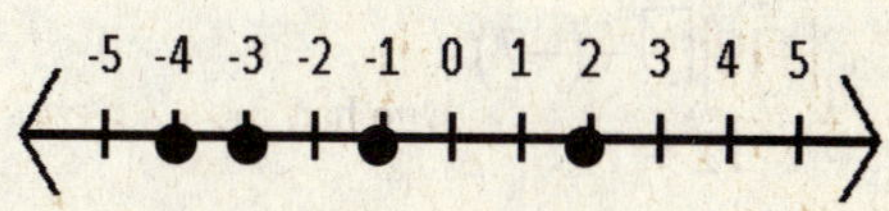

31. $-5 < 5$

33. $-12 < -6$

35. $-10 > -17$

37. $-325 > -532$

39. True

41. True

43. False

45. False

47. $|9| = 9$

49. $|-8| = 8$

51. $|-14| = 14$

53. $|180| = 180$

55. $-(-11) = 11$

57. $-(-4) = 4$

59. $-(-102) = 102$

61. $-(-561) = 561$

63. $-|20| = -20$

65. $-|6| = -6$

67. $-|-253| = -253$

69. $-|-0| = 0$

TRY IT YOURSELF

71. $|-12| \square -(-7)$

$12 > 7$

73. $-|-71| \square -|-65|$

$-71 < -65$

75. $-(-343) \square -(-161)$

$343 > 161$

77. $-|-30| \square -|-(-8)|$

$-30 < -8$

79. $-52, -22, -12, 12, 52, 82$

81. $5, 3, 1, -1, \boxed{-3}, \boxed{-5}, \boxed{-7}, \ldots$

APPLICATIONS

85.

Time	Position
0 sec	0
1 sec	20
2 sec	5
3 sec	-40
4 sec	-120

87. Peaks: 2, 4, 0 ; Valleys: -3, -5, -2

89. a. -1 : 1 below par

b. -3 : 3 below par

c. Most of the scores are below par.

91. a. $-20°$ to $-10°$

b. $40°$

c. $10°$

93. a. 200 years

b. A.D.

c. B.C.

d. The birth of Jesus

95.

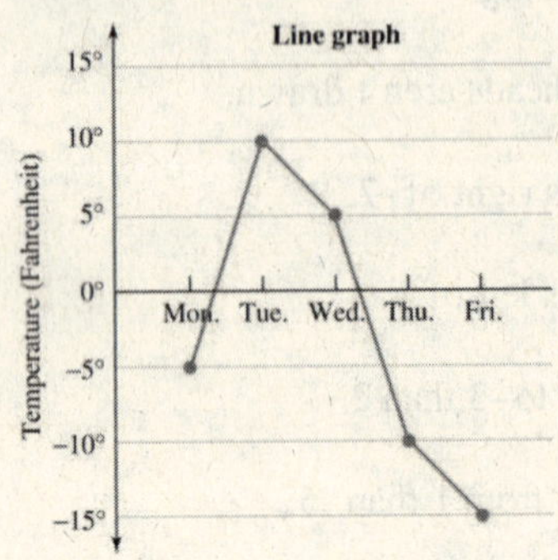

WRITING

97. The opposite of a number is the number that is the same distance from 0 on the number line, but on the other side of 0.

99. Absolute value represents a distance, which cannot be negative.

101. Positive buoyancy would mean you float, negative would mean you sink. Neutral buoyancy would mean that you remain stationary in the water.

103. No it is not: 4 > 3 but -4 < -3.

REVIEW

105. 23,500

107. $2{,}842-2{,}081=761$

109. Associative Property of Multiplication

Section 2.2: Adding Integers

VOCABULARY

1. Two negative integers, as well as two positive integers, are said to have the same or like signs.

3. When 0 is added to a number, the number remains the same. We call 0 the additive identity.

5. Commutative property of addition: The order in which integers are added does not change their sum.

CONCEPTS

7. a. $|10|=10$; $|-12|=12$

b. -12

c. $|-12|-|10|=12-10=2$

9. To add two integers with unlike signs, subtract their absolute values, the smaller from the larger. Then attach to that result the sign of the number with the larger absolute value.

11. a. yes b. yes c. no d. no

13. a. 0 b. 0

NOTATION

15. $-16+(-2)+(-1)=-18+(-1)=-19$

17. $(-3+8)+(-3)=5+(-3)=2$

GUIDED PRACTICE

19. Add the absolute values. Since both numbers are negative, the result is negative.

$6+3=9$ so $-6+(-3)=-9$

21. Add the absolute values. Since both numbers are negative, the result is negative.

$5+5=10$ so $-5+(-5)=-10$

23. Add the absolute values. Since both numbers are negative, the result is negative.

$51+11=62$ so $-51+(-11)=-62$

25. Add the absolute values. Since both numbers are negative, the result is negative.

$69+27=96$ so $-69+(-27)=-96$

27. Add the absolute values. Since both numbers are negative, the result is negative.

$248+131=379$ so $-248+(-131)=-379$

29. Add the absolute values. Since both numbers are negative, the result is negative.

$565+309=874$ so $-565+(-309)=-874$

31. Subtract the smaller absolute value from the larger. Since the negative number has the larger absolute value, the result is negative.

$8-5=3$ so $-8+5=-3$

33. Subtract the smaller absolute value from the larger. Since the positive number has the larger absolute value, the result is positive.

$7-6=1$ so $7+(-6)=1$

35. Subtract the smaller absolute value from the larger. Since the negative number has the larger absolute value, the result is negative.

$42-20=22$ so $20+(-42)=-22$

37. Subtract the smaller absolute value from the larger. Since the positive number has the larger absolute value, the result is positive.

$71-23=48$ so $71+(-23)=48$

39. Subtract the smaller absolute value from the larger. Since the positive number has the larger absolute value, the result is positive.

$479-122=357$ so $479+(-122)=357$

41. Subtract the smaller absolute value from the larger. Since the negative number has the larger absolute value, the result is negative.
$339-279=60$ so $-339+279=-60$

43. $9+(-3)+5+(-4)=6+5+(-4)$
$=11+(-4)=7$

45. $6+(-4)+(-13)+7=2+(-13)+7$
$=-11+7=-4$

47. $[-3+(-4)]+(-5+2)$
$=[-7]+(-3)=-10$

49. $(-1+34)+[16+(-8)]=33+8=41$

51. Look for opposites (additive inverses):
$\underline{23}+\underline{\underline{(-5)}}+3+\underline{\underline{5}}+\underline{(-23)}=0+0+3=3$

53. Look for opposites (additive inverses):
$\underline{-10}+\underline{\underline{(-1)}}+\underline{10}+(-6)+\underline{\underline{1}}=0+0+(-6)=-6$

TRY IT YOURSELF

55. $-2+6+(-1)=4+(-1)=3$

57. $-7+0=-7$

59. Subtract the smaller absolute value from the larger. Since the positive number has the larger absolute value, the result is positive.
$24+(-15)=9$

61. Add the absolute values. Since both numbers are negative, the result is negative.
$-435+(-127)=-562$

63. Subtract the smaller absolute value from the larger. Since the positive number has the larger absolute value, the result is positive.
$-7+9=2$

65. $2+(-2)=0$

67. $-5+(-9+8)+(-2)=-5+(-1)+(-2)$
$=-6+(-2)=-8$

69. $\underline{-9}+\underline{\underline{1}}+(-2)+\underline{\underline{(-1)}}+\underline{9}=0+0+(-2)=-2$

71. $[6+(-4)]+[8+(-11)]=2+[-3]=-1$

73. $(-4+8)+(-11+4)=4+(-7)=-3$

75. $-675+(-456)+99$
$=-1{,}131+99=-1{,}032$

77. $-6+(-7)+(-8)=-13+(-8)=-21$

79. $-2+[789+(-9{,}135)]$
$=-2+[-8{,}346]=-8{,}348$

81. $-45+25=-20$

APPLICATIONS

83. Michigan: $-51°+163°=112°$

Minnesota: $-60°+175°=115°$

85. a. $-15{,}720$ *ft*

b. $-15{,}720+3{,}220=-12{,}500$ *ft*

87. a. -9 *ft*

b. $-9+11=2$ *ft* above flood stage.

89. $-40+200+10+25=160+10+25$
$=170+25=195°$

91. $-4+3+(-3)+3+4+2$
$=0+0+3+2=5$
He has a 4% risk.

93. $-1{,}500+3{,}500+1{,}250$
$=2{,}000+1{,}250=3{,}250$ *m*

95. $5,889+927+(-2,928)+1,645+(-894)+715+(-6,321)$
$=6,816+(-2,928)+1,645+(-894)+715+(-6,321)$
$=3,888+1,645+(-894)+715+(-6,321)$
$=5,533+(-894)+715+(-6,321)$
$=4,639+715+(-6,321)$
$=5,354+(-6,321)$
$=-967$
The entry should be ($967).

WRITING

97. No. If the negative number has a larger absolute value the sum will be negative.

99. Because we are beginning with a negative number and then taking away from it, so the result is still negative.

101. The integers are opposites, like 4 and -4.

105. a. $5+3+5+3=16\,ft.$

b. $5\cdot 3=15\,ft^2$

107. $250=25\cdot 10=5\cdot 5\cdot 5\cdot 2=2\cdot 5^3$

Section 2.3: Subtracting Integers

VOCABULARY

1. -8 is the opposite (or additive inverse) of 8.

3. To evaluate an expression means to find its value.

CONCEPTS

5. To subtract two integers, add the first integer to the opposite (additive inverse) of the integer to be subtracted.

7. Subtracting 3 is the same as adding -3. Subtracting -6 is the same as adding 6.

9. We can find the change in a quantity by subtracting the earlier value from the later value.

11. a. 3 is being subtracted.

b. -12 is being subtracted

13. $3-(-6)=3+6=9$

NOTATION

15. a. $-8-(-4)$

b. $-4-(-8)$

17. $1-3-(-2)=1+(-3)+2$
$=-2+2$
$=0$

19. $(-8-2)-(-6)=[-8+(-2)]-(-6)$
$=-10-(-6)$
$=-10+6$
$=-4$

GUIDED PRACTICE

21. $-4-3=-4+(-3)=-7$

23. $-5-5=-5+(-5)=-10$

25. $8-(-1)=8+1=9$

27. $11-(-7)=11+7=18$

29. $3-21=3+(-21)=-18$

31. $15-65=15+(-65)=-50$

33. a. $-11-(-1)=-11+1=-10$

b. $-1-(-11)=-1+11=10$

35. a. $-16-(-41)=-16+41=25$

b. $-41-(-16)=-41+16=-25$

37. $-4-(-4)-15=-4+4+(-15)$
$=0+(-15)=-15$

39. $10-9-(-8)=10+(-9)+8$
$=1+8=9$

41. $-1-(-3)-4=-1+3+(-4)$
$=2+(-4)=-2$

43. $-5-8-(-3)=-5+(-8)+3$
$=-13+3=-10$

45. $-1-(-4-6)=-1-[-4+(-6)]$
$=-1-(-10)$
$=-1+10=9$

47. $-42-(-16-14)=-42-[-16+(-14)]$
$=-42-(-30)$
$=-42+30=-12$

49. $-9-(6-7)=-9-[6+(-7)]$
$=-9-(-1)$
$=-9+1=-8$

51. $-8-(4-12)=-8-[4+(-12)]$
$=-8-(-8)$
$=-8+8=0$

53. $-(-5)+(-15)-6-(-48)$
$=5+(-15)+(-6)+48$
$=(5+48)+[-15+(-6)]$
$=53+(-21)=32$

55. $-(-3)+(-41)-7-(-19)$
$=3+(-41)+(-7)+19$
$=(3+19)+[-41+(-7)]$
$=22+(-48)=-26$

57. $-1,557-890=-2,447$

59. $-979-(-44,879)=-979+44,879=43,900$

TRY IT YOURSELF

61. $-5+7-(-1)-8=-5+7+1+(-8)$
$=2+1+(-8)$
$=3+(-8)=-5$

63. $7-(-3)=7+3=10$

65. $-2-(-10)=-2+10=8$

67. $0-(-5)=0+5=5$

69. $(6-4)-(1-2)=2-(-1)$
$=2+1=3$

71. $-5-(-4)=-5+4=-1$

73. $-3-3-3=-3+(-3)+(-3)$
$=-6+(-3)=-9$

75. $-(-9)+(-20)-14-(-3)$
$=9+(-20)+(-14)+3$
$=(9+3)+[-20+(-14)]$
$=12+(-34)=-22$

77. $[-4+(-8)]-(-6)+15$
$=-12+6+15=-6+15=9$

79. $-10-(-6)=-10+6=-4$

81. $-3-(-3)=-3+3=0$

83. $-8-[4-(-6)]=-8-[4+6]$
$=-8-10=-8+(-10)=-18$

85. $4-(-4)=4+4=8$

87. $(-6-5)-3+(-11)$
$=-6+(-5)+(-3)+(-11)$
$=-11+(-3)+(-11)$
$=-14+(-11)=-25$

APPLICATIONS

89. $-2,000-200=-2,000+(-200)=-2,200\ ft$

91. $1,348-282=1,348+(-282)=1,066\ ft$

93. $5-7-6=5+(-7)+(-6)$
$=-2+(-6)=-8$

95. $-1+(-6)+(-5)+8$
$=-7+(-5)+8=-12+8$
$=-4\ yds$

97. $15-25-40-30-3(20)$
$=15+(-25)+(-40)+(-30)+(-60)$
$=-10+(-40)+(-30)+(-60)$
$=(-50)+(-30)+(-60)$
$=-80+(-60)=-\$140$

99. Portland (142°), Barrow (135°), Kansas City (131°), Norfolk (108°), Atlantic City (105°),

101. $360-(-110)=360+110=470°F$

103. $-7-(-23)=-7+23=16$ point increase

WRITING

105. It means that we can think of all addition problems as subtraction problems and vice-versa.

107. To check the result of any subtraction problem, the sum of the difference and the subtrahend should be equal to the minuend. In this problem, -11 + 4 = -7, so the result is correct.

REVIEW

109. a. 24,090

b. 6,000

111. $13\cdot 12=156$ oranges needed

Section 2.4: Multiplying Integers

VOCABULARY

1. $-5 \cdot 10 = 50$
factor factor product

3. A positive integer and a negative integer are said to have different signs or unlike signs.

5. Associative property of multiplication: The way in which integers are grouped does not change their product.

CONCEPTS

7. Multiplication of integers is very much like multiplication of whole numbers. The only difference is that we must determine whether the answer is positive or negative.

9. To multiply a positive integer and a negative integer, multiply their absolute values. Then make the final answer negative.

11. The product of two integers with unlike/different signs is negative.

13. The product of any integer and 0 is 0.

15. a. $|-3|=3$ b. $|12|=12$

NOTATION

17. a. base: 8, exponent: 4 b. base -7, exponent: 9

19. $-3(-2)(-4)=6(-4)=-24$

GUIDED PRACTICE

21. Multiply the absolute values. Since one number is negative, the result is negative. $5(-3)=-15$

23. Multiply the absolute values. Since one number is negative, the result is negative. $9(-2)=-18$

25. Multiply the absolute values. Since one number is negative, the result is negative.
$18(-4)=-72$

27. Multiply the absolute values. Since one number is negative, the result is negative.
$21(-6)=-126$

29. Multiply the absolute values. Since one number is negative, the result is negative.
$-45\cdot 37=-1{,}665$

31. Multiply the absolute values. Since one number is negative, the result is negative.
$-94\cdot 1{,}000=-94{,}000$

33. Multiply the absolute values. Since both numbers are negative, the result is positive.
$(-8)(-7)=56$

35. Multiply the absolute values. Since both numbers are negative, the result is positive.

$7\cdot 1=7$ so $-7(-1)=7$

37. Multiply the absolute values. Since both numbers are negative, the result is positive.

$-3(-52)=156$

39. Multiply the absolute values. Since both numbers are negative, the result is positive.

$-6(-46)=276$

41. Multiply the absolute values. Since both numbers are negative, the result is positive.

$-59(-33)=1{,}947$

43. Multiply the absolute values. Since both numbers are negative, the result is positive.

$-60{,}000\cdot(-1{,}200)=72{,}000{,}000$

45. Left to right: $6(-3)(-5)=-18(-5)=90$

47. Left to right: $-5(10)(-3)=-50(-3)=150$

49. First Pair, Last Pair:

$-2(-4)(6)(-8)=8(-48)=-384$

51. First Pair, Last Pair:

$-8(-3)(7)(-2)=24(-14)=-336$

53. Left to right: $-4(-2)(-6)=8(-6)=-48$

55. Left to right: $-3(-9)(-3)=27(-3)=-81$

57. First Pair, Last Pair:

$-1(-3)(-2)(-6)=3(12)=36$

59. First Pair, Last Pair:

$-9(-4)(-1)(-4)=36(4)=144$

61. $(-3)^3=(-3)(-3)(-3)=9(-3)=-27$

63. $(-2)^5=(-2)(-2)(-2)(-2)(-2)$

$=4\cdot 4(-2)=-32$

65. $(-5)^4=(-5)(-5)(-5)(-5)$

$=25(25)=625$

67. $(-1)^8=(-1)(-1)(-1)(-1)(-1)(-1)(-1)(-1)$

$=1\cdot 1\cdot 1\cdot 1=1$

69. $(-7)^2=(-7)(-7)=49$

$-7^2=-7\cdot 7=-49$

71. $(-12)^2=(-12)(-12)=144$

$-12^2=-12\cdot 12=-144$

TRY IT YOURSELF

73. $6(-5)(2)=-30(2)=-60$

75. $-8(0)=0$

77. $(-4)^3=(-4)(-4)(-4)=16(-4)=-64$

79. $(-2)10=-20$

81. $-2(-3)(3)(-1)=6(-3)=-18$

83. $-6(-10)=60$

85. $-6(-4)(-2)=24(-2)=-48$

87. $-42\cdot 200{,}000=-8{,}400{,}000$

89. $-5^4=-5\cdot 5\cdot 5\cdot 5=-25\cdot 25=-625$

91. $-12(-12)=144$

93. $(-1)^6=(-1)(-1)(-1)(-1)(-1)(-1)$

$=1\cdot 1\cdot 1=1$

95. $(-1)(-2)(-3)(-4)(-5)=2(12)(-5)$

$=24(-5)=-120$

APPLICATIONS

97. $8(-250)=-2000\,ft$

99. a. high: 2, low: -3

b. high: $2 \cdot 2 = 4$, low: $2(-3) = -6$

101. a. $6(-67) = -402{,}000$ jobs

b. $9(-47) = -423{,}000$ jobs

c. $7(-83) = -581{,}000$ jobs

d. $6(-88) = -528{,}000$ jobs

103. $4(-81) = -324°F$

105. $6(-200) = -\$1{,}200$

107. $3(-6) = -18\,ft$

109. $74{,}905(-3) = -\$224{,}715$

WRITING

111. Because we are really adding 5 negatives together (in this case) and the sum of any negative numbers is still negative.

113. Multiplying any integer by -1 would simply change the sign of the number, which is exactly what happens when we find the opposite of a number..

REVIEW

115. 2, 3, 5, 7, 11, 13, 17, 19, 23, 29

117.

$$\begin{array}{r} 19 \\ 49\overline{)\ 978} \\ \underline{-49} \\ 488 \\ \underline{-441} \\ 47 \end{array}$$

$978 \div 49 = 19R47$

Section 2.5: Dividing Integers

VOCABULARY

1. 12 ÷ (-4) = -3

dividend divisor quotient

dividend quotient

$$\frac{12}{-4} = -3$$

divisor

3. $\frac{-3}{0}$ is division <u>by</u> 0 and $\frac{0}{-3}$ is division <u>of</u> 0.

CONCEPTS

5. a. $-5(5) = -25$

b. $6(-6) = -36$

c. $0(-15) = 0$

7. a. The quotient of two integers that have the same *(like)* signs is <u>positive</u>.

b. The quotient of two integers that have different *(unlike)* signs is <u>negative</u>.

9. a. If 0 is divided by ay nonzero integer, the quotient is <u>0</u>.

b. Division of any nonzero number by 0 is <u>undefined</u>.

11. a. always true

b. sometimes true

c. always true

GUIDED PRACTICE

13. Divide the absolute values: the signs are opposite so the result is negative.

$$\frac{-14}{2} = -7 \qquad \text{Check: } -7(2) = -14$$

15. Divide the absolute values: the signs are opposite so the result is negative.

$$\frac{-20}{5} = -4 \qquad \text{Check: } -4(5) = -20$$

17. Divide the absolute values: the signs are opposite so the result is negative.

$36 \div (-6) = -6 \quad \text{Check: } (-6)(-6) = 36$

19. Divide the absolute values: the signs are opposite so the result is negative.

$24 \div (-3) = -8 \quad \text{Check: } (-8)(-3) = 24$

21. Divide the absolute values: the signs are opposite so the result is negative.

$\frac{264}{-12} = -22 \quad \text{Check: } (-22)(-12) = 264$

23. Divide the absolute values: the signs are opposite so the result is negative.

$\frac{702}{-18} = -39 \quad \text{Check: } (-39)(-18) = 702$

25. Divide the absolute values: the signs are opposite so the result is negative.

$-9{,}000 \div 300 = -30$

$\text{Check: } (-30)(300) = -9{,}000$

27. Divide the absolute values: the signs are opposite so the result is negative.

$-250{,}000 \div 5{,}000 = -50$

$\text{Check: } (-50)(5000) = -250{,}000$

29. Divide the absolute values: the signs are the same so the result is positive.

$\frac{-8}{-4} = 2 \quad \text{Check: } 2(-4) = -8$

31. Divide the absolute values: the signs are the same so the result is positive.

$\frac{-45}{-9} = 5 \quad \text{Check: } 5(-9) = -45$

33. Divide the absolute values: the signs are the same so the result is positive.

$-63 \div (-7) = 9 \quad \text{Check: } 9(-7) = -63$

35. Divide the absolute values: the signs are the same so the result is positive.

$-32 \div (-8) = 4 \quad \text{Check: } 4(-8) = -32$

37. Divide the absolute values: the signs are the same so the result is positive.

$\frac{-400}{-25} = 16 \quad \text{Check: } 16(-25) = -400$

39. Divide the absolute values: the signs are the same so the result is positive.

$\frac{-651}{-31} = 21 \quad \text{Check: } 21(-31) = -651$

41. Divide the absolute values: the signs are the same so the result is positive.

$-800 \div (-20) = 40$

$\text{Check: } 40(-20) = -800$

43. Divide the absolute values: the signs are the same so the result is positive.

$-15{,}000 \div (-30) = 500$

$\text{Check: } 500(-30) = -15{,}000$

45. a. undefined b. 0

47. a. 0 b. undefined

TRY IT YOURSELF

49. $-36 \div (-12) = 3$

51. $\frac{425}{-25} = -17$

53. $\frac{0}{120} = 0$

55. $-45 \div 9 = -5$

57. $-2{,}500 \div (500) = -5$

59. $\frac{-6}{0}$ is undefined.

61. $\frac{19}{-1} = -19$

63. $-23 \div (-23) = 1$

65. $\frac{40}{-2} = -20$

67. $9 \div (-9) = -1$

69. $\dfrac{-10}{-1} = 10$

71. $\dfrac{-888}{37} = -24$

73. $\dfrac{3{,}000}{-100} = -30$

75. $8 \div (-2) = -4$

77. $\dfrac{-13{,}550}{25} = -542$

79. $\dfrac{27{,}778}{-17} = -1{,}634$

APPLICATIONS

81. $-210 \div 6 = -\$35/wk$

83. $-3{,}030 \div 3 = -1{,}010\,ft$

85. $30 - 72 = -42°$

$-42 \div 6 = -7°/min$

87. $-12 \div 2 = -6$ (6 games behind)

89. $-300 \div 20 = -\$15$ per pair

91. $-4{,}335 \div 255 = -\$17$ per share

WRITING

93. The quotient of two negatives is positive because we are really dividing top and bottom by negative 1.

95. To check that $20 \div 4 = 5$, verify $4 \cdot 5 = 20$.

97. The quotient of two integers that have the same signs is positive and the quotient of two integers that have different signs is negative.

REVIEW

99. $\dfrac{6(32)-72}{3(2^3)} = \dfrac{192-72}{3(8)} = \dfrac{120}{24} = 5$

101. Associative Property of Addition

103. No, because subtraction is not commutative.

Section 2.6: Order of Operations and Estimation

VOCABULARY

1. To evaluate expressions that contain more than one operation, we use the order of operations rule.

3. In the expression $-9-2\left[-5-6(-3-1)\right]$, the parentheses are the inner most grouping symbols and the brackets are the outer most grouping symbols.

CONCEPTS

5. a. square, multiply, subtract

b. multiply, cube, subtract, add

c. grouping symbol, multiply, add

d. square, multiply

NOTATION

7. parentheses, brackets, absolute value symbol, fraction bar

9. $-8-5(-2)^2 = -8-5(4)$

$= -8-20$

$= -8+(-20)$

$= -28$

11. $-9+5(-4\cdot 2+7) = -9+5(-8+7)$

$= -9+5(-1)$

$= -9+(-5)$

$= -14$

GUIDED PRACTICE

13. $-2(-3)^2-(-8) = -2(9)-(-8)$

$= -18+8 = -10$

15. $-5(-4)^2-(-18) = -5(16)-(-18)$

$= -80+18 = -62$

17. $9(7)+(-6)(-2)(-4) = 63+12(-4)$

$= 63-48 = 15$

19. $8(6)+(-2)(-9)(-2) = 48+18(-2)$

$= 48-36 = 12$

21. $30 \div (-5)2 = -6 \cdot 2 = -12$

23. $60 \div (-3)4 = -20 \cdot 4 = -80$

25. $-6^2 - (-6)^2 = -36 - 36 = -72$

27. $-10^2 - (-10)^2 = -100 - 100 = -200$

29. $-14 + 2(-9 + 6 \cdot 3) = -14 + 2(-9 + 18)$
$= -14 + 2 \cdot 9 = -14 + 18 = 4$

31. $-23 + 3(-15 + 8 \cdot 4) = -23 + 3(-15 + 32)$
$= -23 + 3(17) = -23 + 51 = 28$

33. $77 - 2\left[-6 + (3-9)^2\right]$
$= 77 - 2\left[-6 + (-6)^2\right]$
$= 77 - 2[-6 + 36] = 77 - 2 \cdot 30$
$= 77 - 60 = 17$

35. $99 - 4\left[-9 + (6-10)^2\right]$
$= 99 - 4\left[-9 + (-4)^2\right]$
$= 99 - 4[-9 + 16] = 99 - 4 \cdot 7$
$= 99 - 28 = 71$

37. $-\left[4 - \left(3^3 + \frac{22}{-11}\right)\right]$
$= -\left[4 - (27 + (-2))\right]$
$= -[4 - 25] = -(-21) = 21$

39. $-\left[50 - \left(5^3 + \frac{50}{-2}\right)\right]$
$= -\left[50 - (125 + (-25))\right]$
$= -[50 - 100] = -(-50) = 50$

41. $\frac{-24 + 3(-4)}{42 - (-6)^2} = \frac{-24 - 12}{42 - 36}$
$= \frac{-36}{6} = -6$

43. $\frac{-38 + 11(-2)}{69 - (-8)^2} = \frac{-38 - 22}{69 - 64}$
$= \frac{-60}{5} = -12$

45. a. $|-6(2)| = |-12| = 12$

b. $|-12 + 7| = |-5| = 5$

47. a. $|15(-4)| = |-60| = 60$

b. $|16 + (-30)| = |-14| = 14$

49. $16 - 6|-2 - 1| = 16 - 6|-3|$
$= 16 - 6 \cdot 3 = 16 - 18 = -2$

51. $17 - 2|-6 - 4| = 17 - 2|-10|$
$= 17 - 2 \cdot 10 = 17 - 20 = -3$

53. $-379 + (-13) + 287 + (-671)$
$\approx -380 + (-10) + 290 + (-670)$
$= -390 + 290 + (-670)$
$= -100 + (-670) = -770$

55. $-3{,}887 + (-5{,}806) + 4{,}701$
$\approx -3{,}900 + (-5{,}800) + 4{,}700$
$= -9{,}700 + 4{,}700 = -5{,}000$

TRY IT YOURSELF

57. $(-3)^2 - 4^2 = 9 - 16 = -7$

59. $3^2 - 4(-2)(-1) = 9 - 4(-2)(-1) = 9 - 8 = 1$

61. $|-3 \cdot 4 + (-5)| = |-12 + (-5)|$
$= |-17| = 17$

63. $(2-5)(5+2) = (-3)(7) = -21$

65. $6 + \frac{25}{-5} + 6 \cdot 3 = 6 - 5 + 18$
$= 1 + 18 = 19$

67. $$\frac{-6-2^3}{-2-(-4)}=\frac{-6-8}{-2+4}=\frac{-14}{2}=-7$$

69. $$-12\div(-2)2=6\cdot2=12$$

71. $$-16-4\div(-2)=-16+2=-14$$

73. $$-\left|2\cdot7-(-5)^2\right|=-|14-25|=-|-11|=-11$$

75. $$-3|4-7|=-3|-3|=-3(3)=-9$$

77. $$(7-5)^2-(1-4)^2=2^2-(-3)^2$$
$$=4-9=-5$$

79. $$-1\left(2^2-2+1^2\right)=-1(4-2+1)$$
$$=-1(2+1)=-1(3)=-3$$

81. $$\frac{-5-5}{1^4+1^5}=\frac{-10}{1+1}=\frac{-10}{2}=-5$$

83. $$-50-2(-3)^3(4)=-50-2(-27)(4)$$
$$=-50+54\cdot4=-50+216=166$$

85. $$-6^2+6^2=-36+36=0$$

87. $$3\left(\frac{-18}{3}\right)-2(-2)=3(-6)+4$$
$$=-18+4=-14$$

89. $$2|1-8|\cdot|-8|=2|-7|\cdot8$$
$$=2\cdot7\cdot8=112$$

91. $$\frac{2+3[5-(1-10)]}{|2(-8+2)+10|}=\frac{2+3[5-(-9)]}{|2(-6)+10|}$$
$$=\frac{2+3[5+9]}{|-12+10|}=\frac{2+3(14)}{|-2|}=\frac{2+42}{2}$$
$$=\frac{44}{2}=22$$

93. $$-2+\left|6-4^2\right|=-2+|6-16|$$
$$=-2+|-10|=-2+10=8$$

95. $$\frac{-4(-5)-2}{3-3^2}=\frac{20-2}{3-9}=\frac{18}{-6}=-3$$

APPLICATIONS

97. $-70+60-230-30-130=-400$ points

99. $$12(3)+3(-4)+5(-1)$$
$$=36-12-5=19$$

101. $$\frac{-12+(-15)+(-5)+(-16)}{6}=\frac{-48}{6}$$
$=-\$8$ million

103. Over the last five years::
$$\frac{-164+(-107)+(-22)+70+123}{5}=-\$20$$

Over the last four years:
$$\frac{(-107)+(-22)+70+123}{4}=\$16$$

It would be better to refer to the last 4 years because there is a surplus.

105. a. $-30-60=-90\,ft$

b. $10(60)=\$600$ lost

c. $-8{,}000\div20=-400\,ft$

WRITING

107. It is necessary so that everyone gets the same answer from the same problem.

109. a. The division should have been done before the multiplication.

b. The multiplication of the 8 and the absolute value should happen before adding the -1.

REVIEW

111. $$34+y=34$$
$$34+y-34=34-34$$
$$y=0$$

113. yes, yes, yes, yes, yes, no, yes

Section 2.7: Solving Equations That Involve Integers

VOCABULARY

1. To solve an equation means to find all the values of the variable that make the equation true.

3. To check the solution of an equation, we substitute the value for the variable in the original equation and determine whether the result is a true statement.

CONCEPTS

5. a. multiplication by -2

b. addition of -6

c. division by -5

d. subtraction of 4

7. a. add 9 to both sides

b. divide both sides by -8

9. The addition property of equality: Adding the same number to both sides of an equation does not change its solution.

11. * To isolate the variable term $4x$, we undo the addition of 10 by subtracting 10 from both sides.
* To isolate the variable x, we undo the multiplication of 4 by dividing both sides by 4.

NOTATION

13.
$$y+(-7)=-16+3$$
$$y+(-7)=-13$$
$$y+(-7)+7=-13+7$$
$$y=-6$$
Check: $y+(-7)=-16+3$
$$-6+(-7)\stackrel{?}{=}-13$$
$$-13=-13 \text{ True}$$

15. a. $-10x=-10\cdot x$

b. $\dfrac{x}{-8}=x\div(-8)$

GUIDED PRACTICE

17.
$$x+(-3)=-12$$
$$x+(-3)+3=-12+3$$
$$x=-9$$
Check: $x+(-3)=-12$
$$-9+(-3)\stackrel{?}{=}-12$$
$$-12=-12 \text{ True}$$

19.
$$m+(-6)=-1$$
$$m+(-6)+6=-1+6$$
$$m=5$$
Check: $m+(-6)=-1$
$$5+(-6)\stackrel{?}{=}-1$$
$$-1=-1 \text{ True}$$

21.
$$y+20=-4$$
$$y+20-20=-4-20$$
$$y=-24$$
Check: $y+20=-4$
$$-24+20\stackrel{?}{=}-4$$
$$-4=-4 \text{ True}$$

23.
$$t+19=-33$$
$$t+19-19=-33-19$$
$$t=-52$$
Check: $t+19=-33$
$$-52+19\stackrel{?}{=}-33$$
$$-33=-33 \text{ True}$$

25.
$$-7+9=x+5(-3)$$
$$2=x-15$$
$$2+15=x-15+15$$
$$17=x$$
Check: $-7+9=x+5(-3)$
$$-7+9\stackrel{?}{=}17+5(-3)$$
$$-2\stackrel{?}{=}17-15$$
$$-2=-2 \text{ True}$$

27. $-6+3=f+2(-4)$

$-3=f-8$

$-3+8=f-8+8$

$5=f$

Check: $-6+3=f+2(-4)$

$-6+3\stackrel{?}{=}5+2(-4)$

$-3\stackrel{?}{=}5-8$

$-3=-3$ True

29. $-2s=16$

$\frac{-2s}{-2}=\frac{16}{-2}$

$s=-8$

Check: $-2s=16$

$-2(-8)\stackrel{?}{=}16$

$16=16$ True

31. $-150=-6g$

$\frac{-150}{-6}=\frac{-6g}{-6}$

$25=g$

Check: $-150=-6g$

$-150\stackrel{?}{=}-6(25)$

$-150=-150$ True

33. $\frac{t}{-3}=-9$

$-3\cdot\frac{t}{-3}=-3(-9)$

$t=27$

Check: $\frac{t}{-3}=-9$

$\frac{27}{-3}\stackrel{?}{=}-9$

$-9=-9$ True

35. $\frac{x}{-7}=11$

$-7\cdot\frac{x}{-7}=-7\cdot 11$

$x=-77$

Check: $\frac{x}{-7}=11$

$\frac{-77}{-7}\stackrel{?}{=}11$

$11=11$ True

37. $-x=14$

$\frac{-x}{-1}=\frac{14}{-1}$

$x=-14$

Check: $-x=14$

$-(-14)\stackrel{?}{=}14$

$14=14$ True

39. $-y=-58$

$\frac{-y}{-1}=\frac{-58}{-1}$

$y=58$

Check: $-y=-58$

$-58=-58$ True

41. $-5x-9=11$

$-5x-9+9=11+9$

$-5x=20$

$\frac{-5x}{-5}=\frac{20}{-5}$

$x=-4$

Check: $-5x-9=11$

$-5(-4)-9\stackrel{?}{=}11$

$20-9\stackrel{?}{=}11$

$11=11$ True

43.
$$-11y-1=87$$
$$-11y-1+1=87+1$$
$$-11y=88$$
$$\frac{-11y}{-11}=\frac{88}{-11}$$
$$y=-8$$
Check: $-11y-1=87$
$$-11(-8)-1\overset{?}{=}87$$
$$88-1\overset{?}{=}87$$
$$87=87 \quad \text{True}$$

45.
$$-22=8-3x$$
$$-22-8=8-3x-8$$
$$-30=-3x$$
$$\frac{-30}{-3}=\frac{-3x}{-3}$$
$$10=x$$
Check: $-22=8-3x$
$$-22\overset{?}{=}8-3(10)$$
$$-22\overset{?}{=}8-30$$
$$-22=-22 \quad \text{True}$$

47.
$$49=4-5t$$
$$49-4=4-5t-4$$
$$45=-5t$$
$$\frac{45}{-5}=\frac{-5t}{-5}$$
$$-9=t$$
Check: $49=4-5t$
$$49\overset{?}{=}4-5(-9)$$
$$49\overset{?}{=}4+45$$
$$49=49 \quad \text{True}$$

49.
$$\frac{x}{-2}-6=-9$$
$$\frac{x}{-2}-6+6=-9+6$$
$$\frac{x}{-2}=-3$$
$$-2\cdot\frac{x}{-2}=-2(-3)$$
$$x=6$$
Check: $\frac{x}{-2}-6=-9$
$$\frac{6}{-2}-6\overset{?}{=}-9$$
$$-3-6\overset{?}{=}-9$$
$$-9=-9 \quad \text{True}$$

51.
$$\frac{y}{4}+5=-8$$
$$\frac{y}{4}+5-5=-8-5$$
$$\frac{y}{4}=-13$$
$$4\cdot\frac{y}{4}=4(-13)$$
$$y=-52$$
Check: $\frac{y}{4}+5=-8$
$$\frac{-52}{4}+5\overset{?}{=}-8$$
$$-13+5\overset{?}{=}-8$$
$$-8=-8 \quad \text{True}$$

TRY IT YOURSELF

53.
$$-21 = 4h - 5$$
$$-21 + 5 = 4h - 5 + 5$$
$$-16 = 4h$$
$$\frac{-16}{4} = \frac{4h}{4}$$
$$-4 = h$$
Check: $-21 = 4h - 5$
$$-21 \stackrel{?}{=} 4(-4) - 5$$
$$-21 \stackrel{?}{=} -16 - 5$$
$$-21 = -21 \quad \text{True}$$

55.
$$-9h = -3(-3)$$
$$-9h = 9$$
$$\frac{-9h}{-9} = \frac{9}{-9}$$
$$h = -1$$
Check: $-9h = -3(-3)$
$$-9(-1) \stackrel{?}{=} -3(-3)$$
$$9 = 9 \quad \text{True}$$

57.
$$0 = \frac{y}{8}$$
$$8 \cdot 0 = 8 \cdot \frac{y}{8}$$
$$0 = y$$
Check: $0 = \frac{y}{8}$
$$0 \stackrel{?}{=} \frac{0}{8}$$
$$0 = 0 \quad \text{True}$$

59.
$$-5 - 6 - 5x = 4$$
$$-11 - 5x = 4$$
$$-11 - 5x + 11 = 4 + 11$$
$$-5x = 15$$
$$\frac{-5x}{-5} = \frac{15}{-5}$$
$$x = -3$$
Check: $-5 - 6 - 5x = 4$
$$-5 - 6 - 5(-3) \stackrel{?}{=} 4$$
$$-11 + 15 \stackrel{?}{=} 4$$
$$4 = 4 \quad \text{True}$$

61.
$$-15 = -k$$
$$\frac{-15}{-1} = \frac{-k}{-1}$$
$$15 = k$$
Check: $-15 = -k$
$$-15 = -15 \quad \text{True}$$

63.
$$\frac{h}{-6} + 4 = 5$$
$$\frac{h}{-6} + 4 - 4 = 5 - 4$$
$$\frac{h}{-6} = 1$$
$$-6\left(\frac{h}{-6}\right) = -6 \cdot 1$$
$$h = -6$$
Check: $\frac{h}{-6} + 4 = 5$
$$\frac{-6}{-6} + 4 \stackrel{?}{=} 5$$
$$1 + 4 \stackrel{?}{=} 5$$
$$5 = 5 \quad \text{True}$$

65. $r-(-7)=-1-6$

$r+7=-7$

$r+7-7=-7-7$

$r=-14$

Check: $r-(-7)=-1-6$

$(-14)-(-7)\overset{?}{=}-1-6$

$-14+7\overset{?}{=}-7$

$-7=-7$ True

67. $-8=t-6$

$-8+6=t-6+6$

$-2=t$

Check: $-8=t-6$

$-8\overset{?}{=}-2-6$

$-8=-8$ True

69. $2x+3(0)=-6$

$2x=-6$

$\frac{2x}{2}=\frac{-6}{2}$

$x=-3$

Check: $2x+3(0)=-6$

$2(-3)+3(0)\overset{?}{=}-6$

$-6+0\overset{?}{=}-6$

$-6=-6$ True

71. $-x=8$

$\frac{-x}{-1}=\frac{8}{-1}$

$x=-8$

Check: $-x=8$

$-(-8)\overset{?}{=}8$

$8=8$ True

73. $0=y+9$

$0-9=y+9-9$

$-9=y$

Check: $0=y+9$

$0\overset{?}{=}(-9)+9$

$0=0$ True

75. $-1=-8+\frac{h}{-2}$

$-1+8=-8+\frac{h}{-2}+8$

$7=\frac{h}{-2}$

$-2\cdot 7=-2\left(\frac{h}{-2}\right)$

$-14=h$

Check: $-1=-8+\frac{h}{-2}$

$-1\overset{?}{=}-8+\frac{-14}{-2}$

$-1\overset{?}{=}-8+7$

$-1=-1$ True

77. $34=4-5x$

$34-4=4-5x-4$

$30=-5x$

$\frac{30}{-5}=\frac{-5x}{-5}$

$-6=x$

Check: $34=4-5x$

$34\overset{?}{=}4-5(-6)$

$34\overset{?}{=}4+30$

$34=34$ True

79. $t-4=-8-(-2)$

$t-4=-8+2$

$t-4=-6$

$t-4+4=-6+4$

$t=-2$

Check: $t-4=-8-(-2)$

$(-2)-4\overset{?}{=}-8-(-2)$

$-2-4\overset{?}{=}-8+2$

$-6=-6$ True

81. $5-t=500$

$5-t-5=500-5$

$-t=495$

$\frac{-t}{-1}=\frac{495}{-1}$

$t=-495$

Check: $5-t=500$

$5-(-495)\overset{?}{=}500$

$5+495\overset{?}{=}500$

$500=500$ True

83. $4=-3x+(-2)$

$4+2=-3x+(-2)+2$

$6=-3x$

$\frac{6}{-3}=\frac{-3x}{-3}$

$-2=x$

Check: $4=-3x+(-2)$

$4\overset{?}{=}-3(-2)+(-2)$

$4\overset{?}{=}6+(-2)$

$4=4$ True

85. $-2(4)=\frac{t}{-6}+1$

$-8=\frac{t}{-6}+1$

$-8-1=\frac{t}{-6}+1-1$

$-9=\frac{t}{-6}$

$-6(-9)=-6\left(\frac{t}{-6}\right)$

$54=t$

Check: $-2(4)=\frac{t}{-6}+1$

$-2(4)\overset{?}{=}\frac{54}{-6}+1$

$-8\overset{?}{=}-9+1$

$-8=-8$ True

87. $2y+8=-6$

$2y+8-8=-6-8$

$2y=-14$

$\frac{2y}{2}=\frac{-14}{2}$

$y=-7$

Check: $2y+8=-6$

$2(-7)+8\overset{?}{=}-6$

$-14+8\overset{?}{=}-6$

$-6=-6$ True

APPLICATIONS

89. **Analyze**

* The first observations were at a depth of -120 feet.

* The second set of observations were at a depth of -75 feet.

* How many feet was the cage raised?

Assign

Let x = the number of feet the cage was raised between observations.

Form

The key word *raised* suggests addition.

first depth + amount raised = second depth

-120 + x = -75

Solve

$-120+x=-75$

$-120+x+120=-75+120$

$x=45$

State

The shark cage was raised 45 feet.

Check

$-120\,ft+45\,ft=-75\,ft$

91. x = the profit in 2010

$-42+x=26$

$-42+x+42=26+42$

$x=68$

The company had a \$68 million profit in 2010.

93. x = the gain in points by the company

$-43+x=-9$

$-43+x+43=-9+43$

$x=34$

The company gained 34 points of market share.

95. x = the initial balance

$x+95+65=-15$

$x+160=-15$

$x+160-160=-15-160$

$x=-175$

The initial balance was -\$175.

97. x = the rushing yards of the Lions

$189+x=136$

$189+x-189=136-189$

$x=-53$

The Lions rushed for -53 yards that game.

99. x = the length of the Roman Empire

$-27+x=476$

$-27+x+27=476+27$

$x=503$

The Roman Empire lasted 503 years.

101. x = the income for 2011

$76+37+x+(-69)=-30$

$x+44=-30$

$x+44-44=-30-44$

$x=-74$

In 2011, Barnes and Noble lost \$74 million.

WRITING

103. It is not isolated because it still has the negative sign, which can be read as a coefficient of -1.

REVIEW

105. $(-4)(-4)(-4)(-4)(-4)=(-4)^5$

107. $\dfrac{0}{-17}=0$

109.
$$\begin{array}{r} \overset{9}{\cancel{1}}\overset{10}{\cancel{0}},\overset{9}{\overset{10}{\cancel{0}}}\overset{9}{\overset{10}{\cancel{0}}}\overset{10}{\cancel{0}} \\ -\quad 7\ 8\ 2 \\ \hline 9,2\ 1\ 8 \end{array}$$

111.
$$\begin{array}{r} \overset{1}{2}\overset{1}{3},\overset{1}{4}56 \\ 2,345 \\ 234 \\ +\quad 23 \\ \hline 26,058 \end{array}$$

Chapter 2 Review

1. $\{\ldots,-3,-2,-1,0,1,2,3,\ldots\}$

2. a. $-\$1,200$

b. $-10\,\text{sec}$

3. $-33\,ft$

4. a.

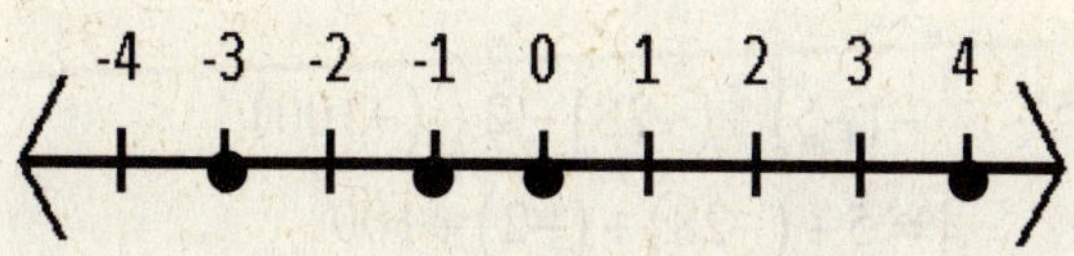

b.

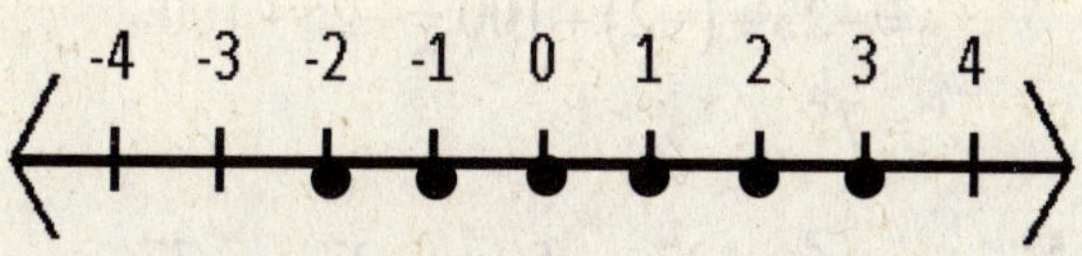

5. a. $0>-7$ b. $-20<-19$

6. a. false b. true

7. a. $|5|=5$ b. $|-43|=43$ c. $|0|=0$

8. a. -8 b. 8 c. 0

9. a. $-|12|=-12$ b. $-(-12)=12$

c. $-0=0$

10. a. negative b. the opposite of

c. negative d. minus

11.

Position	Player	Score to Par
1	Na Yeon Choi	-14
2	So Yeon Ryu	-12
3	Brittany Lincicome	-11
4	Karrie Webb	-10
5	Ai Miyazato	-9
6	Karine Incher	-8

12. a. 1998, \$60 billion

b. 2000, \$230 billion

c. 2009, -\$1,420 billion

13. $-6+(-4)=-10$

14. $-3+(-6)=-9$

15. $-28+60=32$

16. $93+(-20)=73$

17. $-8+8=0$

18. $73+(-73)=0$

19. $-1+(-4)+(-3)=-5+(-3)=-8$

20. $3+(-2)+(-4)=1+(-4)=-3$

21. $[7+(-9)]+(-4+16)$
$=-2+12=10$

22. $(-2+11)+[(-5)+4]$
$=9+(-1)=8$

23. $-4+0=-4$

24. $0+(-20)=-20$

25. $-2+(-1)+(-76)+1+2$
$=-2+2+(-1)+1+(-76)$
$=0+0+(-76)=-76$

26. $-5+(-31)+9+(-9)+5$
$=-5+5+(-31)+9+(-9)$
$=0+(-31)+0=-31$

27. $-102+73+(-345)$
$=-29+(-345)=-374$

28. $-59+3{,}187=3{,}128$

29. a. 11 b. -4

30. a. yes b. yes

c. no d. no

31. a. $-100\,ft$

b. $-100+16+18=-66\,ft$

32. $-129+263=134°F$

33. Subtracting an integer is the same as adding the opposite of that integer.

34. a. $-9-(-1)$

b. $-6-(-10)$

35. $5-8=5+(-8)=-3$

36. $-9-12=-9+(-12)=-21$

37. $-4-(-8)=-4+8=4$

38. $-8-(-2)=-8+2=-6$

39. $-6-106=-6+(-106)=-112$

40. $-7-1=-7+(-1)=-8$

41. $0-37=0+(-37)=-37$

42. $0-(-30)=0+30=30$

43. $12-2-(-6)=12+(-2)+6$
$=10+6=16$

44. $-16-9-(-1)=-16+(-9)+1$
$=-25+1=-24$

45. $-9-7+12=-9+(-7)+12$
$=-16+12=-4$

46. $-5-6+33=-5+(-6)+33$
$=-11+33=22$

47. $1-(2-7)=1-(-5)=1+5=6$

48. $-12-(6-10)=-12-(-4)=-12+4=-8$

49. $-70-[(-6)-2]=-70-(-8)$
$=-70+8=-62$

50. $89-[(-2)-12]=89-[-14]$
$=89+14=103$

51. $-(-5)+(-28)-2-(-100)$
$=5+(-28)+(-2)+100$
$=-23+(-2)+100=-25+100$
$=75$

52. a. $-50-27=-50+(-27)=-77$

b. $27-(-50)=27+50=77$

53. $-150-75=-150+(-75)=-225\,ft$

54. Alaska: $100-(-80)=100+80=180°F$

Virginia: $110-(-30)=110+30=140°F$

55. $28-(-16)=28+16=44$ points

56. $255-300-35$
$=255+(-300)+(-35)$
$=-45+(-35)=-\$80$

57. $7(-2)=-14$

58. $(-8)(47)=-376$

59. $(-23)(-14)=322$

60. $-5(-5)=25$

61. $-1\cdot 25=-25$

62. $(6)(-34)=-204$

63. $-4{,}000(17{,}000)=-68{,}000{,}000$

64. $-100{,}000(-300)=30{,}000{,}000$

65. $(-6)(-2)(-3)=(12)(-3)=-36$

66. $-4(-3)(-3)=12(-3)=-36$

67. $(-3)(4)(2)(-5)=-12(2)(-5)$
$=(-24)(-5)=120$

68. $(-1)(10)(10)(-1)=-10(10)(-1)$
$=-100(-1)=100$

69. $-15(-30)=450$

70. $-16(-3)=48$

71. **Tax Shortfall**

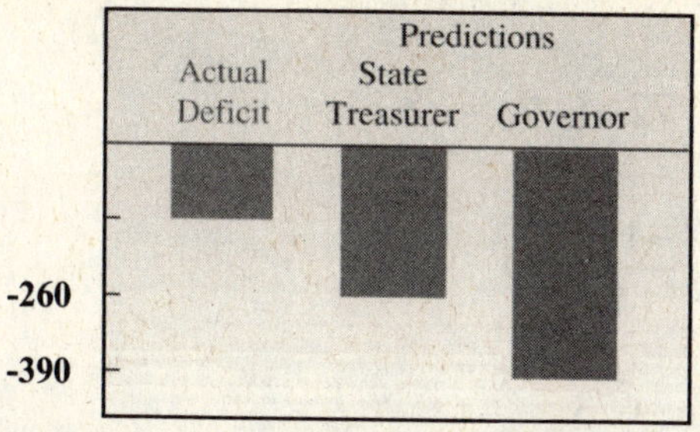

State Treasurer: $2(-130)=-\$260$

Governor: $3(-130)=-\$390$

72. $12(-45)=-540\,ft$

73. $(-5)^3=(-5)(-5)(-5)=25(-5)=-125$

74. $(-2)^5=(-2)(-2)(-2)(-2)(-2)$
$=4(4)(-2)=16(-2)=-32$

75. $(-8)^4=(-8)(-8)(-8)(-8)$
$=64(64)=4{,}096$

76. $(-4)^4=(-4)(-4)(-4)(-4)$
$=16(16)=256$

77. Since the exponent is odd, the result will be negative.

78. In the first instance the base is 9.
$-9^2=-9\cdot 9=-81.$

In the second instance the base is -9.
$(-9)^2=(-9)(-9)=81$

79. We know that $\dfrac{-15}{5}=-3$ because $-3(5)=-15$.

80. $18(-8)=-144\neq -152$ The answer is incorrect.

81. $\dfrac{25}{-5}=-5$

82. $\dfrac{-14}{7}=-2$

83. $(-64)\div(-8)=8$

84. $72\div(-9)=-8$

85. $\dfrac{-10}{-1}=10$

86. $\dfrac{-673}{-673}=1$

87. $-150{,}000\div 3{,}000=-50$

88. $-24{,}000\div(-60)=400$

89. $\dfrac{-1{,}058}{-46}=23$

90. $-272\div 16=-17$

91. $\dfrac{0}{-5}=0$

92. $\dfrac{-4}{0}$ is undefined.

93. $-96\div 3=-32$

94. $-125\div(-25)=5$

95. $-12\div 6=-2\,\text{min}$

96. $-28{,}374\div 6=-4{,}729\,\textit{ft}$

97. $7-8\cdot 6=7-48=-41$

98. $7-(-2)^2+1=7-4+1$
$=3+1=4$

99. $65-8(9)-(-47)=65-72+47$
$=-7+47=40$

100. $-3(-2)^3-16=-3(-8)-16$
$=24-16=8$

101. $-2(5)(-4)+\dfrac{|-9|}{3^2}=-10(-4)+\dfrac{9}{9}$
$=40+1=41$

102. $-4^2+(-4)^2=-16+16=0$

103. $-12-(8-9)^2=-12-(-1)^2$
$=-12-1=-13$

104. $7|-8|-2(3)(4)=7\cdot 8-6(4)$
$=56-24=32$

105. $-4\left(\dfrac{15}{-3}\right)-2^3=-4(-5)-8$
$=20-8=12$

106. $-20+2(12-5\cdot 2)=-20+2(12-10)$
$=-20+2(2)=-20+4=-16$

107. $-20+2\left[12-(-7+5)^2\right]$
$=-20+2\left[12-(-2)^2\right]=-20+2[12-4]$
$=-20+2[8]=-20+16=-4$

108. $8-6|-3\cdot 4+5|=8-6|-12+5|$
$=8-6|-7|=8-6\cdot 7=8-42=-34$

109. $\dfrac{2\cdot 5+(-6)}{-3-1^5}=\dfrac{10+(-6)}{-3-1}=\dfrac{4}{-4}=-1$

110. $\dfrac{3(-6)-11+1}{4^2-3^2}=\dfrac{-18-11+1}{16-9}$
$=\dfrac{-29+1}{7}=\dfrac{-28}{7}=-4$

111. $-\left[1-\left(2^3+\dfrac{100}{-50}\right)\right]$
$=-\left[1-(8+(-2))\right]$
$=-[1-6]=-[-5]=5$

112. $-\left[45-\left(5^3+\dfrac{100}{-4}\right)\right]$
$=-\left[45-(125+(-25))\right]$
$=-\left[45-(100)\right]=-[-55]=55$

113. $-4,500+7,900+2,100+(-3,200)$
$=3,400+2,100+(-3,200)$
$=5,500+(-3,200)=2,300$

114. $\dfrac{-8+4+7+(-11)+2+0+(-6)+(-4)}{8}$
$=\dfrac{-16}{8}=-2$

115. $x+(-16)=-6$
$x+(-16)+16=-6+16$
$x=10$
Check: $x+(-16)=-6$
$10+(-16)\stackrel{?}{=}-6$
$-6=-6$ True

116. $-y=-32$
$\dfrac{-y}{-1}=\dfrac{-32}{-1}$
$y=32$
Check: $-y=-32$
$-32=-32$ True

117. $9x-27-0$
$9x-27+27=0+27$
$9x=27$
$\dfrac{9x}{9}=\dfrac{27}{9}$
$x=3$
Check: $9x-27=0$
$9\cdot3-27\stackrel{?}{=}0$
$27-27\stackrel{?}{=}0$
$0=0$ True

118. $-20+4=a+3(-7)$
$-16=a-21$
$-16+21=a-21+21$
$5=a$
Check: $-20+4=a+3(-7)$
$-20+4\stackrel{?}{=}5+3(-7)$
$-16\stackrel{?}{=}5-21$
$-16=-16$ True

119. $-84=-2t$
$\dfrac{-84}{-2}=\dfrac{-2t}{-2}$
$42=t$
Check: $-84=-2t$
$-84\stackrel{?}{=}-2\cdot42$
$-84=-84$ True

120. $\dfrac{n}{5}+(-2)=-7$
$\dfrac{n}{5}+(-2)+2=-7+2$
$\dfrac{n}{5}=-5$
$5\cdot\dfrac{n}{5}=5(-5)$
$n=-25$
Check: $\dfrac{n}{5}+(-2)=-7$
$\dfrac{-25}{5}+(-2)\stackrel{?}{=}-7$
$-5+(-2)\stackrel{?}{=}-7$
$-7=-7$ True

121. $\frac{s}{3}=-1$

$3\cdot\frac{s}{3}=3(-1)$

$s=-3$

Check: $\frac{s}{3}=-1$

$\frac{-3}{3}\overset{?}{=}-1$

$-1=-1$ True

122. $-11y-10=67$

$-11y-10+10=67+10$

$-11y=77$

$\frac{-11y}{-11}=\frac{77}{-11}$

$y=-7$

Check: $-11y-10=67$

$-11(-7)-10\overset{?}{=}67$

$77-10\overset{?}{=}67$

$67=67$ True

123. $16-6n=22$

$-16+16-6n=-16+22$

$-6n=6$

$\frac{-6n}{-6}=\frac{6}{-6}$

$n=-1$

Check: $16-6n=22$

$16-6(-1)\overset{?}{=}22$

$16+6\overset{?}{=}22$

$22=22$ True

124. $\frac{n}{-2}-27=-27$

$\frac{n}{-2}-27+27=-27+27$

$\frac{n}{-2}=0$

$-2\cdot\frac{n}{-2}=-2\cdot 0$

$n=0$

Check: $\frac{n}{-2}-27=-27$

$\frac{0}{-2}-27\overset{?}{=}-27$

$0-27\overset{?}{=}-27$

$-27=-27$ True

125. $-9-3-20x=-52$

$-12-20x=-52$

$-12+12-20x=-52+12$

$-20x=-40$

$\frac{-20x}{-20}=\frac{-40}{-20}$

$x=2$

Check: $-9-3-20x=-52$

$-9-3-20\cdot 2\overset{?}{=}-52$

$-9-3-40\overset{?}{=}-52$

$-52=-52$ True

126. $15=-13b+(-11)$

$15+11=-13b+(-11)+11$

$26=-13b$

$\frac{26}{-13}=\frac{-13b}{-13}$

$-2=b$

Check: $15=-13b+(-11)$

$15\overset{?}{=}-13(-2)+(-11)$

$15\overset{?}{=}26+(-11)$

$15=15$ True

127. x = the income for 2010

$74+x+33=65$

$x+107=65$

$x+107-107=65-107$

$x=-42$

Dole Food Company lost \$42 million in 2010.

128. x = the gain in points

$-32+x=-3$

$-32+32+x=-3+32$

$x=29$

She gained 29 points in the 8 weeks.

Chapter 2 Test

1. a. $\{\ldots,-5,-4,-3,-2,-1,0,1,2,3,4,5,\ldots\}$ is called the set of integers.

b. The symbols > and < are called inequality symbols.

c. The absolute value of a number is the distance between the number and 0 on the number line.

d. Two numbers that are the same distance from 0 on the number line but on opposite sides of it, are called opposites.

e. In the expression $(-3)^5$, the base is -3 and 5 is the exponent.

f. To solve an equation means to find all the values of the variable that make the equation true.

g. To check the solution of an equation, we substitute the value for the variable in the original equation and determine whether the result is a true statement.

2. a. $-8>-9$ b. $-213<123$ c. $-5<0$

3. a. true b. true c. false d. false e. true

4. Poly will have the greatest shortage.

5.

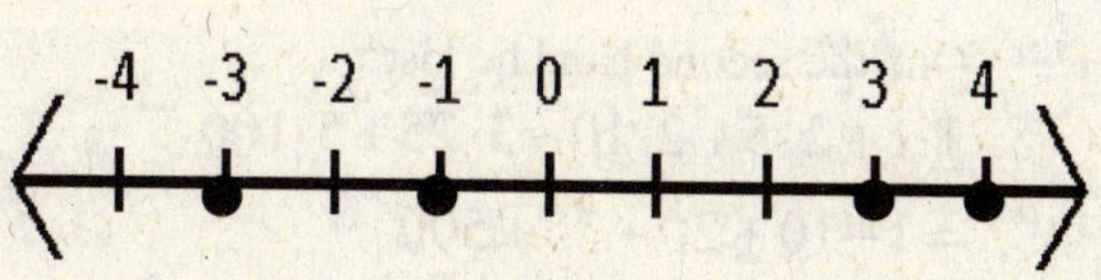

6. a. $-16+14=-2$

b. $-72+(-73)=-145$

c. $8+(-6)+(-9)+5+1$
$=2+(-9)+5+1$
$=-7+5+1$
$=-2+1$
$=-1$

d. $(-31+12)+[3+(-16)]$
$=-19+[-13]$
$=-32$

e. $-24+(-3)+24+(-5)+5$
$=-24+24+(-3)+(-5)+5$
$=0+(-3)+0$
$=-3$

7. a. $-7-6=-7+(-6)=-13$

b. $-7-(-6)=-7+6=-1$

c. $82-(-109)=82+109=191$

d. $0-15=0+(-15)=-15$

e. $-29-(-60)-71=-29+60+(-71)=-40$

8. a. $-10\cdot 7=-70$

b. $(-4)(-8)=32$

c. $-4(2)(-6)=-8(-6)=48$

d. $(-9)(-3)(-1)(-2)=27\cdot 2=54$

e. $-20{,}000(1{,}300)=-26{,}000{,}000$

9. $\frac{-20}{-4}=5 \Leftrightarrow 5(-4)=-20$

10.

a. $\frac{-32}{4}=-8$

$-8\cdot 4=-32$

b. $24\div(-3)=-8$

$-8(-3)=24$

c. $-54\div(-6)=9$

$9(-6)=-54$

d. $\frac{408}{-12}=-34$

$-34(-12)=408$

e. $-560{,}000\div 7{,}000=-80$

$-80\cdot 7{,}000=-560{,}000$

11.

a. $-27+15=-12$

b. $-1-(-19)=-1+19=18$

c. $\frac{-28}{-7}=\frac{28}{7}=4$

d. $10(-8)=-80$

12.

a. Commutative Property of Addition

b. Commutative Property of Multiplication

c. Subtracting is the same as <u>adding</u> the opposite.

13.

a. $\frac{-21}{0}$ is undefined

b. $\frac{-5}{1}=-5$

c. $\frac{0}{-6}=0$

d. $\frac{-18}{-18}=1$

14.

a. $(-4)^2=(-4)(-4)=16$

b. $-4^2=-4\cdot 4=-16$

15. $4-(-3)^2-(-6)=4-9+6$

$=-5+6=1$

16. $-18\div 2\cdot 3=-9\cdot 3=-27$

17. $-3+\left(\frac{-16}{4}\right)-3^3=-3+(-4)-27$

$=-7-27=-34$

18. $94-3\left[-7+(5-8)^2\right]=94-3\left[-7+(-3)^2\right]$

$=94-3[-7+9]=94-3[2]=94-6=88$

19. $\frac{4(-6)-4^2+(-2)}{-3-4\cdot 1^5}=\frac{-24-16+(-2)}{-3-4\cdot 1}$

$=\frac{-40+(-2)}{-3-4}=\frac{-42}{-7}=6$

20. $6(-2\cdot 6+5\cdot 4)=6(-12+20)$

$=6(8)=48$

21. $21-9|-3-4+2|=21-9|-5|$

$=21-9\cdot 5=21-45=-24$

22. $-\left[2-\left(4^3+\frac{20}{-5}\right)\right]=-\left[2-(64+(-4))\right]$

$=-[2-60]=-[-58]=58$

23. $-6\cdot 12=-72°F$

24. On the first hand he won

$3\cdot 1+2\cdot 5+4\cdot 10+2\cdot 25+3\cdot 100$

$=3+10+40+50+300$

$=\$403.$

One the second hand he lost

$1\cdot 1+2\cdot 5+2\cdot 10+3\cdot 25+5\cdot 100$

$=1+10+20+75+500$

$=\$606.$

Overall: $\$403-\$606=-\$203$ (a loss).

25. $436-282=154\,ft$ difference in elevations.

26. $-5{,}250\div 15=-350\,ft$

27. In this hand he has $3+10+9+1=23$ points.

His new score is $8-23=-15$ points.

28. $\dfrac{-60+(-100)+(-20)}{3}=\dfrac{-180}{3}=-60$

Each investor is responsible for \$60 million of debt.

29.
$$\frac{x}{-4}=10$$
$$-4\cdot\frac{x}{-4}=-4\cdot 10$$
$$x=-40$$
Check: $\dfrac{x}{-4}=10$
$$\frac{-40}{-4}\stackrel{?}{=}10$$
$$10=10 \quad \text{True}$$

30.
$$-10=6-x$$
$$-10-6=6-x-6$$
$$-16=-x$$
$$\frac{-16}{-1}=\frac{-x}{-1}$$
$$16=x$$
Check: $-10=6-x$
$$-10\stackrel{?}{=}6-16$$
$$-10=-10 \quad \text{True}$$

31.
$$c-(-7)=-8$$
$$c+7=-8$$
$$c+7-7=-8-7$$
$$c=-15$$
Check: $c-(-7)=-8$
$$-15-(-7)\stackrel{?}{=}-8$$
$$-15+7\stackrel{?}{=}-8$$
$$-8=-8 \quad \text{True}$$

32.
$$-6x=0$$
$$\frac{-6x}{-6}=\frac{0}{-6}$$
$$x=0$$
Check: $-6x=0$
$$-6\cdot 0\stackrel{?}{=}0$$
$$0=0 \quad \text{True}$$

33.
$$3x+(-7)=-11+(-11)$$
$$3x+(-7)+7=-11+(-11)+7$$
$$3x=-15$$
$$\frac{3x}{3}=\frac{-15}{3}$$
$$x=-5$$
Check: $3x+(-7)=-11+(-11)$
$$3(-5)+(-7)\stackrel{?}{=}-11+(-11)$$
$$-15+(-7)\stackrel{?}{=}-11+(-11)$$
$$-22=-22 \quad \text{True}$$

34.
$$-a=38$$
$$\frac{-a}{-1}=\frac{38}{-1}$$
$$a=-38$$
Check: $-a=38$
$$-(-38)\stackrel{?}{=}38$$
$$38=38 \quad \text{True}$$

35.
$$-5=-6s+7$$
$$-5-7=-6s+7-7$$
$$-12=-6s$$
$$\frac{-12}{-6}=\frac{-6s}{-6}$$
$$2=s$$
Check: $-5=-6s+7$
$$-5\stackrel{?}{=}-6\cdot 2+7$$
$$-5\stackrel{?}{=}-12+7$$
$$-5=-5 \quad \text{True}$$

36. $\frac{x}{-2}+3=(-2)(-6)$

$\frac{x}{-2}+3-3=12-3$

$\frac{x}{-2}=9$

$-2\cdot\frac{x}{-2}=-2\cdot 9$

$x=-18$

Check: $\frac{x}{-2}+3=(-2)(-6)$

$\frac{-18}{-2}+3\stackrel{?}{=}(-2)(-6)$

$9+3\stackrel{?}{=}12$

$12=12$ True

37. x = the balance before the deposit

$x+125+100=-19$

$x+225=-19$

$x+225-225=-19-225$

$x=-244$

Her previous balance was -$244.

38. x = the weight of the 2nd floor people

$-165+x=85$

$-165+165+x=85+165$

$x=250$

The weight of the people on the second floor was 250 pounds.

39. x = the storage temperature

$9x=-360$

$\frac{9x}{9}=\frac{-360}{9}$

$x=-40$

Dippin' Dots are stored at $-40°$ F.

40. x = the net income in 2007

$\frac{x}{10}=-3$

$10\cdot\frac{x}{10}=-3\cdot 10$

$x=-30$

The net income for Sprint Nextel in 2007 was -$30 billion

Chapters 1 – 2 Cumulative Review

1. a. 7 millions b. 3

c. 7,326,500 d. 7,330,000

2. The lowest bidder is CRF Cable.

3.

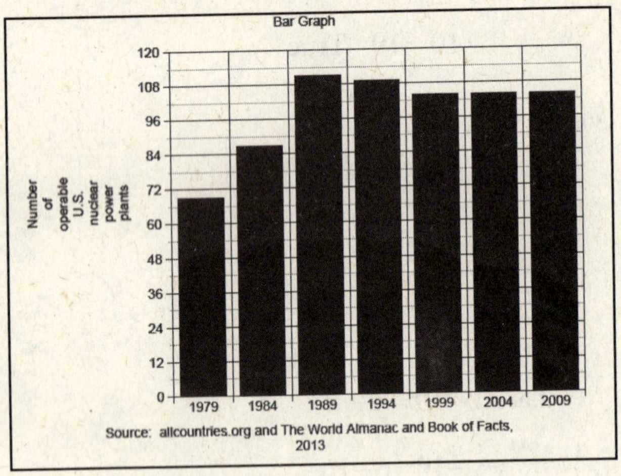

4. $180+180=360$

5.

$$\begin{array}{r} {\scriptstyle 1\,2} \\ 1,237 \\ 549 \\ +\ 68 \\ \hline 1,854 \end{array}$$

6.

$$\begin{array}{r} {\scriptstyle 2\,2\,1\,2} \\ 8,907 \\ 2,345 \\ 7,899 \\ +\ 5,237 \\ \hline 24,388 \end{array}$$

7.
$$\begin{array}{r} {\scriptstyle 5\ \ 12\ 6\ 15} \\ \not{6},\not{3}\not{7}\not{5} \\ -2,569 \\ \hline 3,806 \end{array}$$

8.
$$\begin{array}{r} {\scriptstyle 12} \\ {\scriptstyle 4\ \ 13\ 16} \\ \not{5},\not{3}\not{6}9 \\ -\ 685 \\ \hline 4,684 \end{array}$$

9.
$$\begin{array}{r} {\scriptstyle 14\ \ 9} \\ {\scriptstyle 8\ \ 15\ 10\ 16} \\ 3\not{9},\not{5}\not{0}\not{6} \\ -1,729 \\ \hline 37,777 \end{array}$$

10.
$$\begin{array}{r} 1,736 \\ -304 \\ \hline 1,432 \end{array}$$

11. $132+237=369$ It is incorrect.

12.
$$\begin{aligned} &147-27-55 \\ &=120-55 \\ &=65 \end{aligned}$$

65 wooden chairs

13.
$$\begin{array}{r} 435 \\ \times 27 \\ \hline 3,045 \\ +8,700 \\ \hline 11,745 \end{array}$$

14.
$$\begin{array}{r} 9,183 \\ \times 602 \\ \hline 18,366 \\ 00 \\ +5,509,800 \\ \hline 5,528,166 \end{array}$$

15. $31\cdot 7=217$

attaching 5 zeros gives $21,700,000$

16. $3\cdot 24\cdot 12=72\cdot 12=864$ tennis balls

17.
$$\begin{aligned} P&=17+35+17+35 \\ &=52+52=104\,ft \\ A&=17\cdot 35=595\,ft^2 \end{aligned}$$

18.

a. $5\cdot 5=25$ pixels

b. $12\cdot 12=144$ pixels

c. $100\cdot 100=10,000$ pixels

19.
$$\begin{array}{r} 87 \\ 8\overline{)701} \\ \underline{-64} \\ 61 \\ \underline{-56} \\ 5 \end{array}$$

$87R5$

20.
$$\begin{array}{r} 13 \\ 97\overline{)1261} \\ \underline{-97} \\ 291 \\ \underline{-291} \\ 0 \end{array}$$

21.
$$\begin{array}{r} 467 \\ 38\overline{)17746} \\ \underline{-152} \\ 254 \\ \underline{-228} \\ 266 \\ \underline{-266} \\ 0 \end{array}$$

22. Start by dividing out a zero from the divisor and dividend.

$$\begin{array}{r} 28 \\ 35\overline{)980} \\ \underline{-70} \\ 280 \\ \underline{-280} \\ 0 \end{array}$$

23. $218 \cdot 9 = 1{,}962$ It is correct.

24.

$$\begin{array}{r} 10 \\ 30\overline{)320} \\ \underline{-30} \\ 20 \end{array}$$

$10R20$

It can be filled 10 times, with 20 ounces left over.

25. $15 \cdot 16 = 240\ oz$

$240 \div 4 = 60$

He can make 60 dinner rolls.

26. 1, 2, 3, 6, 9, 18

27. a. prime, odd

b. composite, even

c. neither, even

d. neither, odd

28. $504 = 72 \cdot 7 = 8 \cdot 9 \cdot 7$

$= 2 \cdot 4 \cdot 3 \cdot 3 \cdot 7 = 2 \cdot 2 \cdot 2 \cdot 3 \cdot 3 \cdot 7$

$= 2^3 \cdot 3^2 \cdot 7$

29. $11 \cdot 11 \cdot 11 \cdot 11 = 11^4$

30. $5^2 \cdot 7 = 25 \cdot 7 = 175$

31. $8 = 2 \cdot 2 \cdot 2$

$12 = 2 \cdot 2 \cdot 3$

$LCM(8,12) = 2 \cdot 2 \cdot 2 \cdot 3 = 24$

32. $3 = 3$

$6 = 2 \cdot 3$

$15 = 3 \cdot 5$

$LCM(3,6,15) = 2 \cdot 3 \cdot 5 = 30$

33. $30 = 2 \cdot 3 \cdot 5$

$48 = 2 \cdot 2 \cdot 2 \cdot 2 \cdot 3$

$GCF(30,48) = 2 \cdot 3 = 6$

34. $81 = 3 \cdot 3 \cdot 3 \cdot 3$

$108 = 2 \cdot 2 \cdot 3 \cdot 3 \cdot 3$

$162 = 2 \cdot 3 \cdot 3 \cdot 3 \cdot 3$

$GCF(81,108,162) = 3 \cdot 3 \cdot 3 = 27$

35. $4 - 2[26 + 2(5-3)] = 4 - 2[26 + 2 \cdot 2]$

$= 4 - 2[26 + 4] = 4 - 2 \cdot 30$

$= 4 - 60 = -56$

36. $264 \div 4 - 7(4)2 = 66 - 28 \cdot 2$

$= 66 - 56 = 10$

37. $\dfrac{4^2 - 2 \cdot 3}{2 + (3^2 - 3 \cdot 2)} = \dfrac{16-6}{2+9-6} = \dfrac{10}{5} = 2$

38. $\dfrac{38 + 42 + 36 + 38 + 48 + 44}{6} = \dfrac{246}{6}$

$= 41 mph$

39. $x - 2 = 4$

$6 - 2 \stackrel{?}{=} 4$

$4 = 4$

4 is a solution.

40. a. no b. yes c. no d. no

41. $50 = x + 37$

$50 - 37 = x + 37 - 37$

$13 = x$

Check: $50 = x + 37$

$50 \stackrel{?}{=} 13 + 37$

$50 = 50$

The solution checks.

42. $a-12=41$

$a-12+12=41+12$

$a=53$

Check: $a-12=41$

$53-12\overset{?}{=}41$

$41=41$ True

43. $5p=135$

$\frac{5p}{5}=\frac{135}{5}$

$p=27$

Check: $5p=135$

$5\cdot 27\overset{?}{=}135$

$135=135$ True

44. $\frac{y}{8}=3$

$8\cdot\frac{y}{8}=8\cdot 3$

$y=24$

Check: $\frac{y}{8}=3$

$\frac{24}{8}\overset{?}{=}3$

$3=3$ True

45. x = the number of Dunkin' Donuts shops

$x+26{,}999=37{,}199$

$x+26{,}999-26{,}999=37{,}199-26{,}999$

$x=10{,}200$

There are 10,200 Dunkin' Donuts shops.

46. x = the capacity of Sun Life Stadium

$2x=150{,}000$

$\frac{2x}{2}=\frac{150{,}000}{2}$

$x=75{,}000$

Sun Life Stadium has a capacity of 75,000.

47. a.

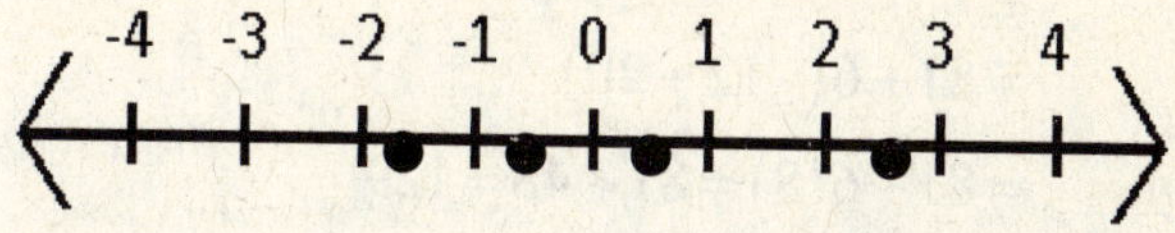

b.

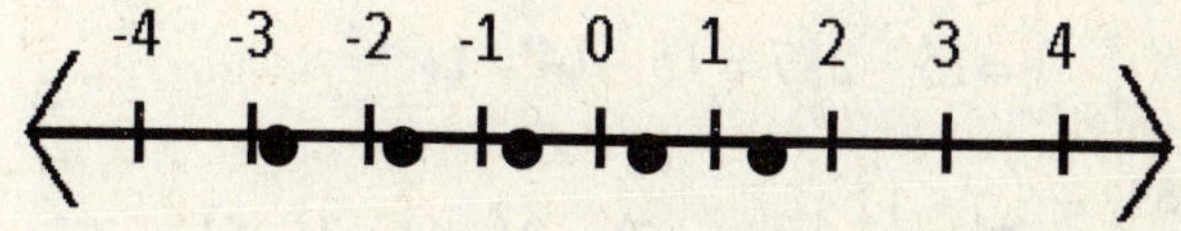

48. $-11+20+(-13)+1=9+(-13)+1$

$=-4+1=-3$

49. $3-(-18)=3+18=21$

50. $48-105-22=48+(-105)+(-22)$

$=-57+(-22)=-79$

The new account balance will be -\$79.

51. $7(-39)=-273$

The melting point of helium is -273 $°C$.

52. $-660{,}000\div 12=-55{,}000$

$\frac{-660{,}000}{12}=-55{,}000$

Each investor was responsible for \$55,000 of debt.

53. $5+(-3)(-7)(-2)=5+21(-2)$

$=5+(-42)=-37$

54. $-2\left[-6\left(5\cdot 1^3\right)-5\right]=-2\left[-6(5)-5\right]$

$=-2[-30-5]=-2[-35]=70$

55. $\frac{10-(-5)}{1-2\cdot 3}=\frac{10+5}{1-6}=\frac{15}{-5}=-3$

56. $\frac{3(-6)-10}{3^2-4^2}=\frac{-18-10}{9-16}$

$=\frac{-28}{-7}=4$

57. $3^4+6(-12+5\cdot 4)$

$=81+6(-12+20)$

$=81+6(8)=81+48=129$

58. $15-2|-3-4|=15-2|-7|$

$=15-2\cdot 7=15-14=1$

59. $2\left(\frac{-12}{3}\right)+3(-5)=2(-4)+3(-5)$

$=-8+(-15)=-23$

60. $-9^2+(-9)^2=-9\cdot 9+(-9)(-9)$

$=-81+81=0$

61. $-\left|\frac{45}{-9}-(-9)\right|=-|-5+9|$

$=-|4|=-4$

62. $\frac{-4(-5)-2}{3-3^2}=\frac{20-2}{3-9}=\frac{18}{-6}=-3$

63. $\frac{-1,200}{12}\approx -100\,ft$

64. $20,000\cdot(-\$200)=\$4,000,000$ loss

65. $-1-h=-8$

$-1-h+1=-8+1$

$-h=-7$

$\frac{-h}{-1}=\frac{-7}{-1}$

$h=7$

Check: $-1-h=-8$

$-1-7\stackrel{?}{=}-8$

$-8=-8$ True

66. $8q-7=25$

$8q-7+7=25+7$

$8q=32$

$\frac{8q}{8}=\frac{32}{8}$

$q=4$

Check: $8q-7=25$

$8\cdot 4-7\stackrel{?}{=}25$

$32-7\stackrel{?}{=}25$

$25=25$ True

67. $-1+7=x+2(-9)$

$6=x-18$

$6+18=x-18+18$

$24=x$

Check: $-1+7=x+2(-9)$

$-1+7\stackrel{?}{=}24+2(-9)$

$6\stackrel{?}{=}24-18$

$6=6$ True

68. $\frac{r}{2}+5=-13$

$\frac{r}{2}+5-5=-13-5$

$\frac{r}{2}=-18$

$2\cdot\frac{r}{2}=2(-18)$

$r=-36$

Check: $\frac{r}{2}+5=-13$

$\frac{-36}{2}+5\stackrel{?}{=}-13$

$-18+5\stackrel{?}{=}-13$

$-13=-13$ True

69. x = the original balance

$x+255+395=-85$

$x+650=-85$

$x+650-650=-85-650$

$x=-735$

The original balance was -$735.

70. et x = the amount it must be heated

$-173+x=173$

$-173+173+x=173+173$

$x=346$

It must be heated $346°F$.

CHAPTER 3 The Language of Algebra

Section 3.1: Algebraic Expressions

VOCABULARY

1. A variable is a letter (or symbol) that stands for a number.

3. Phrases such as *increased by* and *more than* indicate the operation of addition. Phrases such as *decreased by* and *less than* indicate the operation of subtraction.

CONCEPTS

5. a. $10+x$

b. $2-3t$

7. a. ii

b. iii

c. iv

d. i

9. $(12-h)in$

11.

Number of Decades	Number of Years
1	10
2	20
3	30
d	$10d$

We multiply the number of decades by 10 to find the number of years.

NOTATION

13. a. $8x$

b. $5t$

c. $\frac{10}{g}$

GUIDED PRACTICE

15. $l+15$

17. $50x$

19. $\frac{w}{l}$

21. $P+\frac{2}{3}p$

23. $k^2-2{,}005$

25. $2a-1$

27. $\frac{1{,}000}{n}$

29. $2p+90$

31. $3(35+h+300)$

33. $p-680$

35. $4d-15$

37. $2(200+t)$

39. $|a-2|$

41. $\frac{1}{10}d$ or $0.1d$

43. $f-2$

45. $6s$

47. $\frac{p}{15}$

49. $t+2$

51. $\frac{h}{4}$

53. $450-x$

55. w = the width of the rectangle (in inches)
$w+6$ = the length of the rectangle (in inches)

57. q = the number of quarts of coolant originally
$q-3$ = the number of quarts of coolant remaining

59. v = the area of Vermont (in square miles)
$50v + 380$ = the area of Alaska (in square miles)

61. c = number of calories in a scoop of ice cream
$2c + 100$ = number of calories in a slice of pie

63. a. $b+30$

b. $e-30$

65. a. $s+11$

b. $w-11$

67. x = the age of the ATM
$x + 11$ = the age of the digital clock
$x - 15$ = the age of the camcorder

69. E = the age of the Empire State Building
$E + 18$ = the age of the Woolworth Building
$E - 21$ = the age of the United Nations Building

71. $60m$ sec

73. $12f$ *in*

75. $\dfrac{y}{100}$ centuries

77. $\dfrac{e}{12}$ dozens

TRY IT YOURSELF

79. three-fourths of r

81. 50 less than t

83. the product of $x, y,$ and z

85. twice m, increased by 5

87. a. $7x$ hours

b. $365x$ hours

89. a. $\dfrac{s}{12}$ dollars b. $\dfrac{s}{52}$ dollars

APPLICATIONS

91. n = the number of votes for Nixon
$n + 118,500$ = the number of votes for Kennedy

93. x = the age of Apple
$x + 80$ = IBM's age
$x - 9$ = Dell's age

95.

Wind Conditions	Speed of Jet (mph)
In still air	500
With the tail wind	$500 + x$
Against the head wind	$500 - x$

97. $5,000 - x$

WRITING

99. Variables are used to represent an unknown quantity.

101. The variable should represent the height of the building.

REVIEW

103. $-5+(-6)+1=-11+1=-10$

105. $-x = 4$

$$\frac{-x}{-1} = \frac{4}{-1}$$

$x = -4$

107. $(-1-2^2\cdot5)3^2 = (-1-4\cdot5)3^2$
$= (-1-20)3^2 = (-21)(9) = -189$

109. $-3-2=-3+(-2)=-5$

Section 3.2: Evaluating Algebraic Expressions and Formulas

VOCABULARY

1. An algebraic expression is a combination of variables, numbers, and the operation symbols for addition, subtraction, multiplication, and division.

3. To evaluate $a^2 + 10a + 1$ for $a = -3$ we substitute -3 for a and apply the order of operations rule.

5. Temperature can be measured using the Fahrenheit or Celsius scale.

CONCEPTS

7. a. $s = p - d$

b. $p = r - c$

c. $r = c + m$

NOTATION

9. $$\begin{aligned} 9a - a^2 &= 9(5) - (5)^2 \\ &= 9(5) - 25 \\ &= 45 - 25 \\ &= 20 \end{aligned}$$

GUIDED PRACTICE

11. a. x = the length of part 1
$x - 40$ = the length of part 2
$x + 16$ = the length of part 3

b $x = 60$ inches (part 1)
$x - 40 = 60 - 40 = 20$ inches (part 2)
$x + 16 = 60 + 16 = 76$ inches (part 3)

13. a. x = the weight of the Nissan Leaf
$2x - 340$ = the weight of the H2
$x - 1{,}720$ = the weight of a Smart Car

b. $x = 3{,}370$ lbs (Nissan Leaf)
$2x - 340 = 2(3{,}370) - 340 = 6400$ lbs (H2)
$x - 1{,}720 = 3{,}370 - 1{,}720 = 1{,}650$ lbs(Smart Car)

15. $10x - 3 = 10(3) - 3 = 30 - 3 = 27$

17. $\dfrac{-n-1}{3} = \dfrac{-11-1}{3} = \dfrac{-12}{3} = -4$

19. $$\begin{aligned} 3x^2 - 2x &= 3(-2)^2 - 2(-2) = 3 \cdot 4 + 4 \\ &= 12 + 4 = 16 \end{aligned}$$

21. $$\begin{aligned} -y + 3(1+y) &= -(-10) + 3[1 + (-10)] \\ &= 10 + 3(-9) = 10 - 27 = -17 \end{aligned}$$

23. $h^3 - 24 = (-3)^3 - 24 = -27 - 24 = -51$

25. $n^4 + n^2 = (-1)^4 + (-1)^2 = 1 + 1 = 2$

27. $$\begin{aligned} (2ab + 4b)^2 &= (2(-5)(2) + 4(2))^2 \\ &= (-20 + 8)^2 = (-12)^2 = 144 \end{aligned}$$

29. $$\begin{aligned} |-6r - 8s| &= |-6(-11) - 8(9)| \\ &= |66 - 72| = |-6| = 6 \end{aligned}$$

31. $r - c = p$
$\$200 - \$35 = \$165$
The sale price of the skis is \$165.

33. $c + m = r$
$\$125 + \$65 = \$190$
The retail price of the pants is \$190.

35. $r - c = p$
$\$445 - \$295 = \$150$
The profit is \$150.

37. $r - c = p$
$\$13{,}500 - \$5{,}300 = \$8{,}200$
The profit is \$8,200.

39. $d = r \cdot t = 550 \cdot 3 = 1{,}650 mi$
The jet covers 1,650 miles.

41. $d = r \cdot t = 12 \cdot 8 = 96 mi$
The hiker travels 96 miles.

43. $$C = \frac{5(F-32)}{9} = \frac{5(59-32)}{9} = \frac{5(27)}{9}$$
$$= 5 \cdot 3 = 15°C$$

45. $C=\dfrac{5(F-32)}{9}=\dfrac{5(-4-32)}{9}=\dfrac{5(-36)}{9}$
$=5(-4)=-20°C$

47. $d=16t^2=16\cdot 2^2=16\cdot 4=64\,ft$
The ball has fallen 64 feet.

49. $d=16t^2=16\cdot 4^2=16\cdot 16=256\,ft$
He was working 256 feet above the ground.

51. $\dfrac{254+225+238}{3}=\dfrac{717}{3}=239$
The bowler's mean score is 239.

53. $\dfrac{23+18+\cdots 25}{8}=\dfrac{200}{8}=25lb$
The mean weight of the fish is 25 pounds.

TRY IT FOR YOURSELF

55. $\dfrac{-x-8}{2}=\dfrac{-4-8}{2}=\dfrac{-12}{2}=-6$

57. $-p=-(-4)=4$

59. $2(p+9)+2p$
$=2((-12)+9)+2(-12)$
$=2(-3)-24$
$=-6-24=-30$

61. $x^2-x-7=(-5)^2-(-5)-7$
$=25+5-7=23$

63. $\dfrac{x-y}{a-b}=\dfrac{-1-8}{6-3}=\dfrac{-9}{3}=-3$

65. $\dfrac{-b^2+3b}{2b+1}=\dfrac{-(-4)^2+3(-4)}{2(-4)+1}$
$=\dfrac{-16-12}{-8+1}=\dfrac{-28}{-7}=4$

67. $\dfrac{24+k}{3k}=\dfrac{24+3}{3\cdot 3}=\dfrac{27}{9}=3$

69. $(x-a)^2+(y-b)^2=(-2-5)^2+(1-(-3))^2$
$=(-7)^2+4^2=49+16=65$

71. $|6-x|=|6-50|=|-44|=44$

73. $-2|x|-7=-2|-7|-7=-2\cdot 7-7$
$=-14-7=-21$

75. $2-[10-x(5h-1)]=2-[10-(-2)(5(2)-1)]$
$=2-[10-(-2)(10-1)]=2-[10-(-2)\cdot 9]$
$=2-[10+18]=2-28=-26$

77. $b^2-4ac=(-3)^2-4(4)(-1)$
$=9-(-16)=9+16=25$

79. $\dfrac{x}{y+10}=\dfrac{30}{-10+10}=\dfrac{30}{0}$ which is undefined

81. $\dfrac{50-6s}{-t}=\dfrac{50-6\cdot 5}{-4}=\dfrac{50-30}{-4}$
$=\dfrac{20}{-4}=-5$

83. $5rs^2t=5\cdot 2\cdot(-3)^2(-3)$
$=10\cdot 9\cdot(-3)=-270$

85. $\dfrac{|a^2-b^2|}{2a-b}=\dfrac{|(-2)^2-(-5)^2|}{2(-2)-(-5)}=\dfrac{|4-25|}{-4+5}$
$=\dfrac{|-21|}{1}=21$

APPLICATIONS

87. $2010: 10,900-3,850=7,050$
$2011: 11,300-3,910=7,390$

89. The speedometer measures rate, the odometer measures distance, and the clock measures time. The formula relating these is $d=rt$.

93.

Time	Distance Traveled (ft)	Time Intervals
1 sec	$16 \cdot 1^2$ $= 16\,ft$	Distance traveled from 0 sec to 1 sec is 16 feet
2 sec	$16 \cdot 2^2$ $= 64\,ft$	$64 - 16 = 48\,ft$ Distance traveled from 1 sec to 2 sec is 48 feet
3 sec	$16 \cdot 3^2$ $= 144\,ft$	$144 - 64 = 80\,ft$ Distance traveled from 2 sec to 3 sec is 80 feet
4 sec	$16 \cdot 4^2$ $= 256\,ft$	$256 - 144 = 112\,ft$ Distance traveled from 3 sec to 4 sec is 256 feet

95. $\dfrac{39+41+\cdots 35+47}{12} = \dfrac{384}{12} = 32$ therms

WRITING

97. The student needed to put the first 6 in parentheses as a (-6).

99. Revenue is the amount taken in from the consumer. Markup is the amount over production cost the consumer pays. Profit is the difference between revenue and cost.

101. In the following sentence, substitute is used as both a noun (sugar substitute) and a verb (substitute apple sauce).To cook healthier food, substitute apple sauce for oil and use a sugar substitute.

103. Because there were no parentheses around the -8, the student subtracted 8 instead of multiplying by -8.

105. a. doctor b. mechanic c. archaeologist

d. realtor e. nurse f. economist

REVIEW

107. 17, 37, and 41

109. $|-2+(-5)| = |-7| = 7$

111. Division by 3

113. $-3-(-6) = -3+6 = 3$

Section 3.3: Simplifying Algebraic Expressions and the Distributive Property

VOCABULARY

1. To simplify the expression $5(6x)$ means to write it in simpler form: $5(6x) = 30x$.

3. In the expression $2(x+8)$ there are two terms within the parentheses, x and 8.

5. When we use the distributive property to write a product, such as $7(4y+3)$, as the sum $28y+21$, we say we have removed or *cleared* the parentheses.

CONCEPTS

7. a. $4(9t) = (4 \cdot 9)t = 36t$

b. Associative Property of Multiplication

9. $x(y+z) = xy + xz$

11. Distributing multiplication by -1 changes the sign of each term within the parentheses.

$-(x+10) = -1(x+10) = -x-10$

13. a. $6(4x) = (6 \cdot 4)x = 24x$

b. $6(4+x) = 6 \cdot 4 + 6 \cdot x = 24 + 6x$

NOTATION

15. $-5 \cdot 7n = (-5 \cdot 7)n = -35n$

17. $9(5y-4) = 9(5y) - 9(4) = 45y - 36$

19. a. $-(-x) = x$

b. $x-(-5) = x+5$

c. $10y + (-15) = 10y - 15$

d. $5 \cdot x = 5x$

GUIDED PRACTICE

21. $2\cdot 6x=(2\cdot 6)x=12x$

23. $5\cdot 8y=(5\cdot 8)y=40y$

25. $-10t(-10)=-10(-10)t=100t$

27. $-15a(-3)=-15(-3)a=45a$

29. $-7x\cdot 9y=-7\cdot 9\cdot x\cdot y=-63xy$

31. $-4r\cdot 4s=-4\cdot 4\cdot r\cdot s=-16rs$

33. $2(-5x)(3y)=[2(-5)3]xy=-30xy$

35. $5r(2)(-3b)=[5\cdot 2(-3)]rb=-30br$

37. $4(x+1)=4\cdot x+4\cdot 1=4x+4$

39. $7(b+2)=7\cdot b+7\cdot 2=7b+14$

41. $9(3e-3)=9\cdot 3e-9\cdot 3=27e-27$

43. $3(2q-7)=3\cdot 2q-3\cdot 7=6q-21$

45. $-2(3h+5)=-2\cdot 3h+(-2)\cdot 5$
$=-6h-10$

47. $-10(4y+6)=-10\cdot 4y+(-10)\cdot 6$
$=-40y-60$

49. $-8(2q-4)=-8\cdot 2q-(-8)4$
$=-16q-(-32)=-16q+32$

51. $-5(7g-1)=-5\cdot 7g-(-5)\cdot 1$
$=-35g-(-5)=-35g+5$

53. $-4(-5s-3)=-4(-5s)-(-4)(3)$
$=20s+12$

55. $-6(-15t-9)=-6(-15t)-(-6)(9)$
$=90t+54$

57. $-1(x-5)=-1\cdot x-(-1)5=-x+5$

59. $-1(5d+8)=-1\cdot 5d+(-1)\cdot 8=-5d-8$

61. $(4d+7)6=4d\cdot 6+7\cdot 6$
$=24d+42$

63. $(3q+20)7=3q\cdot 7+20\cdot 7=21q+140$

65. $(-4-d)6=-4\cdot 6-d\cdot 6=-24-6d$

67. $(t-12)9=t\cdot 9-12\cdot 9=9t-108$

69. $2(4t-3)3=2\cdot 3(4t-3)$
$=6(4t-3)=6\cdot 4t-6\cdot 3=24t-18$

71. $4(3h+1)5=4\cdot 5(3h+1)$
$=20(3h+1)=20\cdot 3h+20\cdot 1=60h+20$

73. $-3(-3z-3x+5)$
$=-3(-3z)-(-3)3x+(-3)5$
$=9z-(-9x)+(-15)=9z+9x-15$

75. $-8(2a+4b-6)$
$=-8(2a)+(-8)(4b)-(-8)(6)$
$=-16a+(-32b)-(-48)=-16a-32b+48$

77. $-(-3w-4)=3w+4$

79. $-(-18x-19)=18x+19$

81. $-(x+3)=-x-3$

83. $-(4t+5)=-4t-5$

TRY IT YOURSELF

85. $-(x-7)=-x+7$

87. $(4s)3=3\cdot 4s=12s$

89. $5(-7q)=5(-7)q=-35q$

91. $6(-6c+7)=6(-6c)+6(7)=-36c+42$

93. $5(9s-12t-3)=5(9s)-5(12t)+5(-3)$
$=45s-60t-15$

95. $-5\cdot 8h=-40h$

97. $5\cdot 8c\cdot 2=2\cdot 5\cdot 8\cdot c=80c$

99. $2(3t+2)8=8\cdot 2(3t+2)$
$=16(3t+2)=16\cdot 3t+16\cdot 2$
$=48t+32$

101. $(-1)(-2e)(-4)=(-1)(-2)(-4)e=-8e$

103. $-(5x-4y+1)=-5x+4y-1$

105. $2(4x)+2(5)=2(4x+5)$

107. $-3(4y)+(-3)(2)=-3(4y+2)$

109. $3(4)-3(7t)-3(5s)=3(4-7t-5s)$

111. $-4(5)-3x(5)=(-4-3x)5$

WRITING

113. It means to rewrite the expression with fewer terms. An example would be $2(x-3)=2x-6$.

115. After getting the shipment of supplies, the quartermaster distributed the new gear appropriately.

117. Every term inside the parentheses must be multiplied by 2.

REVIEW

119. $|-6+1|=|-5|=5$

121. multiplication, division, subtraction, addition

123. $-6>-7$

125. carpeting and painting

Section 3.4: Combining Like Terms

VOCABULARY

1. A term is a number or a product of a number and one or more variables. A single number or variable is also a term.

3. In the term $5t$, 5 is called the coefficient.

5. Terms such as x and y have an implied coefficient of 1.

7. When we write $9x+x$ as $10x$, we say that we have combined like terms.

CONCEPTS

9. The expression $5x-10+8x$ has 3 terms. The second term is -10. The coefficient of the third term is 8.

11. The term $8m$ has a coefficient of 8 because $8m=\underline{8}\cdot\underline{m}$.

13. a. unlike b. unlike

c. unlike d. like

NOTATION

15. $2x+3x=(2+3)x=5x$

17. $2(x-1)+3x=2x-2+3x=5x-2$

19. a. the perimeter of a rectangle

b. 2 times the length

c. 2 times the width

21. a. true b. true c. true d. false

GUIDED PRACTICE

23. $3x^2,-9x,4$

25. $5,5t,-8t,-1$

27. $-35a$

29. $9mn,-6n$

31. a. term b. factor

33. a. factor b. term

35. 5, 1, -12

37. 1, -27

39. 1, -1, 1, 10

41. -1, 6, -1, 5

43. $8x, 2x$

45. none

47. $-3k^3,\ k^3\ ;\ 6k,\ -3k$

49. $12a\ ,\ 15a\ ;\ -8\ ,\ 1$

51. $6t+9t=15t$

53. $20b+30b=50b$

55. $-5x+(-6x)+2x=-11x+2x=-9x$

57. $-5d+(-9d)+10d=-14d+10d=-4d$

59. $5s^2-3s^2=2s^2$

61. $3e^3-17e^3=-14e^3$

63. does not simplify

65. $14z-8z+2z=6z+2z=8z$

67. $53a+6a-21a=59a-21a=38a$

69. does not simplify

71. $10s-9s=s$

73. $40a^2-a^2=39a^2$

75. $43s^3-44s^3=-s^3$

77. $14r+r=15r$

79. $5x^2+19x-3x+6=5x^2+16x+6$

81. $y^2-8y-2y-4=y^2-10y-4$

83. $5(m+2)-8-(3m-1)=5m+10-8-3m+1$
$=2m+3$

85. $4(x-1)-2-(x+5)=4x-4-2-x-5$
$=3x-11$

87. $P=2l+2w=2\cdot16+2\cdot7=32+14=46\,ft$

89. $P=4s=4\cdot37=148\,yd$

TRY IT YOURSELF

91. $-7b^2+27b^2=20b^2$

93. $-4(-4y+5)+4-6(y+2)$
$=16y-20+4-6y-12=10y-28$

95. $6t+9+5t+3=11t+12$

97. $x+x+x+x=4x$

99. $3t-(t-8)=3t-t+8=2t+8$

101. $5(2x)(-5)=-5\cdot5\cdot2\cdot x=-50x$

103. does not simplify

105. $4x^2-3x-7+4x^2-2-x$
$=8x^2-4x-9$

107. $-6s+6s=0s=0$

109. $-4r+8R+2R-3r+R=-7r+11R$

111. $0-2y^3=-2y^3$

113. $5(3-2s)+4(2-3s)+19s$
$=15-10s+8-12s+19s=-3s+23$

APPLICATIONS

115. a. $(d+15)+d=(2d+15)\,mi$
b. $b+(b+30)=(2b+30)\,ft$

117. $2x+2(x+4)=2x+2x+8$
$=(4x+8)\,ft$

119. Front and Back: $2\cdot 60+2\cdot 10=120+20=140\,ft$

Sides: $2\cdot 10+2\cdot 10=20+20=40\,ft$

Total stripping needed:
$2\cdot 140+2\cdot 40=280+80=360\,ft$

$0.8(360)=288$

The pine will cost $288.

121.

Size of Floor (feet)	Perimeter (feet) *P=4s*
9 x 9	4 x 9 = 36 ft
12 x 12	4 x 12 = 48 ft
15 x 15	4 x 15 = 60 ft
18 x 18	4 x 18 = 72 ft
21 x 21	4 x 21 = 84 ft

WRITING

123. Two terms are like terms if they have the same variables with the same exponents on each of them.

125. A term is a product or quotient of numbers and variables. A factor is component of a term, for example, 2 and x are factors of the term $2x$ and 2 and x are terms of the expression $2+x$.

REVIEW

127. $-12y-8=40$

$-12y-8+8=40+8$

$-12y=48$

$\frac{-12y}{-12}=\frac{48}{-12}$

$y=-4$

129. To <u>check</u> the solution of an equation, we substitute the value in the original equation and determine whether the result is a true statement.

Section 3.5: Simplifying Expressions to Solve Equations

VOCABULARY

1. To <u>solve</u> an equation means to find all values of the variable that make the equation a true statement.

3. To <u>check</u> a solution means to substitute that value into the original equation to see whether a true statement results.

5. When solving equations, <u>simplify</u> the expressions that make up the left and right sides of the equation before using properties of equality to isolate the variable.

CONCEPTS

7. a. $\boxed{5x}+\boxed{3x}=8$ $\boxed{5t}=\boxed{3t}+8$

$7=\boxed{5h}+\boxed{3h}-1$

b. $5t=3t+8$

9. a. combine like terms: $6x=36$

b. distribute: $5x+5=15$

c. combine like terms: $11x-5=2x+4$

d. distribute: $-3x+12=2x+2$

11. $6-(d-4)=8$

$6-1(d-4)=8$

$6-d+4=8$

13. a. $2t-8$

b. $3t-t=-8$

$2t=-8$

$\frac{2t}{2}=\frac{-8}{2}$

$t=-4$

c. $3t-t-8=3(-2)-(-2)-8$

$=-6+2-8=-12$

d. no

NOTATION

15.
$$5x-2x=-27$$
$$3x=-27$$
$$\frac{3x}{3}=\frac{-27}{3}$$
$$x=-9$$
Check: $5x-2x=-27$
$$5(-9)-2(-9)\stackrel{?}{=}-27$$
$$-45-(-18)\stackrel{?}{=}-27$$
$$-45+18\stackrel{?}{=}-27$$
$$-27=-27 \text{ True}$$
The solution is -9.

17.
$$5(x-9)=5$$
$$5x-5(9)=5$$
$$5x-45=5$$
$$5x-45+45=5+45$$
$$5x=50$$
$$\frac{5x}{5}=\frac{50}{5}$$
$$x=10$$
Check: $5(x-9)=5$
$$5(10-9)\stackrel{?}{=}5$$
$$5(1)\stackrel{?}{=}5$$
$$5=5 \text{ True}$$
The solution is 10.

GUIDED PRACTICE

19.
$$5f+8=4f+11$$
$$5(3)+8\stackrel{?}{=}4(3)+11$$
$$15+8\stackrel{?}{=}12+11$$
$$23=23 \text{ True}$$
3 is a solution.

21.
$$2(x-1)=33$$
$$2(12-1)\stackrel{?}{=}33$$
$$2(11)\stackrel{?}{=}33$$
$$22=33 \text{ False}$$
12 is not a solution.

23.
$$3x+6x=54$$
$$9x=54$$
$$\frac{9x}{9}=\frac{54}{9}$$
$$x=6$$
Check: $3x+6x=54$
$$3(6)+6(6)\stackrel{?}{=}54$$
$$18+36\stackrel{?}{=}54$$
$$54=54 \text{ True}$$
The solution checks.

25.
$$6x-3x=9$$
$$3x=9$$
$$\frac{3x}{3}=\frac{9}{3}$$
$$x=3$$
Check: $6x-3x=9$
$$6(3)-3(3)\stackrel{?}{=}9$$
$$18-9\stackrel{?}{=}9$$
$$9=9 \text{ True}$$
The solution checks.

27.
$$250+350=m-12+m$$
$$600=2m-12$$
$$600+12=2m-12+12$$
$$612=2m$$
$$\frac{612}{2}=\frac{2m}{2}$$
$$306=m$$
Check: $250+350=m-12+m$
$$250+350\stackrel{?}{=}(306)-12+(306)$$
$$600=600 \text{ True}$$
The solution checks.

29. $255+275=a+16+a$

$530=2a+16$

$530-16=2a+16-16$

$514=2a$

$\frac{514}{2}=\frac{2a}{2}$

$257=a$

Check: $255+275=a+16+a$

$255+275\stackrel{?}{=}(257)+16+(257)$

$530=530$ True

The solution checks.

31. $3s-1=4s+7$

$3s-1-3s=4s+7-3s$

$-1=s+7$

$-1-7=s+7-7$

$-8=s$

Check: $3s-1=4s+7$

$3(-8)-1\stackrel{?}{=}4(-8)+7$

$-24-1\stackrel{?}{=}-32+7$

$-25=-25$ True

The solution checks.

33. $x-14=2x-10$

$x-14-x=2x-10-x$

$-14=x-10$

$-14+10=x-10+10$

$-4=x$

Check: $x-14=2x-10$

$(-4)-14\stackrel{?}{=}2(-4)-10$

$-18\stackrel{?}{=}-8-10$

$-18=-18$ True

The solution checks.

35. $10m-14=11m+13$

$10m-14+14=11m+13+14$

$10m=11m+27$

$10m-11m=11m+27-11m$

$-m=27$

$\frac{-m}{-1}=\frac{27}{-1}$

$m=-27$

Check: $10m-14=11m+13$

$10(-27)-14\stackrel{?}{=}11(-27)+13$

$-270-14\stackrel{?}{=}-297+13$

$-284=-284$ True

The solution checks.

37. $n+5=2n-3$

$n+5-5=2n-3-5$

$n=2n-8$

$n-2n=2n-8-2n$

$-n=-8$

$\frac{-n}{-1}=\frac{-8}{-1}$

$n=8$

Check: $n+5=2n-3$

$8+5\stackrel{?}{=}2(8)-3$

$13\stackrel{?}{=}16-3$

$13=13$ True

The solution checks.

39.

$$-16-3r-5r=2r+4$$
$$-16-8r=2r+4$$
$$-16-8r+8r=2r+4+8r$$
$$-16=10r+4$$
$$-16-4=10r+4-4$$
$$-20=10r$$
$$\frac{-20}{10}=\frac{10r}{10}$$
$$-2=r$$

Check: $-16-3r-5r=2r+4$

$$-16-3(-2)-5(-2)\stackrel{?}{=}2(-2)+4$$
$$-16+6+10\stackrel{?}{=}-4+4$$
$$0=0 \quad \text{True}$$

The solution checks.

41.

$$60-a-6a=2a-3$$
$$60-7a=2a-3$$
$$60-7a+7a=2a-3+7a$$
$$60=9a-3$$
$$60+3=9a-3+3$$
$$63=9a$$
$$\frac{63}{9}=\frac{9a}{9}$$
$$7=a$$

Check: $60-a-6a=2a-3$

$$60-(7)-6(7)\stackrel{?}{=}2(7)-3$$
$$60-7-42\stackrel{?}{=}14-3$$
$$11=11 \quad \text{True}$$

The solution checks.

43.

$$-3x-7+x=6x+20-5x$$
$$-2x-7=x+20$$
$$-2x-7+7=x+20+7$$
$$-2x=x+27$$
$$-2x-x=x+27-x$$
$$-3x=27$$
$$\frac{-3x}{-3}=\frac{27}{-3}$$
$$x=-9$$

Check: $-3x-7+x=6x+20-5x$

$$-3(-9)-7+(-9)\stackrel{?}{=}6(-9)+20-5(-9)$$
$$27-7-9\stackrel{?}{=}-54+20+45$$
$$11=11 \quad \text{True}$$

The solution checks.

45.

$$7+5y+3y=4y-13$$
$$7+8y=4y-13$$
$$7+8y-7=4y-13-7$$
$$8y=4y-20$$
$$8y-4y=4y-20-4y$$
$$4y=-20$$
$$\frac{4y}{4}=\frac{-20}{4}$$
$$y=-5$$

Check: $7+5y+3y=4y-13$

$$7+5(-5)+3(-5)\stackrel{?}{=}4(-5)-13$$
$$7-25-15\stackrel{?}{=}-20-13$$
$$-33=-33 \quad \text{True}$$

The solution checks.

47. $2(6x+7)=14$

$12x+14=14$

$12x+14-14=14-14$

$12x=0$

$\frac{12x}{12}=\frac{0}{12}$

$x=0$

Check: $2(6x+7)=14$

$2[6(0)+7]\stackrel{?}{=}14$

$2(7)\stackrel{?}{=}14$

$14=14$ True

The solution checks.

49. $8(9b+5)=-32$

$72b+40=-32$

$72b+40-40=-32-40$

$72b=-72$

$\frac{72b}{72}=\frac{-72}{72}$

$b=-1$

Check: $8(9b+5)=-32$

$8[9(-1)+5]\stackrel{?}{=}-32$

$8(-9+5)\stackrel{?}{=}-32$

$8(-4)\stackrel{?}{=}-32$

$-32=-32$ True

The solution checks.

51. $-(6-t)=-14$

$-6+t=-14$

$-6+t+6=-14+6$

$t=-8$

Check: $-(6-t)=-14$

$-[6-(-8)]\stackrel{?}{=}-14$

$-(6+8)\stackrel{?}{=}-14$

$-14=-14$ True

The solution checks.

53. $-2(3x+5)=20$

$-6x-10=20$

$-6x-10+10=20+10$

$-6x=30$

$\frac{-6x}{-6}=\frac{30}{-6}$

$x=-5$

Check: $-2(3x+5)=20$

$-2[3(-5)+5]\stackrel{?}{=}20$

$-2(-15+5)\stackrel{?}{=}20$

$-2(-10)\stackrel{?}{=}20$

$20=20$ True

The solution checks.

55. $3(x-4)+3x=2(x+10)$

$3x-12+3x=2x+20$

$6x-12=2x+20$

$6x-12-2x=2x+20-2x$

$4x-12=20$

$4x-12+12=20+12$

$4x=32$

$\frac{4x}{4}=\frac{32}{4}$

$x=8$

Check: $3(x-4)+3x=2(x+10)$

$3[(8)-4]+3(8)\stackrel{?}{=}2[(8)+10]$

$3(4)+3(8)\stackrel{?}{=}2(18)$

$12+24\stackrel{?}{=}36$

$36=36$ True

The solution checks.

57.

$$6(2j+6)+4j=4(j-30)$$
$$12j+36+4j=4j-120$$
$$16j+36=4j-120$$
$$16j+36-4j=4j-120-4j$$
$$12j+36=-120$$
$$12j+36-36=-120-36$$
$$12j=-156$$
$$\frac{12j}{12}=\frac{-156}{12}$$
$$j=-13$$

Check: $6(2j+6)+4j=4(j-30)$

$$6[2(-13)+6]+4(-13)\stackrel{?}{=}4[(-13)-30]$$
$$6(-26+6)-52\stackrel{?}{=}4(-43)$$
$$6(-20)-52\stackrel{?}{=}-172$$
$$-120-52\stackrel{?}{=}-172$$
$$-172=-172 \quad \text{True}$$

The solution checks.

TRY IT YOURSELF

59.

$$-16=2(t+2)$$
$$-16=2t+4$$
$$-16-4=2t+4-4$$
$$-20=2t$$
$$\frac{-20}{2}=\frac{2t}{2}$$
$$-10=t$$

Check: $-16=2(t+2)$

$$-16\stackrel{?}{=}2(-10+2)$$
$$-16\stackrel{?}{=}2(-8)$$
$$-16=-16 \quad \text{True}$$

The solution checks.

61.

$$-7+5r=83-10r$$
$$-7+5r+10r=83-10r+10r$$
$$-7+15r=83$$
$$-7+15r+7=83+7$$
$$15r=90$$
$$\frac{15r}{15}=\frac{90}{15}$$
$$r=6$$

Check: $-7+5r=83-10r$

$$-7+5(6)\stackrel{?}{=}83-10(6)$$
$$-7+30\stackrel{?}{=}83-60$$
$$23=23 \quad \text{True}$$

The solution checks.

63.

$$T+T-17=57$$
$$2T-17=57$$
$$2T-17+17=57+17$$
$$2T=74$$
$$\frac{2T}{2}=\frac{74}{2}$$
$$T=37$$

Check: $T+T-17=57$

$$(37)+(37)-17\stackrel{?}{=}57$$
$$57=57 \quad \text{True}$$

The solution checks.

65.
$$-15 = 5 + 5(2x + 10)$$
$$-15 = 5 + 10x + 50$$
$$-15 = 10x + 55$$
$$-15 - 55 = 10x + 55 - 55$$
$$-70 = 10x$$
$$\frac{-70}{10} = \frac{10x}{10}$$
$$-7 = x$$

Check: $-15 = 5 + 5(2x + 10)$
$$-15 \stackrel{?}{=} 5 + 5[2(-7) + 10]$$
$$-15 \stackrel{?}{=} 5 + 5(-14 + 10)$$
$$-15 \stackrel{?}{=} 5 + 5(-4)$$
$$-15 \stackrel{?}{=} 5 - 20$$
$$-15 = -15 \quad \text{True}$$

The solution checks.

67.
$$60 = 3v - 5v$$
$$60 = -2v$$
$$\frac{60}{-2} = \frac{-2v}{-2}$$
$$-30 = v$$

Check: $60 = 3v - 5v$
$$60 \stackrel{?}{=} 3(-30) - 5(-30)$$
$$60 \stackrel{?}{=} -90 + 150$$
$$60 = 60 \quad \text{True}$$

The solution checks.

69.
$$9q + 3(q - 7) = 18 - q$$
$$9q + 3q - 21 = 18 - q$$
$$12q - 21 = 18 - q$$
$$12q - 21 + q = 18 - q + q$$
$$13q - 21 = 18$$
$$13q - 21 + 21 = 18 + 21$$
$$13q = 39$$
$$\frac{13q}{13} = \frac{39}{13}$$
$$q = 3$$

Check: $9q + 3(q - 7) = 18 - q$
$$9(3) + 3(3 - 7) \stackrel{?}{=} 18 - 3$$
$$27 + 3(-4) \stackrel{?}{=} 15$$
$$27 - 12 \stackrel{?}{=} 15$$
$$15 = 15 \quad \text{True}$$

The solution checks.

71.
$$60r - 50 = 15r - 5$$
$$60r - 50 + 50 = 15r - 5 + 50$$
$$60r = 15r + 45$$
$$60r - 15r = 15r + 45 - 15r$$
$$45r = 45$$
$$r = 1$$

Check: $60r - 50 = 15r - 5$
$$60(1) - 50 \stackrel{?}{=} 15(1) - 5$$
$$60 - 50 \stackrel{?}{=} 15 - 5$$
$$10 = 10 \quad \text{True}$$

The solution checks.

73.

$3-5x-2x=x+27$

$3-7x=x+27$

$3-7x-3=x+27-3$

$-7x=x+24$

$-7x-x=x+24-x$

$-8x=24$

$x=-3$

Check: $3-5x-2x=x+27$

$3-5(-3)-2(-3)\stackrel{?}{=}-3+27$

$3+15+6\stackrel{?}{=}24$

$24=24$ True

The solution checks.

75.

$38+6x-5x=2x+7$

$38+x=2x+7$

$38+x-38=2x+7-38$

$x=2x-31$

$x-2x=2x-31-2x$

$-x=-31$

$x=31$

Check: $38+6x-5x=2x+7$

$38+6(31)-5(31)\stackrel{?}{=}2(31)+7$

$38+186-155\stackrel{?}{=}62+7$

$69=69$ True

The solution checks.

77.

$5(13x+3)=210$

$65x+15=210$

$65x+15-15=210-15$

$65x=195$

$x=3$

Check: $5(13x+3)=210$

$5[13(3)+3]\stackrel{?}{=}210$

$5(39+3)\stackrel{?}{=}210$

$5(42)\stackrel{?}{=}210$

$210=210$ True

The solution checks.

79.

$5-(7-y)=-5$

$5-7+y=-5$

$-2+y=-5$

$-2+y+2=-5+2$

$y=-3$

Check: $5-(7-y)=-5$

$5-[7-(-3)]\stackrel{?}{=}-5$

$5-10\stackrel{?}{=}-5$

$-5=-5$ True

The solution checks.

81.

$50a-1=60a-101$

$50a-1-50a=60a-101-50a$

$-1=-101+10a$

$-1+101=-101+10a+101$

$100=10a$

$\frac{100}{10}=\frac{10a}{10}$

$10=a$

Check: $50a-1=60a-101$

$50(10)-1\stackrel{?}{=}60(10)-101$

$500-1\stackrel{?}{=}600-101$

$499=499$ True

The solution checks.

83.

$-(4-c)=-3$

$-4+c=-3$

$-4+c+4=-3+4$

$c=1$

Check: $-(4-c)=-3$

$-(4-1)\stackrel{?}{=}-3$

$-3=-3$ True

The solution checks.

85. $-20-8=-m+2m$

$-28=m$

Check: $-20-8=-m+2m$

$-20-8\stackrel{?}{=}-(-28)+2(-28)$

$-28\stackrel{?}{=}28-56$

$-28=-28$ True

The solution checks.

87. $x+x+6=90$

$2x+6=90$

$2x+6-6=90-6$

$2x=84$

$\frac{2x}{2}=\frac{84}{2}$

$x=42$

Check: $x+x+6=90$

$(42)+(42)+6\stackrel{?}{=}90$

$90=90$ True

The solution checks.

89. $8+4(2x-2)=-16+4x$

$8+8x-8=-16+4x$

$8x=-16+4x$

$8x-4x=-16+4x-4x$

$4x=-16$

$\frac{4x}{4}=\frac{-16}{4}$

$x=-4$

Check: $8+4(2x-2)=-16+4x$

$8+4[2(-4)-2]\stackrel{?}{=}-16+4(-4)$

$8+4(-8-2)\stackrel{?}{=}-16-16$

$8+4(-10)\stackrel{?}{=}-32$

$8-40\stackrel{?}{=}-32$

$-32=-32$ True

The solution checks

91. $1,500=b+30+b$

$1,500=2b+30$

$1,500-30=2b+30-30$

$1,470=2b$

$\frac{1,470}{2}=\frac{2b}{2}$

$735=b$

Check: $1,500=b+30+b$

$1,500\stackrel{?}{=}(735)+30+(735)$

$1,500=1,500$ True

The solution checks.

93. $7x=3x+8$

$7x-3x=3x+8-3x$

$4x=8$

$\frac{4x}{4}=\frac{8}{4}$

$x=2$

Check: $7x=3x+8$

$7(2)\stackrel{?}{=}3(2)+8$

$14\stackrel{?}{=}6+8$

$14=14$ True

The solution checks.

95. $100-y=100+y$

$100-y-100=100+y-100$

$-y=y$

$-y+y=y+y$

$0=2y$

$\frac{0}{2}=\frac{2y}{2}$

$0=y$

Check: $100-y=100+y$

$100-0\stackrel{?}{=}100+0$

$100=100$ True

The solution checks.

97.
$$2(4y+8)+3y=-3(2-3y)$$
$$8y+16+3y=-6+9y$$
$$11y+16=-6+9y$$
$$11y+16-9y=-6+9y-9y$$
$$2y+16=-6$$
$$2y+16-16=-6-16$$
$$2y=-22$$
$$\frac{2y}{2}=\frac{-22}{2}$$
$$y=-11$$
Check: $2(4y+8)+3y=-3(2-3y)$
$$2[4(-11)+8]+3(-11)\stackrel{?}{=}-3[2-3(-11)]$$
$$2(-44+8)-33\stackrel{?}{=}-3(2+33)$$
$$2(-36)-33\stackrel{?}{=}-3(35)$$
$$-72-33\stackrel{?}{=}-105$$
$$-105=-105 \quad \text{True}$$
The solution checks.

99.
$$t+5t+3t-40=14t$$
$$9t-40=14t$$
$$9t-40-9t=14t-9t$$
$$-40=5t$$
$$\frac{-40}{5}=\frac{5t}{5}$$
$$-8=t$$
Check: $t+5t+3t-40=14t$
$$(-8)+5(-8)+3(-8)-40\stackrel{?}{=}14(-8)$$
$$-8-40-24-40\stackrel{?}{=}-112$$
$$-112=-112 \quad \text{True}$$
The solution checks.

101.
$$-2(9-3s)-(5s+2)=-25$$
$$-18+6s-5s-2=-25$$
$$s-20=-25$$
$$s-20+20=-25+20$$
$$s=-5$$
Check: $-2(9-3s)-(5s+2)=-25$
$$-2[9-3(-5)]-[5(-5)+2]\stackrel{?}{=}-25$$
$$-2(9+15)-(-25+2)\stackrel{?}{=}-25$$
$$-2(24)-(-23)\stackrel{?}{=}-25$$
$$-48+23\stackrel{?}{=}-25$$
$$-25=-25 \quad \text{True}$$
The solution checks.

103.
$$25+4j=9j$$
$$25+4j-4j=9j-4j$$
$$25=5j$$
$$\frac{25}{5}=\frac{5j}{5}$$
$$5=j$$
Check: $25+4j=9j$
$$25+4(5)\stackrel{?}{=}9(5)$$
$$25+20\stackrel{?}{=}45$$
$$45=45 \quad \text{True}$$
The solution checks.

105.
$$-3(3-2w)=-9$$
$$-9+6w=-9$$
$$-9+6w+9=-9+9$$
$$6w=0$$
$$\frac{6w}{6}=\frac{0}{6}$$
$$w=0$$
Check: $-3(3-2w)=-9$
$$-3[3-2(0)]\stackrel{?}{=}-9$$
$$-3(3)\stackrel{?}{=}-9$$
$$-9=-9 \quad \text{True}$$
The solution checks.

107.
$$-4(12)+12t=16t$$
$$-48+12t=16t$$
$$-48+12t-12t=16t-12t$$
$$-48=4t$$
$$\frac{-48}{4}=\frac{4t}{4}$$
$$-12=t$$
Check: $-4(12)+12t=16t$
$$-4(12)+12(-12)\stackrel{?}{=}16(-12)$$
$$-48-144\stackrel{?}{=}-192$$
$$-192=-192 \quad \text{True}$$
The solution checks.

109.
$$-5g-40=-15g$$
$$-5g-40+5g=-15g+5g$$
$$-40=-10g$$
$$\frac{-40}{-10}=\frac{-10g}{-10}$$
$$4=g$$
Check: $-5g-40=-15g$
$$-5(4)-40\stackrel{?}{=}-15(4)$$
$$-20-40\stackrel{?}{=}-60$$
$$-60=-60 \quad \text{True}$$
The solution checks.

111.
$$4(p-2)=0$$
$$4p-8=0$$
$$4p-8+8=0+8$$
$$4p=8$$
$$\frac{4p}{4}=\frac{8}{4}$$
$$p=2$$
Check: $4(p-2)=0$
$$4[(2)-2]\stackrel{?}{=}0$$
$$4(0)\stackrel{?}{=}0$$
$$0=0 \quad \text{True}$$
The solution checks.

113.
$$16-(x+3)=-13$$
$$16-x-3=-13$$
$$13-x=-13$$
$$13-x-13=-13-13$$
$$-x=-26$$
$$\frac{-x}{-1}=\frac{-26}{-1}$$
$$x=26$$
Check: $16-(x+3)=-13$
$$16-[(26)+3]\stackrel{?}{=}-13$$
$$16-29\stackrel{?}{=}-13$$
$$-13=-13 \quad \text{True}$$
The solution checks.

WRITING

115. The student must combine the like terms, not just cancel an x.

117. In the second step, every term should have been divided by 2.

119. It means to find all values of the variable that make the equation true.

REVIEW

121. $-7-9=-7+(-9)=-16$

123. $2x-8$

125. $-(-5)=5$

127. positive

Section 3.6: Using Equations to Solve Application Problems

VOCABULARY

1. The six-step problem-solving strategy is:

- Analyze the problem
- Assign a variable
- Form an equation
- Solve the equation
- State the conclusion
- Check the result

3. Phrases such as *distributed equally* and *sectioned off uniformly* indicate the operation of division.

5. Words such as *extended* and *reclaimed* indicate the operation of addition.

7. Number·value = total value

9. $5x$

11. $g-100$

13. $3m$

15. $2w$

17. a. 9

b. $9-d$

19. **Analyze**

* There are 88 seats on the plane.

* There are 10 times as many economy seats as first-class seats.

* How many first-class seats are there?

Assign

Let x = the number of first-class seats.

Key phrase: ten times as many

Translation: multiply by 10.

So, $10x$ = the number of economy seats.

Form

Now we translate the words of the problem into an equation.

The number of first-class seats	plus	the number of economy	is	88.
x	$+$	$10x$	$=$	88

Solve

$$x+10x=88$$
$$11x=88$$
$$\frac{11x}{11}=\frac{88}{11}$$
$$x=8$$

State

There are 8 first-class seats.

Check

The number of economy seats, 80, is 10 times the number of first-class seats, 8. Also, if we add the numbers of seats, we get 88. The result checks.

GUIDED PRACTICE

21. x = the number of months to reach his goal

$$15+5x=100$$
$$15+5x-15=100-15$$
$$5x=85$$
$$\frac{5x}{5}=\frac{85}{5}$$
$$x=17$$

It will take the salesman 17 months to reach his goal.

23. x = the number of scholarships awarded last year

$$x+(x+6)=20$$
$$2x+6=20$$
$$2x+6-6=20-6$$
$$2x=14$$
$$\frac{2x}{2}=\frac{14}{2}$$
$$x=7$$

Last year there were 7 scholarships awarded. This year there were 7 + 6 = 13 scholarships awarded.

25. n = the number of nickels
$25-n$ = the number of dimes

$$20(1)+n(5)+(25-n)(10)=220$$
$$20+5n+250-10n=220$$
$$270-5n=220$$
$$270-5n-270=220-270$$
$$-5n=-50$$
$$\frac{-5n}{-5}=\frac{-50}{-5}$$
$$n=10$$

There were 10 nickels and 25 – 10 = 15 dimes in the piggy bank.

27.

Type of Shoe	Number Sold	x	Commission per Pair	=	Total Commission
Dress	10	x	3	=	30
Athletic	12	x	2	=	24
Child's	x	x	5	=	$5x$
Sandal	$9-x$	x	4	=	$4(9-x)$

29. Let e = the number of hours spent tutoring English.

Type of Tutoring	Number of Hours	x	Hourly Rate	=	Total Income
English	h	x	18	=	$18h$
Math	$40-h$	x	20	=	$20(40-h)$

TRY IT YOURSELF

31.
$$x+8=2x$$
$$x+8-x=2x-x$$
$$8=x$$

The number is 8.

33.
$$5x-10=x+6$$
$$5x-10-x=x+6-x$$
$$4x-10=6$$
$$4x-10+10=6+10$$
$$4x=16$$
$$\frac{4x}{4}=\frac{16}{4}$$
$$x=4$$

The number is 4.

APPLICATIONS

35. x = the number of hours she needs

$$24+6x=48$$
$$24+6x-24=48-24$$
$$6x=24$$
$$\frac{6x}{6}=\frac{24}{6}$$
$$x=4$$

She must take 4 more sessions.

37. x = the number of payments made

$$420+30x=600$$
$$420+30x-420=600-420$$
$$30x=180$$
$$\frac{30x}{30}=\frac{180}{30}$$
$$x=6$$

She has made 6 payments.

39. x = the number of months to the goal

$$465+5x=510$$
$$465+5x-465=510-465$$
$$5x=45$$
$$\frac{5x}{5}=\frac{45}{5}$$
$$x=9$$

It will take 9 months to reach the goal.

41. x = the distance of the freighter from the port

$x+3x=84$

$4x=84$

$\frac{4x}{4}=\frac{84}{4}$

$x=21$

The freighter was 21 miles from the port.

43. x = the amount each inherited

$x-5,575=78,525$

$x-5,575+5,575=78,525+5,575$

$x=84,100$

Each son received \$84,100, so a total of $5\cdot 84,100=\$420,500$ was inherited.

45. x = the monthly rent for each person

$x+100=425$

$x+100-100=425-100$

$x=325$

They each pay \$325 / month, so the monthly rent is $3\cdot 325=\$975$.

47. x = the amount in the premium tank

$x+(x-100)=700$

$2x-100=700$

$2x-100+100=700+100$

$2x=800$

$\frac{2x}{2}=\frac{800}{2}$

$x=400$

The premium tank holds 400 gallons.

49. x = the time allotted to commercials

$x+(x+18)=30$

$2x+18=30$

$2x+18-18=30-18$

$2x=12$

$\frac{2x}{2}=\frac{12}{2}$

$x=6$

There were 6 minutes of commercials.

51. x = the width of the room

$2(2x)+2x=60$

$4x+2x=60$

$6x=60$

$\frac{6x}{6}=\frac{60}{6}$

$x=10$

The room is 10 feet wide.

53. x = the width of the court

$2x+2(x+51)=210$

$2x+2x+102=210$

$4x+102=210$

$4x=210-102$

$4x=108$

$\frac{4x}{4}=\frac{108}{4}$

$x=27$

The width is 27 feet and the length is 27 + 51 = 78 feet.

55.

	Number Sold	Commission	Total Paid
Dress Shoes	d	3	$3d$
Athletic Shoes	$9-d$	2	$2(9-d)$
Total			24

$3d+2(9-d)=24$

$3d+18-2d=24$

$d+18=24$

$d+18-18=24-18$

$d=6$

He sold 6 pairs of dress shoes and 9 – 6 = 3 pairs of athletic shoes.

57.

	Number Hours	Cost per Hour	Total Cost
Regular	r	60	$60r$
Stairs	$20-r$	90	$90(20-r)$
Total			1,380

$60r+90(20-r)=1,380$

$60r+1,800-90r=1,380$

$-30r+1,800=1,380$

$-30r+1,800-1,800=1,380-1,800$

$-30r=-420$

$\frac{-30r}{-30}=\frac{-420}{-30}$

$x=14$

He worked 14 regular hours and 20 – 14 = 6 hours going up and down stairs.

59.

	Number Owned	Value Each	Total Value
Movie	m	200	$200m$
Television	$m-6$	75	$75(m-6)$
Total			4,500

$200m+75(m-6)=4,500$

$200m+75m-450=4,500$

$275m-450=4,500$

$275m-450+450=4,500+450$

$275m=4,950$

$\frac{275m}{275}=\frac{4,950}{275}$

$m=18$

He has 18 movie star autographs and 18 – 6 = 12 TV celebrity autographs.

WRITING

61. The six-step problem-solving strategy is:

- Analyze the problem- figure out what we are looking for.
- Assign a variable- give the unknown a name.
- Form an equation – write the problem mathematically.
- Solve the equation – isolate the variable from the previous step.
- State the conclusion – be sure to answer the original problem.
- Check the result – make sure the answer is correct and logical.

63. Tom had his son 20 years ago. If their combined age is 50, how old is Tom?

REVIEW

65. Associative Property of Addition

67. $-10^2=-10\cdot 10=-100$

69. Subtraction of a number is the same as <u>addition</u> of the opposite of that number.

71. $2\cdot 2\cdot 2\cdot 5\cdot 5=2^3\cdot 5^2$

Chapter 3 Review

1. Brandon is closer by 250 miles.

2. $h+7$

3. $n-5$

4. $7x$

5. $\frac{6}{p}$

6. $s+(-15)$

7. $2l$

8. $D-100$

9. $r+2$

10. $\frac{45}{x}$

11. $100-2s$

12. $\left|2-a^2\right|$

13. m reduced by 500

14. a. $(n+4)in$

b. $(b-4)in$

15. $\dfrac{c}{6}$

16. $1{,}000-x$

17. $x+1$

18. $\dfrac{p}{8}$

19. x = the number of hours the wife drove
$2x$ = the number of hours the husband drove

20. w = the width
$w + 3$ = the length

21. v = the weight of the volleyball (in ounces)
$2v + 2$ = the weight of the basketball (in ounces)

22. M = the age of *To Kill a Mockingbird*
$M+6$ = the age of *The Lord of the Rings*
$M-9$ = the age of *The Godfather*

23. $12x$

24. $\dfrac{d}{7}$

25. a. h = the height of the wall
$h - 5$ = the length of the upper base
$2h - 3$ = the length of the lower base

b. $h = 10$ feet
$h - 5 = 10 - 5 = 5$ feet
$2h - 3 = 2(10) - 3 = 17$ feet
The length of the upper base is 5 feet and the length of the lower base is 17 feet.

26. $20{,}000-3(7{,}000)=20{,}000-21{,}000=-1{,}000$

This means that there is not enough sod left in the field to fill the order. Another 1000 square feet are required.

27. $-2x+6=-2(-3)+6=6+6=12$

28. $\dfrac{6-a}{1+a}=\dfrac{6-(-2)}{1+(-2)}=\dfrac{8}{-1}=-8$

29. $(-y-40)^2=[-(-50)-40]^2$
$=(50-40)^2=10^2=100$

30. $\left|x^3-8x^2\right|=\left|4^3-8\cdot4^2\right|=\left|64-8\cdot16\right|$
$=\left|64-128\right|=\left|-64\right|=64$

31. $b^2-4ac=6^2-4\cdot4(-4)$
$=36+64=100$

32. $(x+y)^2-|z+y|=(3+2)^2-|-4+2|$
$=5^2-|-2|=25-2=23$

33. $r-d=s$
$315-37=278$
The sale price is \$278.

34. $c+m=r$
$14{,}505+725=15{,}230$
The retail price is \$15,230.

35. $r-c=p$
$52{,}895-47{,}980=4{,}915$
The profit was \$4,915.

36.

	Rate (mph)	Time (hr)	Distance (mi)
Monorail	65	2	$65\cdot2=130$
Subway	38	3	$38\cdot3=114$
Train	x	6	$x\cdot6=6x$
Bus	55	$t+1$	$55(t+1)$

37. $C=\dfrac{5(F-32)}{9}=\dfrac{5(77-32)}{9}$

$=\dfrac{5(45)}{9}=5(5)=25°C$

The pool is $2°C$ warmer.

38. $d=16t^2$

$d=16\cdot 3^2=16\cdot 9=144$

The wrench will fall 144 feet.

39. $\dfrac{2\cdot 40+2\cdot 18+4}{5}=\dfrac{80+36+4}{5}$

$=\dfrac{120}{5}=24yr$

40. $\dfrac{1\cdot 0+2\cdot 3+3\cdot 3+4\cdot 5+5\cdot 9}{20}$

$=\dfrac{6+9+20+45}{20}=\dfrac{80}{20}=4$

41. $-2\cdot 5x=(-2\cdot 5)x=-10x$

42. $-7x(-6y)=[-7(-6)]xy=42xy$

43. $4d\cdot 3e\cdot 5=4\cdot 3\cdot 5\cdot de=60de$

44. $(4s)8=4\cdot 8\cdot s=32s$

45. $-1(-e)2=-1(2)(-1)e=2e$

46. $7x\cdot 7y=7\cdot 7\cdot x\cdot y=49xy$

47. $4\cdot 3k\cdot 7=4\cdot 3\cdot 7\cdot k=84k$

48. $(-10t)(-10)=(-10)(-10)t=100t$

49. $4(y+5)=4\cdot y+4\cdot 5=4y+20$

50. $-5(6t+9)=(-5)6t+(-5)9=-30t-45$

51. $(-3-3x)7=-3\cdot 7-3x\cdot 7=-21-21x$

52. $5(9s-12t-2)=5(9s)-5(12t)-5(2)$

$=45s-60t-10$

53. $4(6w-3)2=4\cdot 2(6w-3)$

$=8(6w-3)=48w-24$

54. $-9(x+1)4=-9\cdot 4(x+1)$

$=-36(x+1)=-36x-36$

55. $-(6t-4)=-6t+4$

56. $-(5+x)=-5-x$

57. $-(6t-3s+1)=-6t+3s-1$

58. $-(-5a-3)=5a+3$

59. $8x^2$, $-7x$, 9

60. $-15y$

61. $16ab$, $-6b$

62. $4x$, -3 , $5x$, -7

63. 5 , -4 , 8

64. 7 , 3 , 1 , -1

65. 1 , 1 , -1 , 6

66. -5 , 125

67. factor

68. term

69. term

70. factor

71. yes

72. no

73. yes

74. no

75. $3x+4x=(3+4)x=7x$

76. does not simplify

77. $2z+(-5z)=2z-5z=-3z$

78. $6x-x=5x$

79. $-6y-7y-(-y)=-13y+y=-12y$

80. $5w^2-8-4w^2+3=w^2-5$

81. $-45d-2a+4a-d=-46d+2a$

82. $5y+8h-3+7h+5y+2$
$=10y+15h-1$

83. $10a^2+6a-17a+6=10a^2-11a+6$

84. $28w+w=29w$

85. $7(y+6)+3(2y+2)$
$=7y+42+6y+6=13y+48$

86. $-4(t-7)-(t+6)$
$=-4t+28-t-6=-5t+22$

87. $5x-4-2(x-6)$
$=5x-4-2x+12=3x+8$

88. $6f+(-11)+7(12-8f)$
$=6f-11+84-56f=-50f+73$

89. $x+(x+4)+(x-1)=(3x+3)$ feet

90. Perimeter of the house:
$2\cdot 35+2\cdot 42=70+84=154\,ft$
Each window: $4\cdot 5=20\,ft$
$154+20+20=194\,ft$
194 feet of lights will be used.

91. $6a+(-7)=5a-9$
$6(-1)+(-7)\overset{?}{=}5(-1)-9$
$-6-7\overset{?}{=}-5-9$
$-13=-14$ False
-1 is not a solution.

92. $4(8-3t)=32-8(t+2)$
$4[8-3(4)]\overset{?}{=}32-8[(4)+2]$
$4(8-12)\overset{?}{=}32-8(6)$
$4(-4)\overset{?}{=}32-48$
$-16=-16$ True
4 is a solution.

93. $5a-3a=-36$
$2a=-36$
$\frac{2a}{2}=\frac{-36}{2}$
$a=-18$
Check: $5a-3a=-36$
$5(-18)-3(-18)\overset{?}{=}-36$
$-90+54\overset{?}{=}-36$
$-36=-36$ True
The solution checks.

94. $3x-4x=-8$
$-x=-8$
$\frac{-x}{-1}=\frac{-8}{-1}$
$x=8$
Check: $3x-4x=-8$
$3(8)-4(8)\overset{?}{=}-8$
$24-32\overset{?}{=}-8$
$-8=-8$ True
The solution checks.

95. $250+350=x-10+x$

$600=2x-10$

$600+10=2x-10+10$

$610=2x$

$\frac{610}{2}=\frac{2x}{2}$

$305=x$

Check: $250+350=x-10+x$

$250+350\overset{?}{=}(305)-10+(305)$

$600=600$ True

The solution checks.

96. $7x+1=3x-11$

$7x+1-3x=3x-11-3x$

$4x+1=-11$

$4x+1-1=-11-1$

$4x=-12$

$\frac{4x}{4}=\frac{-12}{4}$

$x=-3$

Check: $7x+1=3x-11$

$7(-3)+1\overset{?}{=}3(-3)-11$

$-21+1\overset{?}{=}-9-11$

$-20=-20$ True

The solution checks.

97. $5(y-15)=0$

$5y-75=0$

$5y-75+75=0+75$

$5y=75$

$\frac{5y}{5}=\frac{75}{5}$

$y=15$

Check: $5(y-15)=0$

$5[(15)-15]\overset{?}{=}0$

$5(0)\overset{?}{=}0$

$0=0$ True

The solution checks.

98. $3a-(2a-1)=-2$

$3a-2a+1=-2$

$a+1=-2$

$a+1-1=-2-1$

$a=-3$

Check: $3a-(2a-1)=-2$

$3(-3)-[2(-3)-1]\overset{?}{=}-2$

$-9-(-6-1)\overset{?}{=}-2$

$-9-(-7)\overset{?}{=}-2$

$-9+7\overset{?}{=}-2$

$-2=-2$ True

The solution checks.

99. $15+b=5b+1+3b$

$15+b=8b+1$

$15+b-b=8b+1-b$

$15=7b+1$

$15-1=7b+1-1$

$14=7b$

$\frac{14}{7}=\frac{7b}{7}$

$2=b$

Check: $15+b=5b+1+3b$

$15+(2)\overset{?}{=}5(2)+1+3(2)$

$15+2\overset{?}{=}10+1+6$

$17=17$ True

The solution checks.

100. $-6(2x+3)=-(5x-10)$

$-12x-18=-5x+10$

$-12x-18+12x=-5x+10+12x$

$-18=7x+10$

$-18-10=7x+10-10$

$-28=7x$

$\frac{-28}{7}=\frac{7x}{7}$

$-4=x$

Check: $-6[2(-4)+3]=-[5(-4)-10]$

$-6(-8+3)\overset{?}{=}-(-20-10)$

$-6(-5)\overset{?}{=}-(-30)$

$30=30$ True

The solution checks.

101. $5(2x-4)-5x=0$

$10x-20-5x=0$

$5x-20=0$

$5x-20+20=0+20$

$5x=20$

$x=4$

Check: $5(2x-4)-5x=0$

$5(2\cdot 4-4)-5(4)\overset{?}{=}0$

$5(8-4)-20\overset{?}{=}0$

$5(4)-20\overset{?}{=}0$

$20-20\overset{?}{=}0$

$0=0$ True

The solution checks.

102. $4(9d+2)=4(d-18)-8d$

$36d+8=4d-72-8d$

$36d+8=-4d-72$

$36d+8+4d=-4d-72+4d$

$40d+8=-72$

$40d+8-8=-72-8$

$40d=-80$

$\frac{40d}{40}=\frac{-80}{40}$

$d=-2$

Check: $4(9d+2)=4(d-18)-8d$

$4[9(-2)+2]\overset{?}{=}4[(-2)-18]-8(-2)$

$4(-18+2)\overset{?}{=}4(-20)+16$

$4(-16)\overset{?}{=}-80+16$

$-64=-64$ True

The solution checks.

103. x = the number of hours they can rent

$2,250+150x=3,300$

$2,250+150x-2,250=3,300-2,250$

$150x=1,050$

$\frac{150x}{150}=\frac{1,050}{150}$

$x=7$

They can rent the hall for 7 hours.

104. t = the time to cool in hours

$71-7t=29$

$71-7t-71=29-71$

$-7t=-42$

$\frac{-7t}{-7}=\frac{-42}{-7}$

$t=6$

It will take 6 hours to cool.

105. $t =$ the cost to rent the trailer

$$\frac{1}{2}t+4=20$$

$$\frac{1}{2}t+4-4=20-4$$

$$\frac{1}{2}t=16$$

$$2\cdot\frac{1}{2}t=2\cdot16$$

$$t=32$$

The trailer cost \$32 to rent.

106. $r =$ the distance she runs
r – 3 = the distance she walks

$$r+(r-3)=15$$

$$2r-3=15$$

$$2r-3+3=15+3$$

$$2r=18$$

$$\frac{2r}{2}=\frac{18}{2}$$

$$r=9$$

She runs 9 miles and walks 9 – 3 = 6 miles.

107. $f =$ the attendance on the first day

$$f+2f=6,600$$

$$3f=6,600$$

$$\frac{3f}{3}=\frac{6,600}{3}$$

$$f=2,200$$

The first day there were 2,200 people. The second day there were 2(2,200) = 4,400 people.

108. $w =$ the width of the parking lot
$4w =$ the length of the parking lot

$$P=2l+2w$$

$$250=2(4w)+2w$$

$$250=8w+2w$$

$$250=10w$$

$$\frac{250}{10}=\frac{10w}{10}$$

$$25=w$$

The width is 25 feet and the length is 4(25) = 100 feet.

109.

Type of coin	Number	Value (cents)	Total Value (cents)
Dime	6	10	$6\cdot10=60$
Quarter	7	25	$7\cdot25=175$
Penny	x	1	$x\cdot1=x$
Nickel	$n+25$	5	$(n+25)5$

110. $t =$ the number of three dollar drinks sold
$50 - t =$ the number of four dollar drinks sold

$$3t+4(50-t)=185$$

$$3t+200-4t=185$$

$$-t+200=185$$

$$-t+200-200=185-200$$

$$-t=-15$$

$$\frac{-t}{-1}=\frac{-15}{-1}$$

$$t=15$$

There were 15 three dollar drinks sold and 50 – 15 = 35 four dollar drinks sold.

Chapter 3 Test

1. a. Variables are letters (or symbols) that stand for numbers.

b. To perform the multiplication $3(x+4)$, we use the distributive property.

c. Terms such as $7x^2$ and $5x^2$, which have the same variables raised to exactly the same power, are called like terms.

d. To solve an equation means to find all values of the variable that make the equation true.

e. The coefficient of the term $9y$ is 9.

f. Variables and/or numbers can be combined with the operations of addition, subtraction, multiplication, and division to create algebraic expressions.

g. To evaluate y^2+9y-3 for $y=-5$, we substitute -5 for y and apply the order of operations rule.

h. An equation is a statement indicating that two expressions are equal.

i. When we write $4x+x$ as $5x$, we say we have combined like terms.

j. To check the solution of an equation, we substitute the value for the variable in the original equation and determine whether the result is a true statement.

2. a. $2h-1{,}000$

b. $2\cdot 2{,}350-1{,}000=4{,}700-1{,}000=\$3{,}700$

3. $56-c$

4. t = the trout length
$t+10$ = the salmon length

s = the salmon length
$s-10$ = the trout length

5. a. $r-2$ b. $3xy$ c. $x+100$ d. $\left|\frac{x}{-9}\right|$

6. $10d$

7. a. $x-16=4-16=-12$

b. $2t^2-3(t-s)=2(-2)^2-3(-2-4)$
$=2\cdot 4-3(-6)=8+18=26$

c. $-a^2+10=-(-3)^2+10=-9+10=1$

d. $\left|\frac{-10d+f^3}{-f}\right|=\left|\frac{-10(-1)+(-5)^3}{-(-5)}\right|$
$=\left|\frac{10-125}{5}\right|=\left|\frac{-115}{5}\right|=|-23|=23$

8. $d=r\cdot t=55\cdot 3=165$
The distance traveled is 165 miles.

9. $r=40{,}000+15{,}000=\$55{,}000$
$c=13{,}000+5{,}000=\$18{,}000$
$p=r-c=\$55{,}000-\$18{,}000=\$37{,}000$
The profit is \$37,000.

10. $h=16t^2=16\cdot 3^2=16\cdot 9=144$ *feet*
It has not hit the ground yet.
There are 200-144=56 feet left to go.

11. $\frac{4+2+(-4)+(-4)+4+2+(-4)+2+4+4}{10}$
$=\frac{10}{10}=1$

12. Perimeter of the outside:
$P=4s=4\cdot 40=160\,ft$
Perimeter of the inside:
$P=2\cdot 20+2\cdot 25=40+50=90\,ft$
Total edging needed: $160+90=250\,ft$

13. $C=\dfrac{5(F-32)}{9}=\dfrac{5(59-32)}{9}$

$=\dfrac{5(27)}{9}=5\cdot 3=15$

The temperature was 15°C.

14. a. $5(5x+1)=5\cdot 5x+5\cdot 1=25x+5$

b. $-6(7-x)=-42+6x$

c. $-(6y+4)=-6y-4$

d. $3(2a+3b-7)=6a+9b-21$

e. $(a-15)8=8a-120$

f. $2(6r+9)3=6(6r+9)=36r+54$

15. a. factor b. term

16. a. $7x+4x=11x$

b. $3\cdot 4e=12e$

c. $6x^2-x^2=5x^2$

d. $-5y(-6)=30y$

e. $0-7x=-7x$

f. $0+9y=9y$

g. $-8a\cdot 9b=-72ab$

h. $8(-7m)5=-280m$

17. a. $8x^2$, $-x$, -6

b. 8 , −1 , −6

18. a. $-20y+6-8y+4=-28y+10$

b. $-t-t-t=-3t$

c. $4(y+3)-5(2y+3)$
$=4y+12-10y-15$
$=-6y-3$

d. $-m^4+3m^3+10m^4+20m^3$
$=9m^4+23m^3$

19. a. $10k$ ¢ b. $20(p+2)$ dollars

20. $6x-8=12(x-3)$

$6(-5)-8\overset{?}{=}12[(-5)-3]$

$-30-8\overset{?}{=}12(-8)$

$-38=-96$

-5 is not a solution.

21. $5x-3x=-18$

$2x=-18$

$\dfrac{2x}{2}=\dfrac{-18}{2}$

$x=-9$

Check: $5x-3x=-18$

$5(-9)-3(-9)\overset{?}{=}-18$

$-45+27\overset{?}{=}-18$

$-18=-18$ True

The solution checks.

22. $6r=r-12+r$

$6r=2r-12$

$6r-2r=2r-12-2r$

$4r=-12$

$r=-3$

Check: $6r=r-12+r$

$6(-3)\overset{?}{=}(-3)-12+(-3)$

$-18=-18$ True

The solution checks.

23.
$$-55+10=3(1-4t)$$
$$-45=3-12t$$
$$-45-3=3-12t-3$$
$$-48=-12t$$
$$\frac{-48}{-12}=\frac{-12t}{-12}$$
$$4=t$$
Check: $-55+10=3(1-4t)$
$$-55+10\stackrel{?}{=}3[1-4(4)]$$
$$-45\stackrel{?}{=}3(1-16)$$
$$-45\stackrel{?}{=}3(-15)$$
$$-45=-45 \quad \text{True}$$
The solution checks.

24.
$$6-(y-3)=19$$
$$6-y+3=19$$
$$9-y=19$$
$$9-y-9=19-9$$
$$-y=10$$
$$\frac{-y}{-1}=\frac{10}{-1}$$
$$y=-10$$
Check: $6-(y-3)=19$
$$6-[(-10)-3]\stackrel{?}{=}19$$
$$6-(-13)\stackrel{?}{=}19$$
$$6+13\stackrel{?}{=}19$$
$$19=19 \quad \text{True}$$
The solution checks.

25.
$$8+2(3x-4)=-60$$
$$8+6x-8=-60$$
$$6x=-60$$
$$\frac{6x}{6}=\frac{-60}{6}$$
$$x=-10$$
Check: $6-(y-3)=19$
$$6-(-10-3)\stackrel{?}{=}19$$
$$6-(-13)\stackrel{?}{=}19$$
$$6+13\stackrel{?}{=}19$$
$$19=19 \quad \text{True}$$
The solution checks.

26.
$$-23-n-6n=13+2n$$
$$-23-7n=13+2n$$
$$-23-7n+7n=13+2n+7n$$
$$-23=13+9n$$
$$-23-13=9n$$
$$-36=9n$$
$$\frac{-36}{9}=\frac{9n}{9}$$
$$-4=n$$
Check: $-23-n-6n=13+2n$
$$-23-(-4)-6(-4)\stackrel{?}{=}13+2(-4)$$
$$-23+4+24\stackrel{?}{=}13-8$$
$$5=5 \quad \text{True}$$
The solution checks.

27. $5(x+7)-7x=9(x-1)$

$5x+35-7x=9x-9$

$-2x+35=9x-9$

$-2x+35+2x=9x-9+2x$

$35=11x-9$

$35+9=11x-9+9$

$44=11x$

$\frac{44}{11}=\frac{11x}{11}$

$4=x$

Check: $5(x+7)-7x=9(x-1)$

$5[(4)+7]-7(4)\stackrel{?}{=}9[(4)-1]$

$5(11)-28\stackrel{?}{=}9(3)$

$55-28\stackrel{?}{=}27$

$27=27$ True

The solution checks.

28. $-80+y=-80-y$

$-80+y+80=-80-y+80$

$y=-y$

$y+y=-y+y$

$2y=0$

$\frac{2y}{2}=\frac{0}{2}$

$y=0$

Check: $-80+y=-80-y$

$-80+0\stackrel{?}{=}-80-0$

$-80=-80$ True

The solution checks.

29. x = the length of a classroom section

$6x+2=20$

$6x+2-2=20-2$

$6x=18$

$\frac{6x}{6}=\frac{18}{6}$

$x=3$

Each classroom session is 3 hours long.

30. l = the length of time for local shows

$l+8$ = the length of time for national shows

$l+(l+8)=24$

$2l+8=24$

$2l+8-8=24-8$

$2l=16$

$\frac{2l}{2}=\frac{16}{2}$

$l=8$

Each day there were 8 hours of local shows and 8 + 8 = 16 hours of national shows.

31. d = the amount donated

$\frac{1}{2}d+4+18=d$

$\frac{1}{2}d+22=d$

$\frac{1}{2}d+22-\frac{1}{2}d=d-\frac{1}{2}d$

$22=\frac{1}{2}d$

$2\cdot 22=2\cdot\frac{1}{2}d$

$44=d$

The developer donated 44 acres of land to the city.

32. w = the width

$2w$ = the length

$P=2l+2w$

$144=2(2w)+2w$

$144=4w+2w$

$144=6w$

$\frac{144}{6}=\frac{6w}{6}$

$24=w$

The width of the frame is 24 inches and the length is $2\cdot 24=48$ inches.

33. They do not mean the same thing. We simplify expressions and we solve equations.

34. We cannot simplify $5x+2$ because $5x$ and 2 are not like terms and therefore cannot be combined.

Chapters 1 – 3 Cumulative Review

1. 3,194,754,000

2. approximately 50,000

3.
$$\begin{array}{r} {}^{1\,1\;\,1\,1}\\ 38,908 \\ +15,696 \\ \hline 54,604 \end{array}$$

4.
$$\begin{array}{r} 9,700 \\ -5,491 \\ \hline 4,209 \end{array}$$

5.
$$\begin{array}{r} 345 \\ \times\quad 67 \\ \hline 2,415 \\ +20,700 \\ \hline 23,115 \end{array}$$

6.
$$\begin{array}{r} 87 \\ 23\overline{)2001} \\ -184 \\ 161 \\ -161 \\ 0 \end{array}$$

7. a. add upwards: $683+459=1,142$

b. $\frac{999}{0}$; $\frac{0}{222}$. Division by 0 is undefined.

8. Keep adding 12: $2010+12=2022$.

9. $4\cdot5=5+5+5+5=20$

10. Each side of plywood: $A=lw=22\cdot62=1,364in^2$

8 sides: $8\cdot1,364=10,912in^2$

11. $\frac{558,000}{3,000}=\frac{558}{3}=186$ oil changes

12. a. 1, 2, 3, 6, 9, 18

b. Because it has factors besides 1 and itself.

c. $18=2\cdot9=2\cdot3\cdot3=2\cdot3^2$

13. 2, 3, 5, 7, 11, 13, 17, 19, 23, 29

14. a. $35=5\cdot7$

$45=3\cdot3\cdot5$

$LCM(35,45)=3\cdot3\cdot5\cdot7=315$

b. $12=2\cdot2\cdot3$

$68=2\cdot2\cdot17$

$92=2\cdot2\cdot23$

$GCF(12,68,92)=2\cdot2=4$

15. a. $(9-2)^2-3^3=7^2-27$

$=49-27=22$

b. $\frac{80-2[12-(5+4)]}{8\div8\cdot2}$

$=\frac{80-2[12-9]}{1\cdot2}$

$=\frac{80-2[3]}{2}=\frac{80-6}{2}$

$=\frac{74}{2}=37$

16. Addition Property of Equality

17. $250=\frac{y}{2}$

$2\cdot250=2\cdot\frac{y}{2}$

$500=y$

Check: $250=\frac{y}{2}$

$250\stackrel{?}{=}\frac{500}{2}$

$250=250$ True

The solution checks.

18. a. $-(-6)=6$

b. $|-5|=5$

c. false

19.

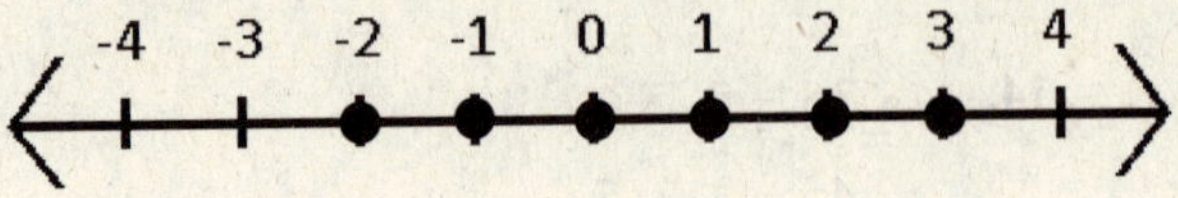

20. $-21-(-73)$

21. a. $-25+5=-20$

b. $25-(-5)=25+5=30$

c. $-25(5)(-1)=-125(-1)=125$

d. $\frac{-25}{-5}=\frac{25}{5}=5$

22. Team 1: $1,705-75=1,630$

Team 2: $695-120=575$

23. $810°-(-290°)=810°+290°=1,100°$

24. a. $-3^2=-(3\cdot 3)=-9$; $(-3)^2=(-3)(-3)=9$

b. Commutative Property of Multiplication

25. $\frac{(-6)^2-1^5}{-4-3}=\frac{36-1}{-7}=\frac{35}{-7}=-5$

26. $-3+3(-4-4\cdot 2)^2=-3+3(-4-8)^2$

$=-3+3(-12)^2=-3+3(144)$

$=-3+432=429$

27. $-4x+4=-24$

$-4x+4-4=-24-4$

$-4x=-28$

$\frac{-4x}{-4}=\frac{-28}{-4}$

$x=7$

Check: $-4x+4=-24$

$-4(7)+4\overset{?}{=}-24$

$-28+4\overset{?}{=}-24$

$-24=-24$ True

The result checks.

28. $-y=10$

$\frac{-y}{-1}=\frac{10}{-1}$

$y=-10$

Check: $-y=10$

$-(-10)\overset{?}{=}10$

$10=10$ True

The result checks.

29. $\frac{m}{-2}-6=-9$

$\frac{m}{-2}-6+6=-9+6$

$\frac{m}{-2}=-3$

$-2\left(\frac{m}{-2}\right)=-2(-3)$

$m=6$

Check: $\frac{m}{-2}-6=-9$

$\frac{6}{-2}-6\overset{?}{=}-9$

$-3-6\overset{?}{=}-9$

$-9=-9$ True

The result checks.

30. $-7+9=a+5(-3)$

$2=a-15$

$2+15=a-15+15$

$17=a$

Check: $-7+9=a+5(-3)$

$-7+9\stackrel{?}{=}17+5(-3)$

$2\stackrel{?}{=}17-15$

$2=2$ True

The result checks.

31. $-90=x+(-3)$

$-90+3=x+(-3)+3$

$-87=x$

Check: $-90=x+(-3)$

$-90\stackrel{?}{=}(-87)+(-3)$

$-90=-90$ True

The result checks.

32. $16=-7-5-7x$

$16=-12-7x$

$16+12=-12-7x+12$

$28=-7x$

$\frac{28}{-7}=\frac{-7x}{-7}$

$-4=x$

Check: $16=-7-5-7x$

$16\stackrel{?}{=}-7-5-7(-4)$

$16\stackrel{?}{=}-12+28$

$16=16$ True

The result checks.

33. a. $h+12$

b. $w-4$

c. $\frac{1{,}000}{x}$

34. a. $(26-x)in$

b. $25q¢$

35. $12f\ in$

36. $x^2-2x+1=(-5)^2-2(-5)+1$

$=25+10+1=36$

37.

	Rate (mph)	Time (hr)	Distance Traveled (mi)
Truck	55	4	$55\cdot 4=220$

38. a. $5(2x-7)=5\cdot 2x-5\cdot 7=10x-35$

b. $-(5t-7)=-5t+7$

39. a. $(-6)(-4t)=[-6(-4)]t=24t$

b. $4(-3y)(4z)=-12y\cdot 4z=-48yz$

40.

Term	Coefficient
$4a$	4
$-2y^2$	-2
x	1
$-m$	-1

41. Term: $x-11{,}139$

Factor: $17x$

42. a. $5b+8-6b-7=-b+1$

b. $4(x+5)-5-(2x-4)$

$=4x+20-5-2x+4$

$=2x+19$

43. $8p+2p-1=-11$

$10p-1=-11$

$10p-1+1=-11+1$

$10p=-10$

$\frac{10p}{10}=\frac{-10}{10}$

$p=-1$

Check: $8p+2p-1=-11$

$8(-1)+2(-1)-1\stackrel{?}{=}-11$

$-8-2-1\stackrel{?}{=}-11$

$-11=-11$ True

The result checks.

44. $7+2x=2-(4x+7)$

$7+2x=2-4x-7$

$7+2x=-5-4x$

$7+2x+4x=-5-4x+4x$

$7+6x=-5$

$7+6x-7=-5-7$

$6x=-12$

$\frac{6x}{6}=\frac{-12}{6}$

$x=-2$

Check: $7+2x=2-(4x+7)$

$7+2(-2)\stackrel{?}{=}2-[4(-2)+7]$

$7-4\stackrel{?}{=}2-(-8+7)$

$3\stackrel{?}{=}2-(-1)$

$3\stackrel{?}{=}2+1$

$3=3$ True

The result checks.

45. $5(2x-4)-5x=0$

$10x-20-5x=0$

$5x-20=0$

$5x-20+20=0+20$

$5x=20$

$x=4$

Check: $5(2x-4)-5x=0$

$5(2\cdot4-4)-5(4)\stackrel{?}{=}0$

$5(8-4)-20\stackrel{?}{=}0$

$5(4)-20\stackrel{?}{=}0$

$20-20\stackrel{?}{=}0$

$0=0$ True

The solution checks.

46. $4(m-30)-4m=6(2m+6)$

$4m-120-4m=12m+36$

$-120=12m+36$

$-120-36=12m+36-36$

$-156=12m$

$\frac{-156}{12}=\frac{12m}{12}$

$m=-13$

Check: $4(m-30)-4m=6(2m+6)$

$4[(-13)-30]-4(-13)\stackrel{?}{=}6[2(-13)+6]$

$4(-43)+52\stackrel{?}{=}6(-26+6)$

$-172+52\stackrel{?}{=}6(-20)$

$-120=-120$ True

The result checks.

47. l = the time they spend in the lab

$$l+(l+50)=300$$
$$2l+50=300$$
$$2l+50-50=300-50$$
$$2l=250$$
$$\frac{2l}{2}=\frac{250}{2}$$
$$l=125$$

The students spend 125 minutes in the lab and 125 + 50 = 175 minutes in lecture per week.

48. w = the width
$5w$ = the length

$$P=2l+2w$$
$$120=2(5w)+2w$$
$$120=10w+2w$$
$$120=12w$$
$$\frac{120}{12}=\frac{12w}{12}$$
$$10=w$$

The width is 10 feet and the length is $5\cdot 10=50$ feet.

CHAPTER 4 Fractions and Mixed Numbers

Section 4.1: An Introduction to Fractions

VOCABULARY

1. A fraction describes the number of equal parts of a whole.

3. If the numerator of a fraction is less than its denominator, the fraction is called a proper fraction. If the numerator of a fraction is greater than or equal to its denominator it is called an improper fraction.

5. Two fractions are equivalent if they represent the same number.

7. A fraction is in simplest form, or lowest terms, when the numerator and denominator have no common factors other than 1.

CONCEPTS

9. Equivalent fractions: $\frac{2}{6}=\frac{1}{3}$

11. a. improper fraction b. proper fraction
c. proper fraction d. improper fraction

13. The common factor of 10 and 15 is 5.

15. Multiplying fractions: To multiply two fractions, multiply the numerators and multiply the denominators.

NOTATION

17. $\frac{-7}{8}, -\frac{7}{8}$

19. $\frac{1}{6}=\frac{1}{6}\cdot\frac{3}{3}=\frac{1\cdot3}{6\cdot3}=\frac{3}{18}$

GUIDED PRACTICE

21. numerator 4; denominator 5

23. numerator 17; denominator 10

25. $\frac{3}{4}$ shaded, $\frac{1}{4}$ not shaded

27. $\frac{5}{8}$ shaded, $\frac{3}{8}$ not shaded

29. $\frac{1}{4}$ shaded, $\frac{3}{4}$ not shaded

31. $\frac{7}{12}$ shaded, $\frac{5}{12}$ not shaded

33. a. $\frac{4}{1}=4$ b. $\frac{8}{8}=1$

c. $\frac{0}{12}=0$ d. $\frac{1}{0}$ undefined

35. a. $\frac{5}{0}$ undefined b. $\frac{0}{50}=0$

c. $\frac{33}{33}=1$ d. $\frac{75}{1}=75$

37. $\frac{7}{8}=\frac{7}{8}\cdot\frac{5}{5}=\frac{7\cdot5}{8\cdot5}=\frac{35}{40}$

39. $\frac{4}{9}=\frac{4}{9}\cdot\frac{3}{3}=\frac{4\cdot3}{9\cdot3}=\frac{12}{27}$

41. $\frac{5}{6}=\frac{5}{6}\cdot\frac{9}{9}=\frac{5\cdot9}{6\cdot9}=\frac{45}{54}$

43. $\frac{2}{7}=\frac{2}{7}\cdot\frac{2}{2}=\frac{2\cdot2}{7\cdot2}=\frac{4}{14}$

45. $4=\frac{4}{1}\cdot\frac{9}{9}=\frac{4\cdot9}{1\cdot9}=\frac{36}{9}$

47. $6=\frac{6}{1}\cdot\frac{8}{8}=\frac{6\cdot8}{1\cdot8}=\frac{48}{8}$

49. $3=\frac{3}{1}\cdot\frac{5}{5}=\frac{3\cdot5}{1\cdot5}=\frac{15}{5}$

51. $14=\frac{14}{1}\cdot\frac{2}{2}=\frac{14\cdot2}{1\cdot2}=\frac{28}{2}$

53. a. No b. Yes

55. a. Yes b. No

57. $\frac{6}{9}=\frac{2\cdot3}{3\cdot3}=\frac{2\cdot\cancel{3}}{3\cdot\cancel{3}}=\frac{2}{3}$

59. $\frac{16}{20}=\frac{2\cdot2\cdot2\cdot2}{2\cdot2\cdot5}=\frac{\cancel{2}\cdot\cancel{2}\cdot2\cdot2}{\cancel{2}\cdot\cancel{2}\cdot5}=\frac{4}{5}$

61. $\frac{5}{15}=\frac{5}{3\cdot5}=\frac{\cancel{5}}{3\cdot\cancel{5}}=\frac{1}{3}$

63. $\frac{2}{48}=\frac{2}{2\cdot2\cdot2\cdot2\cdot3}=\frac{\cancel{2}}{\cancel{2}\cdot2\cdot2\cdot2\cdot3}=\frac{1}{24}$

65. $\frac{16}{17}$ in simplest form

67. $\frac{36}{96}=\frac{3\cdot12}{8\cdot12}=\frac{3\cdot\cancel{12}}{8\cdot\cancel{12}}=\frac{3}{8}$

69. $\frac{55}{62}$ in simplest form

71. $\frac{50}{55}=\frac{2\cdot5\cdot5}{5\cdot11}=\frac{2\cdot\cancel{5}\cdot5}{\cancel{5}\cdot11}=\frac{10}{11}$

73. $\frac{60}{108}=\frac{2\cdot2\cdot3\cdot5}{2\cdot2\cdot3\cdot3\cdot3}=\frac{\cancel{2}\cdot\cancel{2}\cdot\cancel{3}\cdot5}{\cancel{2}\cdot\cancel{2}\cdot\cancel{3}\cdot3\cdot3}=\frac{5}{9}$

75. $\frac{180}{210}=\frac{6\cdot\cancel{30}}{7\cdot\cancel{30}}=\frac{6}{7}$

77. $\frac{306}{234}=\frac{17\cdot\cancel{18}}{13\cdot\cancel{18}}=\frac{17}{13}$

79. $\frac{105}{42}=\frac{3\cdot5\cdot7}{2\cdot3\cdot7}=\frac{\cancel{3}\cdot5\cdot\cancel{7}}{2\cdot\cancel{3}\cdot\cancel{7}}=\frac{5}{2}$

81. $\frac{420}{144}=\frac{\cancel{12}\cdot35}{\cancel{12}\cdot12}=\frac{35}{12}$

83. $-\frac{4}{68}=-\frac{\cancel{4}}{\cancel{4}\cdot17}=-\frac{1}{17}$

85. $-\frac{90}{105}=-\frac{6\cdot\cancel{15}}{7\cdot\cancel{15}}=-\frac{6}{7}$

87. $-\frac{16}{26}=-\frac{\cancel{2}\cdot8}{\cancel{2}\cdot13}=-\frac{8}{13}$

89. $\frac{1}{2}=\frac{1}{2}\cdot\frac{3a}{3a}=\frac{3a}{6a}$

91. $\frac{9}{10}=\frac{9}{10}\cdot\frac{5c}{5c}=\frac{45c}{50c}$

93. $\frac{5}{4n}=\frac{5}{4n}\cdot\frac{11}{11}=\frac{55}{44n}$

95. $\frac{14}{15x}=\frac{14}{15x}\cdot\frac{3}{3}=\frac{42}{45x}$

97. $\frac{7}{14a}=\frac{\cancel{7}}{2\cdot\cancel{7}\cdot a}=\frac{1}{2a}$

99. $\frac{3x}{12}=\frac{\cancel{3}\cdot x}{\cancel{3}\cdot4}=\frac{x}{4}$

101. $\frac{4m^5}{25m^4}=\frac{4\cdot m\cdot\cancel{m}\cdot\cancel{m}\cdot\cancel{m}\cdot\cancel{m}}{25\cdot\cancel{m}\cdot\cancel{m}\cdot\cancel{m}\cdot\cancel{m}}=\frac{4m}{25}$

103. $\frac{6b^4}{9b}=\frac{2\cdot\cancel{3}\cdot\cancel{b}\cdot b\cdot b\cdot b}{\cancel{3}\cdot3\cdot\cancel{b}}=\frac{2b^3}{3}$

105. $\frac{35a^3b^2}{25a^2b^3}=\frac{\cancel{5}\cdot7\cdot\cancel{a}\cdot\cancel{a}\cdot a\cdot\cancel{b}\cdot\cancel{b}}{\cancel{5}\cdot5\cdot\cancel{a}\cdot\cancel{a}\cdot b\cdot\cancel{b}\cdot\cancel{b}}=\frac{7a}{5b}$

107. $-\frac{16n^5p}{24np}=-\frac{2\cdot\cancel{8}\cdot\cancel{n}\cdot n\cdot n\cdot n\cdot n\cdot\cancel{p}}{3\cdot\cancel{8}\cdot\cancel{n}\cdot\cancel{p}}=-\frac{2n^4}{3}$

TRY IT YOURSELF

109.
a. $\frac{2}{14}=\frac{1}{7};\ \frac{6}{36}=\frac{1}{6}$ not equivalent

b. $\frac{4}{30}=\frac{2}{15};\ \frac{12}{90}=\frac{2}{15}$ equivalent

APPLICATIONS

113.
a. 32 b. $\frac{5}{32}$

115.
a. 16 b. $\frac{10}{16}=\frac{\cancel{2}\cdot5}{\cancel{2}\cdot8}=\frac{5}{8}$

117. a. 19 Democratic, 30 Republican, 1 Independent

b. $\frac{19}{50}$

c. $\frac{30}{50}=\frac{\cancel{2}\cdot 3\cdot \cancel{5}}{\cancel{2}\cdot \cancel{5}\cdot 5}=\frac{3}{5}$

d. $\frac{1}{50}$

119.

a. 20 b. $\frac{8}{20}=\frac{2\cdot \cancel{4}}{\cancel{4}\cdot 5}=\frac{2}{5}$ sold

$\frac{12}{20}=\frac{3\cdot \cancel{4}}{\cancel{4}\cdot 5}=\frac{3}{5}$ not sold

121. The second quarter begins on April 1 and ends on June 30.

WRITING

123. Equivalent fractions are those that reduce to the same value. For example, $\frac{3}{6}$ and $\frac{4}{8}$ are equivalent since both reduce to $\frac{1}{2}$.

125. Because the pieces are not the same size.

127. a. Building a fraction: The numerator and denominator must be multiplied by the same number.

b. Reducing a fraction: The same number must be divided out from the numerator and denominator.

REVIEW

129. $\$3{,}575-\$235-\$782$
$-\$148-\$103=\$2{,}307$

Section 4.2: Multiplying Fractions

VOCABULARY

1. When a fraction is followed by the word *of*, such as $\frac{1}{3}$ *of*, it indicates that we are to find a part of some quantity using multiplication.

3. To simplify a fraction we remove the common factors of the numerator and denominator.

5. The area of a triangle is the amount of surface that it encloses.

CONCEPTS

7. To multiply two fractions, multiply the numerators and multiply the denominators. Then simplify if possible.

9. a. negative b. positive

c. positive d. negative

11. a. Area of a triangle = $\frac{1}{2}(\text{base})(\text{height})$ or

$A=\frac{1}{2}bh$

b. Area is measured in square units, such as in^2 and ft^2.

NOTATION

13. a. $4=\frac{4}{1}$ b. $-3=-\frac{3}{1}$ c. $x=\frac{x}{1}$

15.

$$\frac{5}{8}\cdot\frac{7}{15}=\frac{5\cdot 7}{8\cdot 15}$$
$$=\frac{5\cdot 7}{2\cdot 2\cdot 2\cdot 3\cdot 5}$$
$$=\frac{\cancel{5}\cdot 7}{2\cdot 2\cdot 2\cdot 3\cdot \cancel{5}}$$
$$=\frac{7}{24}$$

GUIDED PRACTICE

17. $\frac{1}{4}\cdot\frac{1}{2}=\frac{1}{8}$

19. $\frac{1}{9}\cdot\frac{1}{5}=\frac{1}{45}$

21. $\frac{2}{3}\cdot\frac{7}{9}=\frac{14}{27}$

23. $\frac{8}{11}\cdot\frac{3}{7}=\frac{24}{77}$

25. $-\frac{4}{5}\cdot\frac{1}{3}=-\frac{4}{15}$

27. $\frac{5}{6}\cdot\left(-\frac{7}{12}\right)=-\frac{35}{72}$

29. $\frac{1}{8}\cdot 9=\frac{1}{8}\cdot\frac{9}{1}=\frac{9}{8}$

31. $\frac{1}{2}\cdot 5=\frac{1}{2}\cdot\frac{5}{1}=\frac{5}{2}$

33. $\frac{11}{10}\cdot\frac{5}{11}=\frac{11}{2\cdot5}\cdot\frac{5}{11}=\frac{\cancel{11}}{2\cdot\cancel{5}}\cdot\frac{\cancel{5}}{\cancel{11}}=\frac{1}{2}$

35. $\frac{6}{49}\cdot\frac{7}{6}=\frac{6}{7\cdot7}\cdot\frac{7}{6}=\frac{\cancel{6}}{7\cdot\cancel{7}}\cdot\frac{\cancel{7}}{\cancel{6}}=\frac{1}{7}$

37. $\frac{3}{4}\left(-\frac{8}{35}\right)\left(-\frac{7}{12}\right)=\frac{3}{4}\left(-\frac{2\cdot4}{5\cdot7}\right)\left(-\frac{7}{2\cdot2\cdot3}\right)$

$=\frac{\cancel{3}}{\cancel{4}}\left(-\frac{\cancel{2}\cdot\cancel{4}}{5\cdot\cancel{7}}\right)\left(-\frac{\cancel{7}}{\cancel{2}\cdot2\cdot\cancel{3}}\right)=\frac{1}{10}$

39. $-\frac{5}{8}\left(\frac{16}{27}\right)\left(-\frac{9}{25}\right)=-\frac{5}{8}\left(\frac{2\cdot8}{3\cdot9}\right)\left(-\frac{9}{5\cdot5}\right)$

$=-\frac{\cancel{5}}{\cancel{8}}\left(\frac{2\cdot\cancel{8}}{3\cdot\cancel{9}}\right)\left(-\frac{\cancel{9}}{5\cdot\cancel{5}}\right)=\frac{2}{15}$

41. $\frac{3}{4x}\cdot\frac{2x}{5}=\frac{3\cdot\cancel{2}\cdot\cancel{x}}{2\cdot\cancel{2}\cdot\cancel{x}\cdot5}=\frac{3}{2\cdot5}=\frac{3}{10}$

43. $\frac{6a}{49}\cdot\frac{14}{3a}=\frac{2\cdot\cancel{3}\cdot\cancel{a}\cdot2\cdot\cancel{7}}{\cancel{7}\cdot7\cdot\cancel{3}\cdot\cancel{a}}=\frac{2\cdot2}{7}=\frac{4}{7}$

45. $-\frac{2m^3}{21n}\cdot\frac{27mn}{16m^5}$

$=-\frac{\cancel{2}\cdot\cancel{m}\cdot\cancel{m}\cdot\cancel{m}\cdot\cancel{3}\cdot3\cdot3\cdot\cancel{m}\cdot\cancel{n}}{\cancel{3}\cdot7\cdot\cancel{n}\cdot\cancel{2}\cdot8\cdot\cancel{m}\cdot\cancel{m}\cdot\cancel{m}\cdot\cancel{m}\cdot m}$

$=-\frac{3\cdot3}{7\cdot8\cdot m}=-\frac{9}{56m}$

47. $\left(-\frac{25xy}{8y}\right)\left(-\frac{8x^2}{45x^2y^2}\right)$

$=\frac{\cancel{5}\cdot5\cdot\cancel{x}\cdot\cancel{y}\cdot\cancel{8}\cdot\cancel{x}\cdot x}{\cancel{8}\cdot y\cdot\cancel{5}\cdot9\cdot\cancel{x}\cdot\cancel{x}\cdot\cancel{y}\cdot y}=\frac{5\cdot x}{9\cdot y\cdot y}=\frac{5x}{9y^2}$

49. $\frac{1}{8}(8w)=\frac{1}{8}\cdot\frac{8w}{1}=\frac{1\cdot\cancel{8}\cdot w}{\cancel{8}\cdot1}=w$

51. $7x\cdot\frac{1}{7}=\frac{7x}{1}\cdot\frac{1}{7}=\frac{\cancel{7}\cdot x\cdot1}{1\cdot\cancel{7}}=x$

53. $12a\cdot\frac{1}{6a}=\frac{12a}{1}\cdot\frac{1}{6a}=\frac{2\cdot\cancel{6}\cdot\cancel{a}\cdot1}{1\cdot\cancel{6}\cdot\cancel{a}}=2$

55. $\frac{1}{9}(-36y)=-\frac{1}{9}\cdot\frac{36y}{1}=-\frac{1\cdot4\cdot\cancel{9}\cdot y}{\cancel{9}\cdot1}=-4y$

57.

a. $\left(\frac{3}{5}\right)^2=\frac{3}{5}\cdot\frac{3}{5}=\frac{9}{25}$

b. $\left(-\frac{3}{5}\right)^2=\left(-\frac{3}{5}\right)\left(-\frac{3}{5}\right)=\frac{9}{25}$

59.

a. $-\left(-\frac{1}{6}\right)^2=-\left(-\frac{1}{6}\right)\left(-\frac{1}{6}\right)=-\frac{1}{36}$

b. $\left(-\frac{1}{6}\right)^3=\left(-\frac{1}{6}\right)\left(-\frac{1}{6}\right)\left(-\frac{1}{6}\right)=-\frac{1}{216}$

61. $\left(-\frac{6t}{7}\right)^2=\left(\frac{-6t}{7}\right)\left(-\frac{6t}{7}\right)=\frac{6t\cdot6t}{7\cdot7}=\frac{36t^2}{49}$

63. $\left(-\frac{2a}{5}\right)^3=\left(-\frac{2a}{5}\right)\left(-\frac{2a}{5}\right)\left(-\frac{2a}{5}\right)$

$=-\frac{2a\cdot 2a\cdot 2a}{5\cdot 5\cdot 5}=-\frac{8a^3}{125}$

65. $\frac{3}{4}$ of $\frac{5}{8}=\frac{3}{4}\cdot\frac{5}{8}=\frac{15}{32}$

67. $\frac{1}{6}$ of $54=\frac{1}{6}\cdot\frac{54}{1}=\frac{54}{6}=9$

69. $A=\frac{1}{2}bh=\frac{1}{2}\cdot 10\cdot 3=5\cdot 3=15\,ft^2$

71. $A=\frac{1}{2}bh=\frac{1}{2}\cdot 18\cdot 7=9\cdot 7=63in^2$

73. $A=\frac{1}{2}bh=\frac{1}{2}\cdot 4\cdot 3=2\cdot 3=6m^2$

75. $A=\frac{1}{2}bh=\frac{1}{2}\cdot 24\cdot 5=12\cdot 5=60\,ft^2$

TRY IT YOURSELF

77.

$\cdot$	$\frac{1}{2}$	$\frac{1}{3}$	$\frac{1}{4}$	$\frac{1}{5}$	$\frac{1}{6}$
$\frac{1}{2}$	$\frac{1}{4}$	$\frac{1}{6}$	$\frac{1}{8}$	$\frac{1}{10}$	$\frac{1}{12}$
$\frac{1}{3}$	$\frac{1}{6}$	$\frac{1}{9}$	$\frac{1}{12}$	$\frac{1}{15}$	$\frac{1}{18}$
$\frac{1}{4}$	$\frac{1}{8}$	$\frac{1}{12}$	$\frac{1}{16}$	$\frac{1}{20}$	$\frac{1}{24}$
$\frac{1}{5}$	$\frac{1}{10}$	$\frac{1}{15}$	$\frac{1}{20}$	$\frac{1}{25}$	$\frac{1}{30}$
$\frac{1}{6}$	$\frac{1}{12}$	$\frac{1}{18}$	$\frac{1}{24}$	$\frac{1}{30}$	$\frac{1}{36}$

79. $-\frac{15x^3}{24}\cdot\frac{8}{25x^2}=-\frac{\cancel{3}\cdot\cancel{5}\cdot\cancel{x}\cdot\cancel{x}\cdot x\cdot\cancel{8}}{\cancel{3}\cdot\cancel{8}\cdot\cancel{5}\cdot 5\cdot\cancel{x}\cdot\cancel{x}}=-\frac{x}{5}$

81. $\frac{3}{8}\cdot\frac{7}{16}=\frac{21}{128}$

83. $\frac{1}{2}\left(-\frac{1}{3}\right)\left(-\frac{1}{4}\right)=\frac{1}{24}$

85. $-\frac{5}{6}\cdot 18x=-\frac{5}{6}\cdot\frac{18x}{1}=-\frac{5\cdot 3\cdot\cancel{6}\cdot x}{\cancel{6}\cdot 1}=-15x$

87. $\left(-\frac{3}{4n}\right)^3=\left(-\frac{3}{4n}\right)\left(-\frac{3}{4n}\right)\left(-\frac{3}{4n}\right)$

$=-\frac{3\cdot 3\cdot 3}{4n\cdot 4n\cdot 4n}=-\frac{27}{64n^3}$

89. $\frac{3a^3}{4}\cdot\frac{4}{3a^2}=\frac{\cancel{3}\cdot\cancel{a}\cdot\cancel{a}\cdot a\cdot\cancel{4}}{\cancel{4}\cdot\cancel{3}\cdot\cancel{a}\cdot\cancel{a}}=a$

91. $\frac{5}{3}\left(-\frac{6}{15}\right)(-4)=\frac{5}{3}\left(-\frac{2\cdot 3}{3\cdot 5}\right)(-4)$

$=\frac{\cancel{5}}{\cancel{3}}\left(-\frac{2\cdot\cancel{3}}{3\cdot\cancel{5}}\right)(-4)=\frac{8}{3}$

93. $-\frac{11}{12}\cdot\frac{18}{55}\cdot 5=-\frac{11}{2\cdot 6}\cdot\frac{3\cdot 6}{5\cdot 11}\cdot\frac{5}{1}$

$=-\frac{\cancel{11}}{2\cdot\cancel{6}}\cdot\frac{3\cdot\cancel{6}}{\cancel{5}\cdot\cancel{11}}\cdot\frac{\cancel{5}}{1}=-\frac{3}{2}$

95. $\left(-\frac{11}{21}\right)\left(-\frac{14}{33}\right)=\left(-\frac{11}{3\cdot 7}\right)\left(-\frac{2\cdot 7}{3\cdot 11}\right)$

$=\left(-\frac{\cancel{11}}{3\cdot\cancel{7}}\right)\left(-\frac{2\cdot\cancel{7}}{3\cdot\cancel{11}}\right)=\frac{2}{9}$

97. $-\left(-\frac{5}{9}\right)^2=-\left(-\frac{5}{9}\right)\left(-\frac{5}{9}\right)=-\frac{25}{81}$

99. $\frac{7x^4}{10xy^3}\left(\frac{20y^2}{21x^2}\right)=\frac{\cancel{7}\cdot\cancel{x}\cdot\cancel{x}\cdot\cancel{x}\cdot x\cdot 2\cdot\cancel{10}\cdot\cancel{y}\cdot\cancel{y}}{\cancel{10}\cdot\cancel{x}\cdot\cancel{y}\cdot\cancel{y}\cdot y\cdot 3\cdot\cancel{7}\cdot\cancel{x}\cdot\cancel{x}}$

$=\frac{x\cdot 2}{y\cdot 3}=\frac{2x}{3y}$

101. $\frac{3}{4}\left(\frac{5}{7}\right)\left(\frac{2}{3}\right)\left(\frac{7}{3}\right)=\frac{3}{2\cdot 2}\left(\frac{5}{7}\right)\left(\frac{2}{3}\right)\left(\frac{7}{3}\right)$

$=\frac{\cancel{3}}{\cancel{2}\cdot 2}\left(\frac{5}{\cancel{7}}\right)\left(\frac{\cancel{2}}{\cancel{3}}\right)\left(\frac{\cancel{7}}{3}\right)=\frac{5}{6}$

APPLICATIONS

103. $\frac{3}{5}$ of $100=\frac{3}{5}\cdot 100=\frac{3}{\cancel{5}}\cdot\frac{\cancel{5}\cdot 20}{1}=60$ votes

105. h_1 = height of first bounce

h_2 = height of second bounce

h_3 = height of third bounce

$$h_1 = \frac{1}{3}(54)$$

$$h_1 = \frac{1}{3}\left(\frac{54}{1}\right)$$

$$h_1 = \frac{1}{\cancel{3}}\left(\frac{18\cdot\cancel{3}}{1}\right) = \frac{18}{1} = 18$$

$$h_2 = \frac{1}{3}(18)$$

$$h_2 = \frac{1}{3}\left(\frac{18}{1}\right)$$

$$h_2 = \frac{1}{\cancel{3}}\left(\frac{6\cdot\cancel{3}}{1}\right) = \frac{6}{1} = 6$$

$$h_3 = \frac{1}{3}(6)$$

$$h_3 = \frac{1}{3}\left(\frac{6}{1}\right)$$

$$h_3 = \frac{1}{\cancel{3}}\left(\frac{2\cdot\cancel{3}}{1}\right) = \frac{2}{1} = 2$$

18 inches on first bounce, 6 inches on second bounce, and 2 inches on third bounce.

107. $\frac{1}{2}\cdot\frac{3}{4} = \frac{3}{8}$ cups sugar

$\frac{1}{2}\cdot\frac{1}{3} = \frac{1}{6}$ cups molasses

109.

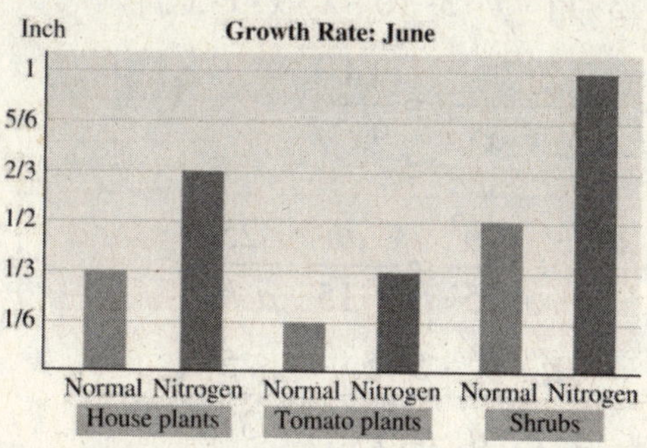

111. $\frac{1}{2}\cdot 6\cdot 9 = 3\cdot 9 = 27\,ft^2$

113. $\frac{1}{2}\cdot 12\cdot 7 = 6\cdot 7 = 42\,ft^2$

115. $\frac{1}{2}\cdot 106\cdot 182 = 53\cdot 182 = 9{,}646\,mi^2$

117. $12\cdot\frac{1}{16} = \frac{3\cdot\cancel{4}}{1}\cdot\frac{1}{4\cdot\cancel{4}} = \frac{3}{4}$ inches

WRITING

119. $\frac{1}{2}$ of a pizza, $\frac{2}{3}$ of an inning, $\frac{1}{4}$ of an hour

121. The student neglected to factor and cancel before multiplying.

REVIEW

123. $\frac{-8}{4} = \frac{-2\cdot\cancel{4}}{\cancel{4}} = -2$

Check: $(-2)(4) = -8$

125. $-36 \div (-9) = \frac{-36}{-9} = \frac{-9\cdot 4}{-9} = 4$

Check: $(4)(-9) = -36$

Section 4.3: Dividing Fractions

VOCABULARY

1. The reciprocal of $\frac{5}{12}$ is $\frac{12}{5}$.

3. The answer to a division is called the quotient.

CONCEPTS

5. a. To divide two fractions, multiply the first fraction by the reciprocal of the second fraction.

b. $\frac{1}{2} \div \frac{2}{3} = \frac{1}{2}\cdot\frac{3}{2}$

7. a. Negative b. Positive

9.

a. $\frac{4}{5}\cdot\frac{5}{4}=\frac{20}{20}=1$

b. $\left(-\frac{3}{5}\right)\left(-\frac{5}{3}\right)=\frac{15}{15}=1$

NOTATION

11. $\frac{4}{9}\div\frac{8}{27}=\frac{4}{9}\cdot\frac{27}{8}$

$=\frac{4\cdot 27}{9\cdot 8}$

$=\frac{\cancel{4}\cdot 3\cdot\cancel{9}}{\cancel{9}\cdot 2\cdot\cancel{4}}$

$=\frac{3}{2}$

GUIDED PRACTICE

13. a. $\frac{7}{6}$ b. $-\frac{8}{15}$ c. $\frac{1}{10}$

15. a. $\frac{8}{11a}$ b. $-14b$ c. $-\frac{1}{63x}$

17. $\frac{1}{8}\div\frac{2}{3}=\frac{1}{8}\cdot\frac{3}{2}=\frac{3}{16}$

19. $\frac{2}{23}\div\frac{1}{7}=\frac{2}{23}\cdot\frac{7}{1}=\frac{14}{23}$

21. $\frac{25}{32}\div\frac{5}{28}=\frac{25}{32}\cdot\frac{28}{5}=\frac{\cancel{5}\cdot 5}{\cancel{4}\cdot 8}\cdot\frac{\cancel{4}\cdot 7}{\cancel{5}}=\frac{35}{8}$

23. $\frac{27}{32}\div\frac{9}{8}=\frac{27}{32}\cdot\frac{8}{9}=\frac{3\cdot\cancel{9}}{4\cdot\cancel{8}}\cdot\frac{\cancel{8}}{\cancel{9}}=\frac{3}{4}$

25. $50\div\frac{10}{9}=\frac{50}{1}\cdot\frac{9}{10}=\frac{5\cdot\cancel{10}}{1}\cdot\frac{9}{\cancel{10}}=45$

27. $150\div\frac{15}{32}=\frac{150}{1}\cdot\frac{32}{15}=\frac{\cancel{15}\cdot 10}{1}\cdot\frac{32}{\cancel{15}}=320$

29. $\frac{1}{8}\div\left(-\frac{1}{32}\right)=\frac{1}{8}\cdot\left(-\frac{32}{1}\right)=\frac{-32}{8}=-4$

31. $\frac{2}{5}\div\left(-\frac{4}{35}\right)=\frac{2}{5}\cdot\left(-\frac{35}{4}\right)=\frac{\cancel{2}}{\cancel{5}}\left(-\frac{\cancel{5}\cdot 7}{\cancel{2}\cdot 2}\right)=-\frac{7}{2}$

33. $-\frac{28}{55}\div(-7)=-\frac{28}{55}\cdot\left(-\frac{1}{7}\right)$

$=-\frac{4\cdot\cancel{7}}{55}\cdot\left(-\frac{1}{\cancel{7}}\right)=\frac{4}{55}$

35. $-\frac{33}{23}\div(-11)=-\frac{33}{23}\cdot\left(-\frac{1}{11}\right)$

$=-\frac{3\cdot\cancel{11}}{23}\cdot\left(-\frac{1}{\cancel{11}}\right)=\frac{3}{23}$

37. $\frac{4a}{5}\div\frac{3}{7}=\frac{4a}{5}\cdot\frac{7}{3}=\frac{28a}{15}$

39. $\frac{6x}{7}\div\frac{5}{11}=\frac{6x}{7}\cdot\frac{11}{5}=\frac{66x}{35}$

41. $\frac{6b^2}{7a}\div\frac{9b}{14a^3}=\frac{6b^2}{7a}\cdot\frac{14a^3}{9b}$

$=\frac{2\cdot\cancel{3}\cdot\cancel{b}\cdot b\cdot 2\cdot\cancel{7}\cdot a\cdot\cancel{a}\cdot a}{\cancel{7}\cdot\cancel{a}\cdot\cancel{3}\cdot 3\cdot\cancel{b}}$

$=\frac{2\cdot b\cdot 2\cdot a\cdot a}{3}=\frac{4a^2b}{3}$

43. $\frac{20x}{33y}\div\frac{30x^4}{77y}=\frac{20x}{33y}\cdot\frac{77y}{30x^4}$

$=\frac{2\cdot\cancel{10}\cdot\cancel{x}\cdot 7\cdot\cancel{11}\cdot\cancel{y}}{3\cdot\cancel{11}\cdot\cancel{y}\cdot 3\cdot\cancel{10}\cdot\cancel{x}\cdot x\cdot x\cdot x}$

$=\frac{2\cdot 7}{3\cdot 3\cdot x\cdot x\cdot x}=\frac{14}{9x^3}$

45. $-\frac{ab^2}{15}\div\frac{a^2b^2}{25}=-\frac{ab^2}{15}\cdot\frac{25}{a^2b^2}$

$=-\frac{\cancel{a}\cdot\cancel{b^2}\cdot\cancel{5}\cdot 5}{3\cdot\cancel{5}\cdot\cancel{a}\cdot a\cdot\cancel{b^2}}=-\frac{5}{3\cdot a}=-\frac{5}{3a}$

47.

$$-\frac{48}{m^{11}} \div \left(-\frac{16}{m^8}\right) = \frac{48}{m^{11}} \cdot \frac{m^8}{16}$$

$$= \frac{3 \cdot \cancel{16} \cdot \cancel{m^8}}{\cancel{m^8} \cdot m^3 \cdot \cancel{16}} = \frac{3}{m^3}$$

TRY IT YOURSELF

49.

$$120 \div \frac{12}{5x} = \frac{120}{1} \cdot \frac{5x}{12} = \frac{10 \cdot \cancel{12} \cdot 5 \cdot x}{\cancel{12}}$$

$$= 10 \cdot 5 \cdot x = 50x$$

51.

$$-8x \div \left(-\frac{4x^3}{9}\right) = \frac{8x}{1} \cdot \frac{9}{4x^3}$$

$$= \frac{2 \cdot \cancel{4} \cdot \cancel{x} \cdot 9}{\cancel{4} \cdot \cancel{x} \cdot x \cdot x} = \frac{2 \cdot 9}{x \cdot x} = \frac{18}{x^2}$$

53.

$$\left(-\frac{7}{4}\right) \div \left(-\frac{21}{8}\right) = \left(-\frac{7}{4}\right)\left(-\frac{8}{21}\right)$$

$$= \left(-\frac{\cancel{7}}{\cancel{4}}\right)\left(-\frac{2 \cdot \cancel{4}}{3 \cdot \cancel{7}}\right) = \frac{2}{3}$$

55. $\frac{4}{5} \div \frac{4}{5} = \frac{4}{5} \cdot \frac{5}{4} = \frac{20}{20} = 1$

57.

$$-\frac{15}{32} \div \frac{3}{4} = -\frac{15}{32} \cdot \frac{4}{3} = -\frac{\cancel{3} \cdot 5}{\cancel{4} \cdot 8} \cdot \frac{\cancel{4}}{\cancel{3}} = -\frac{5}{8}$$

59.

$$3a \div \frac{1}{12a} = 3a \cdot 12a = 36a^2$$

61.

$$-\frac{4}{5} \div (-6) = -\frac{4}{5}\left(-\frac{1}{6}\right)$$

$$= -\frac{2 \cdot \cancel{2}}{5} \cdot \left(-\frac{1}{\cancel{2} \cdot 3}\right) = \frac{2}{15}$$

63. $\frac{15}{16} \div 180 = \frac{15}{16} \cdot \frac{1}{180} = \frac{\cancel{15}}{16} \cdot \frac{1}{12 \cdot \cancel{15}} = \frac{1}{192}$

65.

$$-\frac{9n^5}{10} \div \frac{4n^4}{15} = -\frac{9n^5}{10} \cdot \frac{15}{4n^4}$$

$$= -\frac{9 \cdot n \cdot \cancel{n^4} \cdot 3 \cdot \cancel{5}}{2 \cdot \cancel{5} \cdot 4 \cdot \cancel{n^4}} = -\frac{9 \cdot n \cdot 3}{2 \cdot 4} = -\frac{27n}{8}$$

67.

$$\frac{9a^3}{10b} \div \left(-\frac{3a^4}{25b^3}\right) = -\frac{9a^3}{10b} \cdot \frac{25b^3}{3a^4}$$

$$= -\frac{\cancel{3} \cdot 3 \cdot \cancel{a^3} \cdot \cancel{5} \cdot 5 \cdot \cancel{b} \cdot b^2}{2 \cdot \cancel{5} \cdot \cancel{b} \cdot \cancel{3} \cdot a \cdot \cancel{a^3}}$$

$$= -\frac{3 \cdot 5 \cdot b^2}{2 \cdot a} = -\frac{15b^2}{2a}$$

69.

$$-\frac{x^2}{y^3} \div \frac{x}{y} = -\frac{x^2}{y^3} \cdot \frac{y}{x}$$

$$= -\frac{\cancel{x} \cdot x \cdot \cancel{y}}{\cancel{y} \cdot y^2 \cdot \cancel{x}} = -\frac{x}{y^2}$$

71.

$$-\frac{1}{8} \div 8 = -\frac{1}{8} \cdot \frac{1}{8} = -\frac{1}{64}$$

73.

$$-\frac{7m^3}{6}\left(-\frac{9}{49m^2}\right) = \frac{\cancel{7} \cdot m \cdot \cancel{m^2} \cdot \cancel{3} \cdot 3}{2 \cdot \cancel{3} \cdot \cancel{7} \cdot 7 \cdot \cancel{m^2}}$$

$$= \frac{m \cdot 3}{2 \cdot 7} = \frac{3m}{14}$$

75.

$$-\frac{1}{3} \div \frac{4}{5} = -\frac{1}{3} \cdot \frac{5}{4} = -\frac{5}{12}$$

77. $\frac{13}{16} \div x = \frac{13}{16} \cdot \frac{1}{x} = \frac{13}{16x}$

79.

$$\left(-\frac{11}{21}\right)\left(-\frac{14}{33}\right) = \left(-\frac{\cancel{11}}{3 \cdot \cancel{7}}\right)\left(-\frac{2 \cdot \cancel{7}}{3 \cdot \cancel{11}}\right) = \frac{2}{9}$$

81.

$$-\frac{15x}{32y} \div \frac{5x}{64y^4} = -\frac{15x}{32y} \cdot \frac{64y^4}{5x}$$

$$= -\frac{3 \cdot \cancel{5} \cdot \cancel{x} \cdot 2 \cdot \cancel{32} \cdot \cancel{y} \cdot y^3}{\cancel{32} \cdot \cancel{y} \cdot \cancel{5} \cdot \cancel{x}}$$

$$= -3 \cdot 2 \cdot y^3 = -6y^3$$

83. $11 \cdot \frac{1}{6} = \frac{11}{1} \cdot \frac{1}{6} = \frac{11}{6}$

85. $\frac{3x}{4} \cdot \frac{5}{7} = \frac{15x}{28}$

87. $\frac{25}{7}\div\left(-\frac{30}{21}\right)=\frac{25}{7}\cdot\left(-\frac{21}{30}\right)$

$=\frac{\cancel{5}\cdot 5}{\cancel{7}}\cdot\left(-\frac{\cancel{3}\cdot\cancel{7}}{2\cdot\cancel{3}\cdot\cancel{5}}\right)=-\frac{5}{2}$

APPLICATIONS

89. $\frac{3}{8}\div\frac{3}{32}=\frac{3}{8}\cdot\frac{32}{3}=\frac{\cancel{3}}{\cancel{8}}\cdot\frac{4\cdot\cancel{8}}{\cancel{3}}=4$ applications

91. $\frac{3}{4}\div\frac{1}{8}=\frac{3}{4}\cdot\frac{8}{1}=\frac{24}{4}=6$ one-eighth cups

93.
a. $12\div\frac{2}{5}=\frac{12}{1}\cdot\frac{5}{2}=\frac{6}{1}\cdot\frac{5}{1}=30$ days

b. 7+8=15 miles

c. $15\div\frac{3}{5}=\frac{\overset{5}{\cancel{15}}}{1}\cdot\frac{5}{\underset{1}{\cancel{3}}}=\frac{5}{1}\cdot\frac{5}{1}=25$ days

d. Route 2 will take less time.

95.
a. 16 b. $\frac{3}{4}in$

c. $\frac{3}{4}\div 90=\frac{3}{4}\cdot\frac{1}{90}=\frac{1}{120}in$

97. $6,284\div\frac{4}{5}=\frac{6,284}{1}\cdot\frac{5}{4}$

$=\frac{\cancel{4}\cdot 1571}{1}\cdot\frac{5}{\cancel{4}}=7,855$ sections

WRITING

99. To divide two fractions, multiply the first fraction by the reciprocal of the second fraction.

101. 0 does not have a reciprocal because $\frac{1}{0}$ is undefined.

103. Bill has 10 pizzas. If he evenly slices them so that each guest receives $\frac{1}{5}$ of a pizza, how many guests does Bill have?

REVIEW

105. The symbol < means is less than.

107. Zero is neither positive nor negative.

109. $6=2\cdot 3$

$9=3^2$

$\text{LCM}(6,9)=2\cdot 3^2=2\cdot 9=18$

Section 4.4: Adding and Subtracting Fractions

VOCABULARY

1. Because the denominators of $\frac{3}{8}$ and $\frac{7}{8}$ are the same number, we say that they have a common denominator.

CONCEPTS

3. a. To add (or subtract) fractions that have the same denominator, add (or subtract) their numerators and write the sum (or difference) over the common denominator. Simplify the result, if possible.

b. To add (or subtract) fractions that have different denominators, we express each fraction as an equivalent fraction that has the LCD for its denominator. Then we use the rule for adding (subtracting) fractions that have the same denominator.

5. Since $4\cdot 9=36$, the form of 1 should be $\frac{9}{9}$.

7. a. a 5 appears once

b. a 3 appears twice

c. a 2 appears three times

NOTATION

9. $\frac{2}{5}+\frac{1}{7}=\frac{2}{5}\cdot\frac{7}{7}+\frac{1}{7}\cdot\frac{5}{5}$

$=\frac{14}{35}+\frac{5}{35}=\frac{14+5}{35}=\frac{19}{35}$

GUIDED PRACTICE

11. $\frac{4}{9}+\frac{1}{9}=\frac{5}{9}$

13. $\frac{3}{8}+\frac{1}{8}=\frac{4}{8}=\frac{\cancel{4}}{2\cdot\cancel{4}}=\frac{1}{2}$

15. $\frac{11}{15}-\frac{7}{15}=\frac{4}{15}$

17. $\frac{11}{20}-\frac{3}{20}=\frac{8}{20}=\frac{2\cdot\cancel{4}}{\cancel{4}\cdot 5}=\frac{2}{5}$

19. $-\frac{11}{5}-\left(-\frac{8}{5}\right)=\frac{-11}{5}+\frac{8}{5}=\frac{-3}{5}=-\frac{3}{5}$

21. $-\frac{7}{21}-\left(-\frac{2}{21}\right)=\frac{-7}{21}+\frac{2}{21}=\frac{-5}{21}=-\frac{5}{21}$

23. $\frac{19}{40}-\frac{3}{40}-\frac{1}{40}=\frac{15}{40}=\frac{3\cdot\cancel{5}}{\cancel{5}\cdot 8}=\frac{3}{8}$

25. $\frac{13}{33}+\frac{1}{33}+\frac{7}{33}=\frac{21}{33}=\frac{\cancel{3}\cdot 7}{\cancel{3}\cdot 11}=\frac{7}{11}$

27. $\frac{1}{3}+\frac{1}{7}=\frac{1}{3}\cdot\frac{7}{7}+\frac{1}{7}\cdot\frac{3}{3}=\frac{7}{21}+\frac{3}{21}=\frac{10}{21}$

29. $\frac{2}{5}+\frac{1}{9}=\frac{2}{5}\cdot\frac{9}{9}+\frac{1}{9}\cdot\frac{5}{5}=\frac{18}{45}+\frac{5}{45}=\frac{23}{45}$

31. $\frac{4}{5}-\frac{3}{4}=\frac{4}{5}\cdot\frac{4}{4}-\frac{3}{4}\cdot\frac{5}{5}=\frac{16}{20}-\frac{15}{20}=\frac{1}{20}$

33. $\frac{3}{4}-\frac{2}{7}=\frac{3}{4}\cdot\frac{7}{7}-\frac{2}{7}\cdot\frac{4}{4}=\frac{21}{28}-\frac{8}{28}=\frac{13}{28}$

35. $\frac{11}{12}-\frac{2}{3}=\frac{11}{12}-\frac{2}{3}\cdot\frac{4}{4}=\frac{11}{12}-\frac{8}{12}$

$=\frac{3}{12}=\frac{3}{4\cdot 3}=\frac{1}{4}$

37. $\frac{9}{14}-\frac{1}{7}=\frac{9}{14}-\frac{1}{7}\cdot\frac{2}{2}=\frac{9}{14}-\frac{2}{14}$

$=\frac{7}{14}=\frac{\cancel{7}}{2\cdot\cancel{7}}=\frac{1}{2}$

39. $-2+\frac{5}{9}=\frac{-2}{1}+\frac{5}{9}$

$=\frac{-2}{1}\cdot\frac{9}{9}+\frac{5}{9}=\frac{-18}{9}+\frac{5}{9}=\frac{-13}{9}=-\frac{13}{9}$

41. $-3+\frac{9}{4}=\frac{-3}{1}+\frac{9}{4}$

$=\frac{-3}{1}\cdot\frac{4}{4}+\frac{9}{4}=\frac{-12}{4}+\frac{9}{4}=\frac{-3}{4}=-\frac{3}{4}$

43. $\frac{1}{6}+\frac{5}{8}=\frac{1}{6}\cdot\frac{4}{4}+\frac{5}{8}\cdot\frac{3}{3}=\frac{4}{24}+\frac{15}{24}=\frac{19}{24}$

45. $\frac{4}{9}+\frac{5}{12}=\frac{4}{9}\cdot\frac{4}{4}+\frac{5}{12}\cdot\frac{3}{3}=\frac{16}{36}+\frac{15}{36}=\frac{31}{36}$

47. $\frac{9}{10}-\frac{3}{14}=\frac{9}{10}\cdot\frac{7}{7}-\frac{3}{14}\cdot\frac{5}{5}$

$=\frac{63}{70}-\frac{15}{70}=\frac{48}{70}=\frac{\cancel{2}\cdot 24}{\cancel{2}\cdot 35}=\frac{24}{35}$

49. $\frac{11}{12}-\frac{7}{15}=\frac{11}{12}\cdot\frac{5}{5}-\frac{7}{15}\cdot\frac{4}{4}$

$=\frac{55}{60}-\frac{28}{60}=\frac{27}{60}=\frac{\cancel{3}\cdot 9}{\cancel{3}\cdot 20}=\frac{9}{20}$

51. $\frac{x}{12}+\frac{5x}{12}=\frac{x+5x}{12}=\frac{6x}{12}=\frac{\cancel{6}\cdot x}{2\cdot\cancel{6}}=\frac{x}{2}$

53. $\frac{2c}{7}+\frac{3c}{7}=\frac{2c+3c}{7}=\frac{5c}{7}$

55. $\frac{16}{21m}-\frac{11}{21m}=\frac{16-11}{21m}=\frac{5}{21m}$

57. $\frac{16}{15y}-\frac{7}{15y}=\frac{16-7}{15y}=\frac{9}{15y}=\frac{\cancel{3}\cdot 3}{\cancel{3}\cdot 5\cdot y}=\frac{3}{5y}$

59. $\frac{a}{5}+\frac{2}{3}=\frac{a}{5}\cdot\frac{3}{3}+\frac{2}{3}\cdot\frac{5}{5}=\frac{3a}{15}+\frac{10}{15}=\frac{3a+10}{15}$

61. $\frac{1}{8}+\frac{x}{3}=\frac{1}{8}\cdot\frac{3}{3}+\frac{x}{3}\cdot\frac{8}{8}=\frac{3}{24}+\frac{8x}{24}=\frac{3+8x}{24}$

63. $\frac{3}{n}-\frac{5}{12}=\frac{3}{n}\cdot\frac{12}{12}-\frac{5}{12}\cdot\frac{n}{n}$

$=\frac{36}{12n}-\frac{5n}{12n}=\frac{36-5n}{12n}$

65. $\frac{8}{9}-\frac{11}{d}=\frac{8}{9}\cdot\frac{d}{d}-\frac{11}{d}\cdot\frac{9}{9}=\frac{8d}{9d}-\frac{99}{9d}-\frac{8d-99}{9d}$

67. $\frac{3}{8}\ \square\ \frac{5}{16}$

$\frac{3}{8}\cdot\frac{2}{2}\ \square\ \frac{5}{16}$

$\frac{6}{16}\ \boxed{>}\ \frac{5}{16}$

$\frac{3}{8}$ is larger.

69. $\frac{4}{5}\ \square\ \frac{2}{3}$

$\frac{4}{5}\cdot\frac{3}{3}\ \square\ \frac{2}{3}\cdot\frac{5}{5}$

$\frac{12}{15}\ \boxed{>}\ \frac{10}{15}$

$\frac{4}{5}$ is larger.

71. $\frac{7}{9}\ \square\ \frac{11}{12}$

$\frac{7}{9}\cdot\frac{4}{4}\ \square\ \frac{11}{12}\cdot\frac{3}{3}$

$\frac{28}{36}\ \boxed{<}\ \frac{33}{36}$

$\frac{11}{12}$ is larger.

73. $\frac{23}{20}\ \square\ \frac{7}{6}$

$\frac{23}{20}\cdot\frac{3}{3}\ \square\ \frac{7}{6}\cdot\frac{10}{10}$

$\frac{69}{60}\ \boxed{<}\ \frac{70}{60}$

$\frac{7}{6}$ is larger.

75. $\frac{1}{6}+\frac{5}{18}+\frac{2}{9}=\frac{1}{6}\cdot\frac{3}{3}+\frac{5}{18}+\frac{2}{9}\cdot\frac{2}{2}$

$=\frac{3}{18}+\frac{5}{18}+\frac{4}{18}=\frac{12}{18}=\frac{2\cdot\cancel{6}}{3\cdot\cancel{6}}=\frac{2}{3}$

77. $\frac{4}{15}+\frac{2}{3}+\frac{1}{6}=\frac{4}{15}\cdot\frac{2}{2}+\frac{2}{3}\cdot\frac{10}{10}+\frac{1}{6}\cdot\frac{5}{5}$

$=\frac{8}{30}+\frac{20}{30}+\frac{5}{30}=\frac{33}{30}=\frac{\cancel{3}\cdot 11}{\cancel{3}\cdot 10}=\frac{11}{10}$

TRY IT YOURSELF

79. $-\frac{1}{12}-\left(-\frac{5}{12}\right)=\frac{-1}{12}+\frac{5}{12}=\frac{4}{12}=\frac{\cancel{4}}{3\cdot\cancel{4}}=\frac{1}{3}$

81. $\frac{3n}{4}+\frac{2}{3}=\frac{3n}{4}\cdot\frac{3}{3}+\frac{2}{3}\cdot\frac{4}{4}$

$=\frac{9n}{12}+\frac{8}{12}=\frac{9n+8}{12}$

83. $\frac{12}{25}-\frac{1}{25}-\frac{1}{25}=\frac{10}{25}=\frac{2\cdot\cancel{5}}{5\cdot\cancel{5}}=\frac{2}{5}$

85. $-\frac{7}{20}-\frac{1}{5}=-\frac{7}{20}-\frac{1}{5}\cdot\frac{4}{4}$

$=\frac{-7}{20}-\frac{4}{20}=\frac{-11}{20}=-\frac{11}{20}$

87. $-\frac{7}{16}+\frac{1}{4}=\frac{-7}{16}+\frac{1}{4}\cdot\frac{4}{4}$

$=\frac{-7}{16}+\frac{4}{16}=\frac{-3}{16}=-\frac{3}{16}$

89. $\frac{x}{9}+\frac{2x}{9}=\frac{x+2x}{9}=\frac{3x}{9}=\frac{\cancel{3}\cdot x}{\cancel{3}\cdot 3}=\frac{x}{3}$

91. $\frac{2}{3}+\frac{4}{5}+\frac{5}{6}=\frac{2}{3}\cdot\frac{10}{10}+\frac{4}{5}\cdot\frac{6}{6}+\frac{5}{6}\cdot\frac{5}{5}$

$=\frac{20}{30}+\frac{24}{30}+\frac{25}{30}=\frac{69}{30}=\frac{\cancel{3}\cdot 23}{\cancel{3}\cdot 10}=\frac{23}{10}$

92. $\frac{3}{4}+\frac{2}{5}+\frac{3}{10}=\frac{3}{4}\cdot\frac{5}{5}+\frac{2}{5}\cdot\frac{4}{4}+\frac{3}{10}\cdot\frac{2}{2}$

$=\frac{15}{20}+\frac{8}{20}+\frac{6}{20}=\frac{29}{20}$

93. $\frac{9}{20}-\frac{1}{30}=\frac{9}{20}\cdot\frac{3}{3}-\frac{1}{30}\cdot\frac{2}{2}$
$=\frac{27}{60}-\frac{2}{60}=\frac{25}{60}=\frac{\not{5}\cdot 5}{\not{5}\cdot 12}=\frac{5}{12}$

95. $\frac{27}{50}+\frac{5}{16}=\frac{27}{50}\cdot\frac{8}{8}+\frac{5}{16}\cdot\frac{25}{25}$
$=\frac{216}{400}+\frac{125}{400}=\frac{341}{400}$

97. $\frac{13}{20}-\frac{1}{5}=\frac{13}{20}-\frac{1}{5}\cdot\frac{4}{4}=\frac{13}{20}-\frac{4}{20}=\frac{9}{20}$

99. $-\frac{3}{4}-5=\frac{-3}{4}-\frac{5}{1}\cdot\frac{4}{4}$
$=\frac{-3}{4}-\frac{20}{4}=\frac{-23}{4}=-\frac{23}{4}$

101. $\frac{7}{30}-\frac{19}{75}=\frac{7}{30}\cdot\frac{5}{5}-\frac{19}{75}\cdot\frac{2}{2}$
$=\frac{35}{150}-\frac{38}{150}=\frac{-3}{150}$
$=\frac{-\not{3}}{\not{3}\cdot 50}=\frac{-1}{50}=-\frac{1}{50}$

103. $\frac{13}{d}-\frac{3}{2}=\frac{13}{d}\cdot\frac{2}{2}-\frac{3}{2}\cdot\frac{d}{d}$
$=\frac{26}{2d}-\frac{3d}{2d}=\frac{26-3d}{2d}$

105. $\frac{2}{15}-\frac{5}{12}=\frac{2}{15}\cdot\frac{4}{4}-\frac{5}{12}\cdot\frac{5}{5}$
$=\frac{8}{60}-\frac{25}{60}=\frac{-17}{60}=-\frac{17}{60}$

APPLICATIONS

107.
a. $\frac{21}{32}+\frac{5}{16}=\frac{21}{32}+\frac{10}{32}=\frac{31}{32}in$

b. $\frac{21}{32}-\frac{5}{16}=\frac{21}{32}-\frac{10}{32}=\frac{11}{32}in$

109. $\frac{8}{x}-\frac{1}{2}=\frac{16}{2x}-\frac{x}{2x}=\frac{16-x}{2x}\ cm$

111. $\frac{1}{2}+\frac{1}{3}+\frac{1}{5}=\frac{1}{2}\cdot\frac{15}{15}+\frac{1}{3}\cdot\frac{10}{10}+\frac{1}{5}\cdot\frac{6}{6}$
$=\frac{15}{30}+\frac{10}{30}+\frac{6}{30}=\frac{31}{30}$

Since this is more than one ton, this truck should not be used .

113. $\frac{7}{8}-\frac{15}{64}-\frac{15}{64}=\frac{56}{64}-\frac{15}{64}-\frac{15}{64}=\frac{26}{64}=\frac{13}{32}in$

115.
The scale shows that it is undercharging by $\frac{1}{16}lb$.

117. $\frac{2}{5}+\frac{3}{10}=\frac{4}{10}+\frac{3}{10}=\frac{7}{10}$

119. $\frac{1}{8}+\frac{1}{4}+\frac{1}{16}+\frac{1}{8}+\frac{1}{4}+\frac{1}{16}$
$=\frac{2}{16}+\frac{4}{16}+\frac{1}{16}+\frac{2}{16}+\frac{4}{16}+\frac{1}{16}$
$=\frac{14}{16}=\frac{7}{8}$; No

121.

16/64	20/64
14/64	21/64

a. The right rear has the most tread.

b. The left rear has the least tread.

WRITING

123. Because they do not have the same denominator

125. The same number must be divided out of each term in the numerator and the denominator. There is no 2 to cancel from the $5x$.

REVIEW

127.

a. $\frac{1}{4}+\frac{1}{8}=\frac{1}{4}\cdot\frac{2}{2}+\frac{1}{8}=\frac{2}{8}+\frac{1}{8}=\frac{3}{8}$

b. $\frac{1}{4}-\frac{1}{8}=\frac{1}{4}\cdot\frac{2}{2}-\frac{1}{8}=\frac{2}{8}-\frac{1}{8}=\frac{1}{8}$

c. $\frac{1}{4}\cdot\frac{1}{8}=\frac{1}{32}$

d. $\frac{1}{4}\div\frac{1}{8}=\frac{1}{4}\cdot\frac{8}{1}=\frac{8}{4}=2$

Section 4.5: Multiplying and Dividing Mixed Numbers

VOCABULARY

1. A <u>mixed</u> number, such as $8\frac{4}{5}$, is the sum of a whole number and a proper fraction.

3. The numerator of an <u>improper</u> fraction is greater than or equal to its denominator.

CONCEPTS

5. a. $5\frac{1}{3}°$ b. $-6\frac{7}{8}\,in.$

7. To write a mixed number as an improper fraction:
1. <u>Multiply</u> the denominator of the fraction by the whole-number part.
2. <u>Add</u> the numerator of the fraction to the result from Step 1.
3. Write the sum from Step 2 over the original <u>denominator</u>.

9. $-\frac{4}{5}, -\frac{2}{5}, \frac{1}{5}$

11. To multiply or divide mixed numbers, first change the mixed numbers to <u>improper</u> fractions. Then perform the multiplication or division of the fractions as usual.

13. $4\frac{1}{5}\cdot 2\frac{5}{7}=7\frac{2}{35}$ is not reasonable since

$4\frac{1}{5}\cdot 2\frac{5}{7}\approx 4\cdot 3=12$

NOTATION

15.

a. We read $5\frac{11}{16}$ as "five <u>and</u> eleven-<u>sixteenths</u>."

b. We read $-4\frac{2}{3}$ as "<u>negative</u> four and <u>two</u>-thirds."

17.

$$\begin{aligned}5\frac{1}{4}\cdot 1\frac{1}{7}&=\frac{21}{4}\cdot\frac{8}{7}\\&=\frac{21\cdot 8}{4\cdot 7}\\&=\frac{\cancel{7}\cdot 3\cdot\cancel{4}\cdot 2}{\cancel{4}\cdot\cancel{7}}\\&=\frac{6}{1}\\&=6\end{aligned}$$

GUIDED PRACTICE

19. $\frac{19}{8}; 2\frac{3}{8}$

21. $\frac{34}{25}; 1\frac{9}{25}$

23. $6\frac{1}{2}=\frac{2\cdot 6+1}{2}=\frac{12+1}{2}=\frac{13}{2}$

25. $20\frac{4}{5}=\frac{5\cdot 20+4}{5}=\frac{100+4}{5}=\frac{104}{5}$

27. $-7\frac{5}{9}=-\frac{9\cdot 7+5}{9}=-\frac{63+5}{9}=-\frac{68}{9}$

29. $-8\frac{2}{3}=-\frac{3\cdot 8+2}{3}=-\frac{24+2}{3}=-\frac{26}{3}$

31. $\frac{13}{4}=3\frac{1}{4}$

33. $\frac{28}{5}=5\frac{3}{5}$

35. $\frac{42}{9}=4\frac{6}{9}=4\frac{2\cdot\cancel{3}}{3\cdot\cancel{3}}=4\frac{2}{3}$

37. $\frac{84}{8}=10\frac{4}{8}=10\frac{\cancel{4}}{\cancel{4}\cdot 2}=10\frac{1}{2}$

39. $\frac{52}{13}=4$

41. $\frac{34}{17}=2$

43. $-\frac{58}{7}=-8\frac{2}{7}$

45. $-\frac{20}{6}=-3\frac{2}{6}=-3\frac{\cancel{2}}{\cancel{2}\cdot 3}=-3\frac{1}{3}$

47.

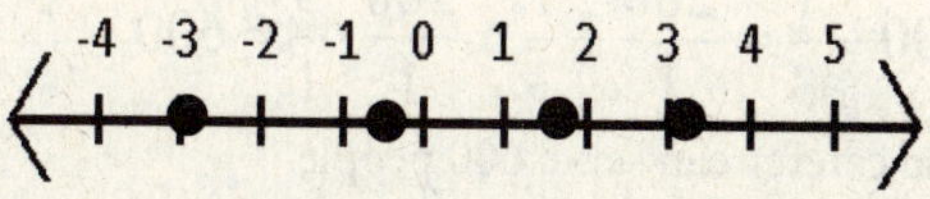

49.

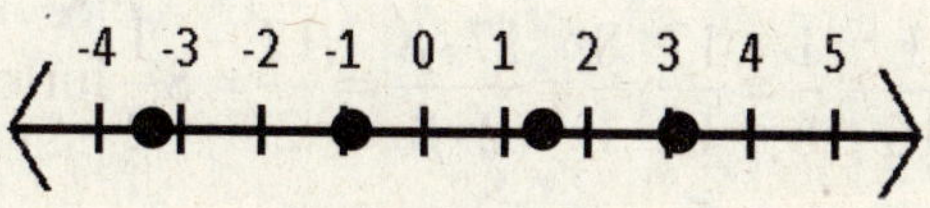

51. $3\frac{1}{2}\cdot 2\frac{1}{3}=\frac{7}{2}\cdot\frac{7}{3}=\frac{49}{6}=8\frac{1}{6}$

53. $2\frac{2}{5}\left(3\frac{1}{12}\right)=\frac{\cancel{12}}{5}\cdot\frac{37}{\cancel{12}}=\frac{37}{5}=7\frac{2}{5}$

55. $6\frac{1}{2}\cdot 1\frac{3}{13}=\frac{\cancel{13}}{2}\cdot\frac{16}{\cancel{13}}=\frac{16}{2}=8$

57. $-2\frac{1}{2}(4)=-\frac{5}{2}\cdot\frac{4}{1}=-\frac{5}{\cancel{2}}\cdot\frac{\cancel{2}\cdot 2}{1}=-10$

59. $-1\frac{13}{15}\div\left(-4\frac{1}{5}\right)=-\frac{28}{15}\div\left(-\frac{21}{5}\right)$

$=-\frac{28}{15}\left(-\frac{5}{21}\right)=-\frac{4\cdot\cancel{7}}{3\cdot\cancel{5}}\left(-\frac{\cancel{5}}{3\cdot\cancel{7}}\right)=\frac{4}{9}$

61. $15\frac{1}{3}\div 2\frac{2}{9}=\frac{46}{3}\div\frac{20}{9}=\frac{46}{3}\cdot\frac{9}{20}$

$=\frac{\cancel{2}\cdot 23}{\cancel{3}}\cdot\frac{\cancel{3}\cdot 3}{\cancel{2}\cdot 10}=\frac{69}{10}=6\frac{9}{10}$

63. $1\frac{3}{4}\div\frac{3}{4}=\frac{7}{4}\div\frac{3}{4}=\frac{7}{\cancel{4}}\cdot\frac{\cancel{4}}{3}=\frac{7}{3}=2\frac{1}{3}$

65. $1\frac{7}{24}\div\frac{7}{8}=\frac{31}{24}\div\frac{7}{8}=\frac{31}{24}\cdot\frac{8}{7}$

$=\frac{31}{3\cdot\cancel{8}}\cdot\frac{\cancel{8}}{7}=\frac{31}{21}=1\frac{10}{21}$

TRY IT YOURSELF

67. $-6\cdot 2\frac{7}{24}=-\frac{6}{1}\cdot\frac{55}{24}$

$=-\frac{\cancel{6}}{1}\cdot\frac{55}{4\cdot\cancel{6}}=-\frac{55}{4}=-13\frac{3}{4}$

69. $-6\frac{3}{5}\div 7\frac{1}{3}=-\frac{33}{5}\div\frac{22}{3}$

$=-\frac{33}{5}\cdot\frac{3}{22}=-\frac{99}{110}=-\frac{9\cdot\cancel{11}}{10\cdot\cancel{11}}=-\frac{9}{10}$

71. $\left(1\frac{2}{3}\right)^2=\left(\frac{5}{3}\right)^2=\left(\frac{5}{3}\right)\left(\frac{5}{3}\right)=\frac{25}{9}=2\frac{7}{9}$

73. $8\div 3\frac{1}{5}=\frac{8}{1}\div\frac{16}{5}=\frac{8}{1}\cdot\frac{5}{16}=\frac{\cancel{8}}{1}\cdot\frac{5}{2\cdot\cancel{8}}$

$=\frac{5}{2}=2\frac{1}{2}$

75. $-20\frac{1}{4}\div\left(-1\frac{11}{16}\right)=-\frac{81}{4}\div\left(-\frac{27}{16}\right)$

$=-\frac{81}{4}\cdot\left(-\frac{16}{27}\right)=-\frac{3\cdot\cancel{27}}{\cancel{4}}\cdot\left(-\frac{\cancel{4}\cdot 4}{\cancel{27}}\right)$

$=12$

77. $3\frac{1}{16}\cdot 4\frac{4}{7}=\frac{49}{16}\cdot\frac{32}{7}=\frac{\cancel{7}\cdot 7}{\cancel{16}}\cdot\frac{2\cdot\cancel{16}}{\cancel{7}}=14$

79.
$$-4\frac{1}{2}\div 2\frac{1}{4}=-\frac{9}{2}\div\frac{9}{4}$$
$$=-\frac{9}{2}\cdot\frac{4}{9}=-\frac{\cancel{9}}{\cancel{2}}\cdot\frac{\cancel{2}\cdot 2}{\cancel{9}}=-2$$

81.
$$2\frac{1}{2}\left(-3\frac{1}{3}\right)=\frac{5}{2}\left(-\frac{10}{3}\right)=-\frac{50}{6}$$
$$=-8\frac{2}{6}=-8\frac{2}{2\cdot 3}=-8\frac{\cancel{2}}{\cancel{2}\cdot 3}=-8\frac{1}{3}$$

83.
$$7\cdot 1\frac{3}{28}=\frac{7}{1}\cdot\frac{31}{28}=\frac{\cancel{7}}{1}\cdot\frac{31}{4\cdot\cancel{7}}=\frac{31}{4}=7\frac{3}{4}$$

85.
$$8\div 3\frac{1}{5}=\frac{8}{1}\div\frac{16}{5}=\frac{8}{1}\cdot\frac{5}{16}=\frac{\cancel{8}}{1}\cdot\frac{5}{2\cdot\cancel{8}}=\frac{5}{2}=2\frac{1}{2}$$

87.
$$1\frac{2}{3}\cdot 6\cdot\left(-\frac{1}{8}\right)=\frac{5}{3}\cdot\frac{6}{1}\cdot\left(-\frac{1}{8}\right)$$
$$=\frac{5}{\cancel{3}}\cdot\frac{\cancel{2}\cdot\cancel{3}}{1}\cdot\left(-\frac{1}{\cancel{2}\cdot 4}\right)=-\frac{5}{4}=-1\frac{1}{4}$$

89.
$$\left(-1\frac{1}{3}\right)^3=\left(-\frac{4}{3}\right)^3=\left(-\frac{4}{3}\right)\left(-\frac{4}{3}\right)\left(-\frac{4}{3}\right)$$
$$=-\frac{64}{27}=-2\frac{10}{27}$$

APPLICATIONS

91.
a. $3\frac{2}{3}$

b. $3\frac{2}{3}=\frac{3\cdot 3+2}{3}=\frac{9+2}{3}=\frac{11}{3}$

93.
Forward $2\frac{1}{2}$ somersaults.

95.
a. $2\frac{2}{3}$

b. $2\frac{2}{3}-\frac{12}{3}=\frac{8}{3}-\frac{12}{3}=\frac{-4}{3}=-1\frac{1}{3}$

97.
Since $59<60\frac{3}{4}<61$ she needs a size 14.

Since $24\frac{1}{4}<24\frac{1}{2}<24\frac{3}{4}$ she needs a slim cut.

99.
$$12\frac{1}{4}\cdot 6\frac{1}{4}=\frac{49}{4}\cdot\frac{25}{4}=\frac{1225}{16}=76\frac{9}{16}in$$

101.
$$\frac{1}{2}\cdot 8\frac{1}{4}\cdot 10\frac{1}{3}=\frac{1}{2}\cdot\frac{\overset{11}{\cancel{33}}}{4}\cdot\frac{31}{\underset{1}{\cancel{3}}}=\frac{341}{8}=42\frac{5}{8}in^2$$

103.
$$3\frac{1}{5}\cdot 20=\frac{16}{5}\cdot\frac{20}{1}=\frac{16}{\cancel{5}}\cdot\frac{4\cdot\cancel{5}}{1}=64 \text{ calories}$$

105.
$$84\cdot 4\frac{1}{4}=\overset{21}{\cancel{84}}\cdot\frac{17}{\underset{1}{\cancel{4}}}=357\text{¢}=\$3.57$$

107.
$$13\frac{3}{4}\div 11=\frac{\overset{5}{\cancel{55}}}{4}\cdot\frac{1}{\underset{1}{\cancel{11}}}=\frac{5}{4}=1\frac{1}{4}\text{ cups}$$

109.
$$200\div\frac{1}{3}=\frac{200}{1}\div\frac{1}{3}=\frac{200}{1}\cdot\frac{3}{1}=600$$
The caterer can serve 600 people.

111.
$$1\frac{1}{16}\div\frac{1}{8}=\frac{17}{16}\cdot\frac{8}{1}=\frac{17\cdot\cancel{8}}{2\cdot\cancel{8}}=\frac{17}{2}=8\frac{1}{2}\text{ furlongs}$$

WRITING

113. $2\frac{3}{4}$ represents $2+\frac{3}{4}$ whereas

$2\left(\frac{3}{4}\right)$ represents $2\times\frac{3}{4}$.

REVIEW

115. $5=5$

$12=2\cdot 2\cdot 3$

$15=3\cdot 4$

$LCM=2\cdot 2\cdot 3\cdot 5=60$

117. $12=2\cdot 2\cdot 3$

$68=2\cdot 2\cdot 17$

$92=2\cdot 2\cdot 23$

$GCF=2\cdot 2=4$

Section 4.6: Adding and Subtracting Mixed Numbers

VOCABULARY

1. A mixed number, such as $1\frac{7}{8}$, contains a whole number part and a fractional part.

3. To add (or subtract) mixed numbers written in vertical form, we add (or subtract) the fractions separately and the whole numbers separately.

5. Consider the following problem:

$$\begin{array}{r} 36\frac{5}{7} \\ +42\frac{4}{7} \\ \hline \end{array}$$

$$78\frac{9}{7} = 78 + 1\frac{2}{7} = 79\frac{2}{7}$$

Since we don't want an improper fraction in the answer we write $\frac{9}{7}$ as $1\frac{2}{7}$, carry the 1 and add it to 78 to get 79.

CONCEPTS

7. a. 76, $\frac{3}{4}$ b. $76\frac{3}{4} = 76 + \frac{3}{4}$

9. a. 12 b. 30 c. 18 d. 24

NOTATION

11.

$$\begin{array}{rcrcr} 6\frac{3}{5} = & & 6\frac{3}{5}\cdot\frac{7}{7} = & & 6\frac{21}{35} \\ +3\frac{2}{7} = & & +3\frac{2}{7}\cdot\frac{5}{5} = & & +3\frac{10}{35} \\ \hline & & & & 9\frac{31}{35} \end{array}$$

GUIDED PRACTICE

13.
$$1\frac{1}{4} + 2\frac{1}{3} = \frac{5}{4} + \frac{7}{3} = \frac{5}{4}\cdot\frac{3}{3} + \frac{7}{3}\cdot\frac{4}{4}$$
$$= \frac{15}{12} + \frac{28}{12} = \frac{43}{12} = 3\frac{7}{12}$$

15.
$$2\frac{1}{3} + 4\frac{2}{5} = \frac{7}{3} + \frac{22}{5} = \frac{7}{3}\cdot\frac{5}{5} + \frac{22}{5}\cdot\frac{3}{3}$$
$$= \frac{35}{15} + \frac{66}{15} = \frac{101}{15} = 6\frac{11}{15}$$

17.
$$-4\frac{1}{8} + 1\frac{3}{4} = \frac{-33}{8} + \frac{7}{4} = \frac{-33}{8} + \frac{7}{4}\cdot\frac{2}{2}$$
$$= \frac{-33}{8} + \frac{14}{8} = \frac{-19}{8} = -2\frac{3}{8}$$

19.
$$-6\frac{5}{6} + 3\frac{2}{3} = \frac{-41}{6} + \frac{11}{3} = \frac{-41}{6} + \frac{11}{3}\cdot\frac{2}{2}$$
$$= \frac{-41}{6} + \frac{22}{6} = \frac{-19}{6} = -3\frac{1}{6}$$

21.
$$334\frac{1}{7} + 42\frac{2}{3} = 334 + 42 + \frac{1}{7} + \frac{2}{3}$$
$$= 376 + \frac{1}{7}\cdot\frac{3}{3} + \frac{2}{3}\cdot\frac{7}{7} = 376 + \frac{3}{21} + \frac{14}{21}$$
$$= 376\frac{17}{21}$$

23.
$$667\frac{1}{5} + 47\frac{3}{4} = 667 + 47 + \frac{1}{5} + \frac{3}{4}$$
$$= 714 + \frac{1}{5}\cdot\frac{4}{4} + \frac{3}{4}\cdot\frac{5}{5}$$
$$= 714 + \frac{4}{20} + \frac{15}{20} = 714\frac{19}{20}$$

25.
$$41\frac{2}{9} + 18\frac{2}{5} = 41 + 18 + \frac{2}{9} + \frac{2}{5}$$
$$= 59 + \frac{2}{9}\cdot\frac{5}{5} + \frac{2}{5}\cdot\frac{9}{9} = 59 + \frac{10}{45} + \frac{18}{45}$$
$$= 59\frac{28}{45}$$

27.
$$89\frac{6}{11} + 43\frac{1}{3} = 89 + 43 + \frac{6}{11} + \frac{1}{3}$$
$$= 132 + \frac{6}{11}\cdot\frac{3}{3} + \frac{1}{3}\cdot\frac{11}{11}$$
$$= 132 + \frac{18}{33} + \frac{11}{33} = 132\frac{29}{33}$$

29.

$$14\frac{1}{4}+29\frac{1}{20}+78\frac{3}{5}$$

$$=14+29+78+\frac{1}{4}+\frac{1}{20}+\frac{3}{5}$$

$$=121+\frac{1}{4}\cdot\frac{5}{5}+\frac{1}{20}+\frac{3}{5}\cdot\frac{4}{4}$$

$$=121+\frac{5}{20}+\frac{1}{20}+\frac{12}{20}=121\frac{18}{20}=121\frac{9}{10}$$

31.

$$106\frac{5}{18}+22\frac{1}{2}+19\frac{1}{9}$$

$$=106+22+19+\frac{5}{18}+\frac{1}{2}+\frac{1}{9}$$

$$=147+\frac{5}{18}+\frac{1}{2}\cdot\frac{9}{9}+\frac{1}{9}\cdot\frac{2}{2}$$

$$=147+\frac{5}{18}+\frac{9}{18}+\frac{2}{18}=147\frac{16}{18}=147\frac{8}{9}$$

33.

$$\begin{array}{rcrcr} 39\frac{5}{8} & = & 39\frac{5}{8}\cdot\frac{3}{3} & = & 39\frac{15}{24} \\ \underline{+62\frac{11}{12}} & = & \underline{+62\frac{11}{12}\cdot\frac{2}{2}} & = & \underline{+62\frac{22}{24}} \\ & & & & 101\frac{37}{24} \end{array}$$

$$=101+1\frac{13}{24}=102\frac{13}{24}$$

35.

$$\begin{array}{rcrcr} 82\frac{8}{9} & = & 82\frac{8}{9}\cdot\frac{5}{5} & = & 82\frac{40}{45} \\ \underline{+46\frac{11}{15}} & = & \underline{+46\frac{11}{15}\cdot\frac{3}{3}} & = & \underline{+46\frac{33}{45}} \\ & & & & 128\frac{73}{45} \end{array}$$

$$=128+1\frac{28}{45}=129\frac{28}{45}$$

37.

$$19\frac{11}{12}-9\frac{2}{3}=19\frac{11}{12}-9\frac{8}{12}=10\frac{3}{12}=10\frac{1}{4}$$

39.

$$21\frac{5}{6}-8\frac{3}{10}=21\frac{25}{30}-8\frac{9}{30}$$

$$=13\frac{16}{30}=13\frac{8}{15}$$

41.

$$47\frac{1}{11}-15\frac{2}{3}=47\frac{3}{33}-15\frac{22}{33}$$

$$=46\frac{36}{33}-15\frac{22}{33}=31\frac{14}{33}$$

43.

$$84\frac{5}{8}-12\frac{6}{7}=84\frac{35}{56}-12\frac{48}{56}$$

$$=83\frac{91}{56}-12\frac{48}{56}=71\frac{43}{56}$$

45.

$$674-94\frac{11}{15}=673\frac{15}{15}-94\frac{11}{15}=579\frac{4}{15}$$

47.

$$112-49\frac{9}{32}=111\frac{32}{32}-49\frac{9}{32}=62\frac{23}{32}$$

TRY IT YOURSELF

49.

$$140\frac{5}{6}-129\frac{4}{5}=140\frac{25}{30}-129\frac{24}{30}=11\frac{1}{30}$$

51.

$$4\frac{1}{6}+1\frac{1}{5}=\frac{25}{6}+\frac{6}{5}=\frac{125}{30}+\frac{36}{30}=\frac{161}{30}=5\frac{11}{30}$$

53.

$$5\frac{1}{2}+3\frac{4}{5}=\frac{11}{2}+\frac{19}{5}=\frac{55}{10}+\frac{38}{10}=\frac{93}{10}=9\frac{3}{10}$$

55.

$$2+1\frac{7}{8}=3\frac{7}{8}$$

57.

$$8\frac{7}{9}-3\frac{1}{9}=5\frac{6}{9}=5\frac{2}{3}$$

59.

$$140\frac{3}{16}-129\frac{3}{4}$$

$$=139\frac{19}{16}-129\frac{12}{16}=10\frac{7}{16}$$

61.

$$380\frac{1}{6}+17\frac{1}{4}=380\frac{4}{24}+17\frac{6}{24}$$

$$=397\frac{10}{24}=397\frac{5}{12}$$

63.

$$-2\frac{5}{6}+1\frac{3}{8}=-\frac{17}{6}+\frac{11}{8}$$

$$=\frac{-68}{24}+\frac{33}{24}=-\frac{35}{24}=-1\frac{11}{24}$$

65. $3\frac{1}{4}+4\frac{1}{4}=7\frac{2}{4}=7\frac{1}{2}$

67. $-3\frac{3}{4}+\left(-1\frac{1}{2}\right)=-\frac{15}{4}-\frac{3}{2}$

$=-\frac{15}{4}-\frac{6}{4}=-\frac{21}{4}=-5\frac{1}{4}$

69. $7-\frac{2}{3}=\frac{21}{3}-\frac{2}{3}=\frac{19}{3}=6\frac{1}{3}$

71. $12\frac{1}{2}+5\frac{3}{4}+35\frac{1}{6}=12\frac{6}{12}+5\frac{9}{12}+35\frac{2}{12}$

$=52\frac{17}{12}=53\frac{5}{12}$

73. $16\frac{1}{4}-13\frac{3}{4}=15\frac{5}{4}-13\frac{3}{4}=2\frac{2}{4}=2\frac{1}{2}$

75. $-4\frac{5}{8}-1\frac{1}{4}=-\frac{37}{8}-\frac{5}{4}$

$=-\frac{37}{8}-\frac{10}{8}=-\frac{47}{8}=-5\frac{7}{8}$

77. $6\frac{5}{8}-3=3\frac{5}{8}$

79. $\frac{7}{3}+2=\frac{7}{3}+\frac{6}{3}=\frac{13}{3}=4\frac{1}{3}$

81. $58\frac{7}{8}+340\frac{1}{2}+61\frac{3}{4}$

$=58\frac{7}{8}+340\frac{4}{8}+61\frac{6}{8}=459\frac{17}{8}=461\frac{1}{8}$

83. $9-8\frac{3}{4}=\frac{36}{4}-\frac{35}{4}=\frac{1}{4}$

85. $167+45=212$

Not Reasonable

87. $616-489=127$

Reasonable

APPLICATIONS

89. $3\frac{3}{4}+1\frac{1}{2}=3\frac{3}{4}+1\frac{2}{4}=4\frac{5}{4}=5\frac{1}{4}\,hrs$

91. $2\frac{3}{4}+\frac{1}{2}+\frac{2}{3}+\frac{1}{3}+2\frac{2}{3}+\frac{1}{4}$

$=\frac{11}{4}+\frac{1}{2}+\frac{2}{3}+\frac{1}{3}+\frac{8}{3}+\frac{1}{4}$

$=\frac{33}{12}+\frac{6}{12}+\frac{8}{12}+\frac{4}{12}+\frac{32}{12}+\frac{3}{12}$

$=\frac{86}{12}=7\frac{2}{12}=7\frac{1}{6}$ cups

93. $2\frac{11}{16}+2\frac{3}{4}+3\frac{1}{4}+2\frac{1}{2}$

$+1\frac{1}{2}+2\frac{3}{4}+1\frac{15}{16}+2\frac{11}{16}$

$=2\frac{11}{16}+2\frac{12}{16}+3\frac{4}{16}+2\frac{8}{16}$

$+1\frac{8}{16}+2\frac{12}{16}+1\frac{15}{16}+2\frac{11}{16}$

$=15+\frac{81}{16}=15+5\frac{1}{16}=20\frac{1}{16}\,lbs$

95. $P=2l+2w$

$P=2\left(24\frac{1}{2}\right)+2\left(29\frac{3}{4}\right)$

$P=\frac{98}{2}+\frac{119}{2}=\frac{217}{2}$

$P=108\frac{1}{2}\,in.$

97. $3\frac{1}{2}-\frac{3}{4}=\frac{7}{2}-\frac{3}{4}$

$=\frac{14}{4}-\frac{3}{4}=\frac{11}{4}=2\frac{3}{4}\,mi$

99. $50-1\frac{1}{2}=\frac{100}{2}-\frac{3}{2}=\frac{97}{2}=48\frac{1}{2}\,ft.$

101. a. 20¢ b. 20¢

103. $5-1\frac{3}{4}=\frac{20}{4}-\frac{7}{4}=\frac{13}{4}=3\frac{1}{4}\,in$

105. $22+10+28=60\ \textit{inches}$ (about)

WRITING

107. Answers will vary.

109. Change $\frac{7}{5}$ into $1\frac{2}{5}$ and add the whole number parts to get $13\frac{2}{5}$.

111. $18\frac{13}{25}$ rounds to 19 because $\frac{13}{25}$ is greater than $\frac{1}{2}$.

REVIEW

113.

a. $3\frac{1}{2}+1\frac{1}{4}=3\frac{2}{4}+1\frac{1}{4}=4\frac{3}{4}$

b. $3\frac{1}{2}-1\frac{1}{4}=3\frac{2}{4}-1\frac{1}{4}=2\frac{1}{4}$

c. $3\frac{1}{2}\cdot 1\frac{1}{4}=\frac{7}{2}\cdot\frac{5}{4}=\frac{35}{8}=4\frac{3}{8}$

d. $3\frac{1}{2}\div 1\frac{1}{4}=\frac{7}{2}\div\frac{5}{4}=\frac{7}{2}\cdot\frac{4}{5}$
$=\frac{28}{10}=\frac{14}{5}=2\frac{4}{5}$

Section 4.7: Order of Operations, Complex Fractions

VOCABULARY

1. We use the order of <u>operations</u> rule to evaluate expressions that contain more than one operation.

3. $\dfrac{\frac{1}{2}}{\frac{3}{4}}$ and $\dfrac{\frac{7}{8}+\frac{2}{5}}{\frac{1}{2}-\frac{1}{3}}$ are examples of <u>complex</u> fractions.

CONCEPTS

5. Raising to a power, multiplication, and addition

7. $\left(\frac{2}{3}-\frac{1}{10}\right)+1\frac{2}{15}$ or $1\frac{2}{15}+\left(\frac{2}{3}-\frac{1}{10}\right)$

9. $\frac{2}{3}\div\frac{1}{5}$

11. $\dfrac{\frac{1}{8}-\frac{3}{16}}{5\frac{3}{4}}=\dfrac{\frac{1}{8}-\frac{3}{16}}{\frac{23}{4}}$

NOTATION

13.
$$\frac{7}{12}-\frac{1}{2}\cdot\frac{1}{3}=\frac{7}{12}-\frac{1\cdot 1}{2\cdot 3}$$
$$=\frac{7}{12}-\frac{1}{6}$$
$$=\frac{7}{12}-\frac{1}{6}\cdot\frac{2}{2}$$
$$=\frac{7}{12}-\frac{2}{12}$$
$$=\frac{5}{12}$$

GUIDED PRACTICE

15.
$$\frac{3}{4}+\frac{2}{5}\left(-\frac{1}{2}\right)^2=\frac{3}{4}+\frac{\overset{1}{\cancel{2}}}{5}\left(\frac{1}{\underset{2}{\cancel{4}}}\right)=\frac{3}{4}+\frac{1}{10}$$
$$=\frac{15}{20}+\frac{2}{20}=\frac{17}{20}$$

17.
$$\frac{1}{6}+\frac{9}{8}\left(-\frac{2}{3}\right)^3=\frac{1}{6}+\frac{\overset{1}{\cancel{9}}}{\underset{1}{\cancel{8}}}\left(-\frac{\overset{1}{\cancel{8}}}{\underset{3}{\cancel{27}}}\right)$$
$$=\frac{1}{6}-\frac{1}{3}=\frac{1}{6}-\frac{2}{6}=-\frac{1}{6}$$

19.
$$\left(\frac{3}{4}-\frac{1}{6}\right)\div\left(-2\frac{1}{6}\right)=\left(\frac{9}{12}-\frac{2}{12}\right)\div\left(-\frac{13}{6}\right)$$
$$=\left(\frac{7}{\underset{2}{\cancel{12}}}\right)\cdot\left(-\frac{\overset{1}{\cancel{6}}}{13}\right)=-\frac{7}{26}$$

21. $\left(\frac{15}{16}-\frac{1}{8}\right)\div\left(-9\frac{3}{4}\right)=\left(\frac{15}{16}-\frac{2}{16}\right)\div\left(-\frac{39}{4}\right)$

$=\left(\frac{\overset{1}{\cancel{13}}}{\underset{4}{\cancel{16}}}\right)\cdot\left(-\frac{\overset{1}{\cancel{4}}}{\underset{3}{\cancel{39}}}\right)=-\frac{1}{12}$

23. $\left(\frac{5}{6}-\frac{2}{3}\right)+5\frac{4}{15}=\left(\frac{5}{6}-\frac{4}{6}\right)+\frac{79}{15}$

$=\frac{1}{6}+\frac{79}{15}=\frac{5}{30}+\frac{158}{30}=\frac{163}{30}=5\frac{13}{30}$

25. $\left(\frac{7}{9}-\frac{1}{2}\right)+2\frac{7}{18}=\left(\frac{14}{18}-\frac{9}{18}\right)+\frac{43}{18}$

$=\frac{5}{18}+\frac{43}{18}=\frac{48}{18}=2\frac{12}{18}=2\frac{2}{3}$

27. $A=\frac{1}{2}h(a+b)$

$A=\frac{1}{2}\left(5\frac{1}{4}\right)\left(2\frac{1}{2}+7\frac{1}{2}\right)$

$A=\frac{1}{2}\left(\frac{21}{4}\right)(10)=\frac{210}{8}$

$=26\frac{2}{8}=26\frac{1}{4}$

29. $A=\frac{1}{2}h(a+b)$

$A=\frac{1}{2}\left(4\frac{1}{2}\right)\left(1\frac{1}{4}+6\frac{3}{4}\right)$

$A=\frac{1}{2}\left(\frac{9}{2}\right)(8)=\frac{72}{4}=18$

31. $\frac{\frac{1}{16}}{\frac{2}{5}}=\frac{1}{16}\cdot\frac{5}{2}=\frac{5}{32}$

33. $\frac{\frac{5}{8}}{\frac{3}{4}}=\frac{5}{8}\cdot\frac{4}{3}=\frac{20}{24}=\frac{5}{6}$

35. $\frac{-\frac{1}{4}+\frac{2}{3}}{\frac{5}{6}+\frac{2}{3}}=\frac{-\frac{3}{12}+\frac{8}{12}}{\frac{5}{6}+\frac{4}{6}}=\frac{\frac{5}{12}}{\frac{9}{6}}=\frac{5}{\underset{2}{\cancel{12}}}\cdot\frac{\overset{1}{\cancel{6}}}{9}=\frac{5}{18}$

37. $\frac{\frac{1}{3}-\frac{3}{4}}{\frac{1}{6}+\frac{2}{3}}=\frac{\frac{4}{12}-\frac{9}{12}}{\frac{1}{6}+\frac{4}{6}}=\frac{-\frac{5}{12}}{\frac{5}{6}}=-\frac{\overset{1}{\cancel{5}}}{\underset{2}{\cancel{12}}}\cdot\frac{\overset{1}{\cancel{6}}}{\underset{1}{\cancel{5}}}=-\frac{1}{2}$

39. $\frac{5-\frac{5}{6}}{1\frac{1}{12}}=\frac{\frac{25}{6}}{\frac{13}{12}}=\frac{25}{\underset{1}{\cancel{6}}}\cdot\frac{\overset{2}{\cancel{12}}}{13}=\frac{50}{13}=3\frac{11}{13}$

41. $\frac{4-\frac{7}{8}}{3\frac{1}{4}}=\frac{\frac{25}{8}}{\frac{13}{4}}=\frac{25}{\underset{2}{\cancel{8}}}\cdot\frac{\overset{1}{\cancel{4}}}{13}=\frac{25}{26}$

TRY IT YOURSELF

43. $\frac{7}{8}-\left(\frac{4}{5}+1\frac{3}{4}\right)=\frac{7}{8}-\left(\frac{4}{5}+\frac{7}{4}\right)$

$=\frac{7}{8}-\frac{51}{20}=-\frac{67}{40}=-1\frac{27}{40}$

45. $\frac{-\frac{14}{15}}{\frac{7}{10}}=-\frac{\overset{2}{\cancel{14}}}{\underset{3}{\cancel{15}}}\cdot\frac{\overset{2}{\cancel{10}}}{\underset{1}{\cancel{7}}}=-\frac{4}{3}=-1\frac{1}{3}$

47. $A=\frac{1}{2}bh$

$A=\frac{1}{2}(10)\left(7\frac{1}{5}\right)=\frac{1}{\cancel{2}}\cdot\frac{\cancel{10}}{1}\cdot\frac{36}{\cancel{5}}$

$A=36$

49. $\frac{2}{3}\left(-\frac{1}{4}\right)+\frac{1}{2}=-\frac{1}{6}+\frac{1}{2}=-\frac{1}{6}+\frac{3}{6}=\frac{1}{3}$

51. $\frac{1}{2}\left(\frac{1}{8}\right)+\left(-\frac{1}{4}\right)^2=\frac{1}{16}+\frac{1}{16}=\frac{2}{16}=\frac{1}{8}$

53. $\frac{\frac{3}{8}+\frac{1}{4}}{\frac{3}{8}-\frac{1}{4}}=\frac{\frac{5}{8}}{\frac{1}{8}}=\frac{5}{\underset{1}{\cancel{8}}}\cdot\frac{\overset{1}{\cancel{8}}}{1}=5$

55. $$\left(5\frac{1}{6}-3\frac{7}{8}\right)+12\frac{11}{12}=\left(\frac{31}{6}-\frac{31}{8}\right)+\frac{155}{12}$$
$$=\left(\frac{124}{24}-\frac{93}{24}\right)+\frac{310}{24}=\frac{31}{24}+\frac{310}{24}$$
$$=\frac{341}{24}=14\frac{5}{24}$$

57. $$\frac{5\frac{1}{2}}{-\frac{1}{4}+\frac{3}{4}}=\frac{\frac{11}{2}}{\frac{1}{2}}=\frac{11}{\cancel{2}_1}\cdot\frac{\cancel{2}^1}{1}=11$$

59. $$\left|\frac{2}{3}-\frac{9}{10}\right|\div\left(-\frac{1}{5}\right)=\left|-\frac{7}{30}\right|\cdot\left(-\frac{5}{1}\right)$$
$$=\frac{7}{\cancel{30}_6}\cdot\left(\frac{-\cancel{5}^1}{1}\right)=-\frac{7}{6}=-1\frac{1}{6}$$

61. $$\frac{\frac{1}{5}-\left(-\frac{1}{4}\right)}{\frac{1}{4}+\frac{4}{5}}=\frac{\frac{4}{20}+\frac{5}{20}}{\frac{5}{20}+\frac{16}{20}}$$
$$=\frac{\frac{9}{20}}{\frac{21}{20}}=\frac{\cancel{9}^3}{\cancel{20}_1}\cdot\frac{\cancel{20}^1}{\cancel{21}_7}=\frac{3}{7}$$

63. $$1\frac{3}{5}\left(\frac{1}{2}\right)^2\left(\frac{3}{4}\right)=\frac{\cancel{8}^{\cancel{2}^1}}{5}\cdot\frac{1}{\cancel{4}_1}\cdot\frac{3}{\cancel{4}_2}=\frac{3}{10}$$

65. $$A=l\cdot w$$
$$A=5\frac{5}{6}\cdot 7\frac{3}{5}=\frac{\cancel{35}^7}{\cancel{6}_3}\cdot\frac{\cancel{38}^{19}}{\cancel{5}_1}=\frac{133}{3}=44\frac{1}{3}$$

67. $$\left(2-\frac{1}{2}\right)^2+\left(2+\frac{1}{2}\right)^2=\left(\frac{3}{2}\right)^2+\left(\frac{5}{2}\right)^2$$
$$=\frac{9}{4}+\frac{25}{4}=\frac{34}{4}=8\frac{1}{2}$$

69. $$\frac{-\frac{5}{6}}{-1\frac{7}{8}}=\frac{-\frac{5}{6}}{-\frac{15}{8}}=\left(-\frac{\cancel{5}^1}{\cancel{6}_3}\right)\cdot\left(-\frac{\cancel{8}^4}{\cancel{15}_3}\right)=\frac{4}{9}$$

71. $$\left(7\frac{3}{7}+3\frac{1}{5}\right)-9\frac{1}{10}=\left(\frac{52}{7}+\frac{16}{5}\right)-\frac{91}{10}$$
$$=\left(\frac{520}{70}+\frac{224}{70}\right)-\frac{637}{700}=\frac{744}{70}-\frac{637}{70}$$
$$=\frac{107}{70}=1\frac{37}{70}$$

73. $$\frac{\frac{3}{5}-2}{\frac{2}{5}-2}=\frac{\frac{3}{5}-\frac{10}{5}}{\frac{2}{5}-\frac{10}{5}}=\frac{-\frac{7}{5}}{-\frac{8}{5}}=\left(-\frac{7}{5}\right)\div\left(-\frac{8}{5}\right)$$
$$=\left(-\frac{7}{\cancel{5}}\right)\cdot\left(-\frac{\cancel{5}}{8}\right)=\frac{7}{8}$$

75. $$\left(\frac{8}{5}-1\frac{1}{3}\right)-\left(-\frac{4}{\cancel{5}_1}\cdot\cancel{10}^2\right)=\left(\frac{8}{5}-\frac{4}{3}\right)-(-8)$$
$$=\left(\frac{24}{15}-\frac{20}{15}\right)+8=\frac{4}{15}+8=8\frac{4}{15}$$

APPLICATIONS

77. $$20\left(4\frac{1}{2}+\frac{1}{16}\right)=20\left(4\frac{8}{16}+\frac{1}{16}\right)$$
$$=20\left(4\frac{9}{16}\right)=\cancel{20}^5\cdot\frac{73}{\cancel{16}_4}=\frac{365}{4}=91\frac{1}{4}\,in$$

79. $$\frac{1}{16}+3\cdot\frac{1}{16}+\frac{5}{8}=\frac{1}{16}+\frac{3}{16}+\frac{10}{16}$$
$$=\frac{14}{16}=\frac{7}{8}\,oz$$

It can be mailed for the 1-ounce rate.

81. $$5\left(\frac{1}{4}\right)+4\left(\frac{1}{2}\right)=\frac{5}{4}+2=1\frac{1}{4}+2=3\frac{1}{4}\,hr$$

83.

Total distance: $2\frac{4}{5}+1\frac{2}{5}+1\frac{4}{5}=4\frac{10}{5}=6mi$

$6\div\frac{2}{3}=6\cdot\frac{3}{2}=\frac{18}{2}=9$ parts

85.

$\frac{1}{2}+\frac{2}{3}+\frac{3}{4}=\frac{6}{12}+\frac{8}{12}+\frac{9}{12}=\frac{23}{12}$

$\frac{23}{12}\div\frac{1}{4}=\frac{23}{\cancel{12}_3}\cdot\frac{\cancel{4}^1}{1}=\frac{23}{3}$

7 full tubes were used, $\frac{2}{3}$ of a tube is left over.

87.

$A=\frac{1}{2}h(b_1+b_2)=\frac{1}{2}\cdot\left(2\frac{1}{3}\right)\left(2\frac{1}{2}+3\frac{1}{2}\right)$

$=\frac{1}{\cancel{2}}\cdot\frac{7}{\cancel{3}}\cdot\cancel{6}=7yd^2$

89.

$\frac{1}{\frac{1}{10}+\frac{1}{15}}=\frac{1}{\frac{3}{30}+\frac{2}{30}}=\frac{1}{\frac{5}{30}}=\frac{30}{5}=6\text{sec}.$

WRITING

91. The order of operations agreement is necessary to insure that everyone will get the same answer if done correctly.

93. A complex is a fraction whose the numerator and / or denominator has a fraction, decimal, or mixed number.

REVIEW

95.

$a-5=66$

$a-5+5=66+5$

$a=71$

97.

$2x+5=17$

$2x+5-5=17-5$

$2x=12$

$\frac{2x}{2}=\frac{12}{2}$

$x=6$

99. 1, 2, 3, 4, 6, 8, 12, 24

Section 4.8: Solving Equations That Involve Fractions

VOCABULARY

1. To solve an equation, we find all the values of the variable that make the equation true.

3. To find the reciprocal of a fraction, we invert the numerator and denominator.

CONCEPTS

5.

$\frac{5}{8}\cdot 40=\frac{5}{\cancel{8}}\cdot\frac{\cancel{40}^5}{1}=5\cdot 5=25$ so 40 is a solution.

7. When a fraction is multiplied by its reciprocal the result is 1.

9. a. $\frac{4}{5}p$ b. $\frac{1}{4}t$

11. a. the LCD of 2 and 3, which is 6

b. the LCD of 8, 3, and 2, which is 24

NOTATION

13.

$\frac{7}{8}x=21$

$\frac{8}{7}\left(\frac{7}{8}x\right)=\frac{8}{7}(21)$

$x=24$

The solution is 24.

15. a. true b. false c. true d. true

GUIDED PRACTICE

17.

$$y-\frac{1}{20}=\frac{13}{20}$$

$$y-\frac{1}{20}+\frac{1}{20}=\frac{13}{20}+\frac{1}{20}$$

$$y=\frac{14}{20}$$

$$y=\frac{7}{10}$$

Check: $y-\frac{1}{20}=\frac{13}{20}$

$$\frac{7}{10}-\frac{1}{20}\stackrel{?}{=}\frac{13}{20}$$

$$\frac{14}{20}-\frac{1}{20}\stackrel{?}{=}\frac{13}{20}$$

$$\frac{13}{20}=\frac{13}{20}$$

19.

$$x-\frac{1}{15}=\frac{11}{15}$$

$$x-\frac{1}{15}+\frac{1}{15}=\frac{11}{15}+\frac{1}{15}$$

$$x=\frac{12}{15}$$

$$x=\frac{4}{5}$$

Check: $x-\frac{1}{15}=\frac{11}{15}$

$$\frac{4}{5}-\frac{1}{15}\stackrel{?}{=}\frac{11}{15}$$

$$\frac{12}{15}-\frac{1}{15}\stackrel{?}{=}\frac{11}{15}$$

$$\frac{11}{15}=\frac{11}{15}$$

21.

$$\frac{5}{6}=x+\frac{1}{9}$$

$$\frac{5}{6}-\frac{1}{9}=x+\frac{1}{9}-\frac{1}{9}$$

$$\frac{15}{18}-\frac{2}{18}=x$$

$$\frac{13}{18}=x$$

Check:

$$\frac{5}{6}=x+\frac{1}{9}$$

$$\frac{5}{6}\stackrel{?}{=}\frac{13}{18}+\frac{1}{9}$$

$$\frac{5}{6}\stackrel{?}{=}\frac{13}{18}+\frac{2}{18}$$

$$\frac{5}{6}\stackrel{?}{=}\frac{15}{18}$$

$$\frac{5}{6}=\frac{5}{6}$$

23.

$$\frac{7}{10}=a+\frac{1}{4}$$

$$\frac{7}{10}-\frac{1}{4}=a+\frac{1}{4}-\frac{1}{4}$$

$$\frac{14}{20}-\frac{5}{20}=a$$

$$\frac{9}{20}=a$$

Check:

$$\frac{7}{10}=a+\frac{1}{4}$$

$$\frac{7}{10}\stackrel{?}{=}\frac{9}{20}+\frac{1}{4}$$

$$\frac{7}{10}\stackrel{?}{=}\frac{9}{20}+\frac{5}{20}$$

$$\frac{7}{10}\stackrel{?}{=}\frac{14}{20}$$

$$\frac{7}{10}=\frac{7}{10}$$

25.

$$\frac{4}{9}x = 12$$
$$\frac{9}{4}\left(\frac{4}{9}x\right) = \frac{9}{4}\cdot 12$$
$$1x = \frac{9}{\cancel{4}}\cdot \overset{3}{\cancel{12}}$$
$$x = 27$$

Check: $\frac{4}{9}x = 12$
$$\frac{4}{\cancel{9}}\cdot \overset{3}{\cancel{27}} \overset{?}{=} 12$$
$$12 = 12$$

27.

$$\frac{3}{7}w = 30$$
$$\frac{7}{3}\left(\frac{3}{7}w\right) = \frac{7}{3}\cdot 30$$
$$1w = \frac{7}{\cancel{3}}\cdot \overset{10}{\cancel{30}}$$
$$w = 70$$

Check: $\frac{3}{7}w = 30$
$$\frac{3}{\cancel{7}}\cdot \overset{10}{\cancel{70}} \overset{?}{=} 30$$
$$30 = 30$$

29.

$$-\frac{8}{3}w = 7$$
$$-\frac{3}{8}\left(-\frac{8}{3}w\right) = -\frac{3}{8}\cdot 7$$
$$1w = -\frac{3}{8}\cdot 7$$
$$w = -\frac{21}{8}$$

Check: $-\frac{8}{3}w = 7$
$$-\frac{\cancel{8}}{\cancel{3}}\left(-\frac{\overset{7}{\cancel{21}}}{\cancel{8}}\right) \overset{?}{=} 7$$
$$7 = 7$$

31.

$$-\frac{16}{9}w = 5$$
$$-\frac{9}{16}\left(-\frac{16}{9}w\right) = -\frac{9}{16}\cdot 5$$
$$1w = -\frac{9}{16}\cdot 5$$
$$w = -\frac{45}{16}$$

Check: $-\frac{16}{5}w = 5$
$$-\frac{\cancel{16}}{\cancel{9}}\left(-\frac{\overset{5}{\cancel{45}}}{\cancel{16}}\right) \overset{?}{=} 5$$
$$5 = 5$$

33.

$$\frac{4}{9}x = 12$$
$$9\left(\frac{4}{9}x\right) = 9(12)$$
$$4x = 108$$
$$\frac{4x}{4} = \frac{108}{4}$$
$$x = 27$$

35.

$$\frac{3}{7}w = 30$$
$$7\left(\frac{3}{7}w\right) = 7(30)$$
$$3w = 210$$
$$\frac{3w}{3} = \frac{210}{3}$$
$$w = 70$$

37.

$$\frac{1}{8}x = \frac{25}{72}$$

$$8 \cdot \frac{1}{8}x = 8 \cdot \frac{25}{72}$$

$$1x = \cancel{8} \cdot \frac{25}{\cancel{72}_{9}}$$

$$x = \frac{25}{9}$$

Check:

$$\frac{1}{8}x = \frac{25}{72}$$

$$\frac{1}{8}\left(\frac{25}{9}\right) \stackrel{?}{=} \frac{25}{72}$$

$$\frac{25}{72} = \frac{25}{72}$$

39.

$$\frac{1}{5}x = \frac{43}{55}$$

$$5 \cdot \frac{1}{5}x = 5 \cdot \frac{43}{55}$$

$$1x = \cancel{5} \cdot \frac{43}{\cancel{55}_{11}}$$

$$x = \frac{43}{11}$$

Check:

$$\frac{1}{5}x = \frac{43}{55}$$

$$\frac{1}{5}\left(\frac{43}{11}\right) \stackrel{?}{=} \frac{43}{55}$$

$$\frac{43}{55} = \frac{43}{55}$$

41.

$$17x = -\frac{15}{4}$$

$$\frac{1}{17} \cdot 17x = \frac{1}{17}\left(-\frac{15}{4}\right)$$

$$x = -\frac{15}{68}$$

Check:

$$17x = -\frac{15}{4}$$

$$\cancel{17} \cdot \left(-\frac{15}{\cancel{68}_{4}}\right) \stackrel{?}{=} -\frac{15}{4}$$

$$-\frac{15}{4} = -\frac{15}{4}$$

43.

$$29x = -\frac{13}{3}$$

$$\frac{1}{29} \cdot 29x = \frac{1}{29}\left(-\frac{13}{3}\right)$$

$$x = -\frac{13}{87}$$

Check:

$$29x = -\frac{13}{3}$$

$$\cancel{29} \cdot \left(-\frac{13}{\cancel{87}_{3}}\right) \stackrel{?}{=} -\frac{13}{3}$$

$$-\frac{13}{3} = -\frac{13}{3}$$

45.

$$\frac{5}{6}k-5=-15$$
$$\frac{5}{6}k-5+5=-15+5$$
$$\frac{5}{6}k=-10$$
$$\frac{6}{5}\cdot\frac{5}{6}k=\frac{6}{5}(-10)$$
$$1k=6(-2)$$
$$k=-12$$

Check: $\frac{5}{6}k-5=-15$
$$\frac{5}{\not 6}\left(-\overset{2}{\not 12}\right)-5\overset{?}{=}-15$$
$$-10-5\overset{?}{=}-15$$
$$-15=-15$$

47.

$$\frac{7}{16}h+28=21$$
$$\frac{7}{16}h+28-28=21-28$$
$$\frac{7}{16}h=-7$$
$$\frac{16}{7}\cdot\frac{7}{16}h=\frac{16}{7}(-7)$$
$$1h=16(-1)$$
$$h=-16$$

Check: $\frac{7}{16}h+28=21$
$$\frac{7}{\not 16}\left(-\overset{1}{\not 16}\right)+28\overset{?}{=}21$$
$$-7+28\overset{?}{=}21$$
$$21=21$$

49.

$$\frac{2}{9}=r-\frac{1}{6}$$
$$(18)\frac{2}{9}=(18)r-(18)\frac{1}{6}$$
$$\left(\overset{2}{\not 18}\right)\frac{2}{\not 9}=(18)r-\left(\overset{3}{\not 18}\right)\frac{1}{\not 6}$$
$$4=18r-3$$
$$4+3=18r-3+3$$
$$7=18r$$
$$\frac{7}{18}=r$$

Check:
$$\frac{2}{9}=r-\frac{1}{6}$$
$$\frac{2}{9}\overset{?}{=}\frac{7}{18}-\frac{1}{6}$$
$$\frac{2}{9}\overset{?}{=}\frac{7}{18}-\frac{3}{18}$$
$$\frac{2}{9}\overset{?}{=}\frac{4}{18}$$
$$\frac{2}{9}=\frac{2}{9}$$

51.

$$\frac{1}{4}=x-\frac{2}{9}$$
$$(36)\frac{1}{4}=(36)x-(36)\frac{2}{9}$$
$$\left(\cancel{36}^{9}\right)\frac{1}{\cancel{4}}=(36)x-\left(\cancel{36}^{4}\right)\frac{2}{\cancel{9}}$$
$$9=36x-8$$
$$9+8=36x-8+8$$
$$17=36x$$
$$\frac{17}{36}=\frac{36}{36}x$$
$$\frac{17}{36}=x$$

Check:

$$\frac{1}{4}=x-\frac{2}{9}$$
$$\frac{1}{4}\stackrel{?}{=}\frac{17}{36}-\frac{2}{9}$$
$$\frac{1}{4}\stackrel{?}{=}\frac{17}{36}-\frac{8}{36}$$
$$\frac{1}{4}\stackrel{?}{=}\frac{9}{36}$$
$$\frac{1}{4}=\frac{1}{4}$$

53.

$$\frac{a}{4}-\frac{a}{7}=-6$$
$$28\cdot\frac{a}{4}-28\cdot\frac{a}{7}=28(-6)$$
$$\cancel{28}^{7}\cdot\frac{a}{\cancel{4}}-\cancel{28}^{4}\cdot\frac{a}{\cancel{7}}=28(-6)$$
$$7a-4a=-168$$
$$3a=-168$$
$$\frac{3a}{3}=\frac{-168}{3}$$
$$a=-56$$

Check: $\frac{a}{4}-\frac{a}{7}=-6$

$$\frac{-56}{4}-\left(\frac{-56}{7}\right)\stackrel{?}{=}-6$$
$$-14-(-8)\stackrel{?}{=}-6$$
$$-6=-6$$

55.

$$\frac{k}{7}-\frac{k}{3}=-24$$
$$21\cdot\frac{k}{7}-21\cdot\frac{k}{3}=21(-24)$$
$$\cancel{21}^{3}\cdot\frac{k}{\cancel{7}}-\cancel{21}^{7}\cdot\frac{k}{\cancel{3}}=21(-24)$$
$$3k-7k=-504$$
$$-4k=-504$$
$$\frac{-4k}{-4}=\frac{-504}{-4}$$
$$k=126$$

Check: $\frac{k}{7}-\frac{k}{3}=-24$

$$\frac{126}{7}-\left(\frac{126}{3}\right)\stackrel{?}{=}-24$$
$$18-42\stackrel{?}{=}-24$$
$$-24=-24$$

57.

$$\frac{3}{5}d-\frac{4}{5}=\frac{1}{3}d$$
$$15\cdot\frac{3}{5}d-15\cdot\frac{4}{5}=15\cdot\frac{1}{3}d$$
$$\overset{3}{\cancel{15}}\cdot\frac{3}{\cancel{5}}d-\overset{3}{\cancel{15}}\cdot\frac{4}{\cancel{5}}=\overset{5}{\cancel{15}}\cdot\frac{1}{\cancel{3}}d$$
$$9d-12=5d$$
$$9d-12-9d=5d-9d$$
$$-12=-4d$$
$$\frac{-12}{-4}=\frac{-4d}{-4}$$
$$3=d$$

Check: $\frac{3}{5}d-\frac{4}{5}=\frac{1}{3}d$

$$\frac{3}{5}(3)-\frac{4}{5}\overset{?}{=}\frac{1}{3}(3)$$
$$\frac{9}{5}-\frac{4}{5}\overset{?}{=}1$$
$$\frac{5}{5}\overset{?}{=}1$$
$$1=1$$

59.

$$\frac{7}{8}w+\frac{1}{4}=\frac{3}{4}w$$
$$8\cdot\frac{7}{8}w+8\cdot\frac{1}{4}=8\cdot\frac{3}{4}w$$
$$\overset{1}{\cancel{8}}\cdot\frac{7}{\cancel{8}}w+\overset{2}{\cancel{8}}\cdot\frac{1}{\cancel{4}}=\overset{2}{\cancel{8}}\cdot\frac{3}{\cancel{4}}w$$
$$7w+2=6w$$
$$7w+2-7w=6w-7w$$
$$2=-w$$
$$\frac{2}{-1}=\frac{-w}{-1}$$
$$-2=w$$

Check: $\frac{7}{8}w+\frac{1}{4}=\frac{3}{4}w$

$$\frac{7}{\underset{4}{\cancel{8}}}(-\cancel{2})+\frac{1}{4}\overset{?}{=}\frac{3}{\underset{2}{\cancel{4}}}(-\cancel{2})$$
$$-\frac{7}{4}+\frac{1}{4}\overset{?}{=}-\frac{3}{2}$$
$$-\frac{6}{4}\overset{?}{=}-\frac{3}{2}$$
$$-\frac{3}{2}=-\frac{3}{2}$$

TRY IT YOURSELF

61.

$$\frac{1}{2}x-\frac{1}{9}=\frac{1}{3}$$
$$18\cdot\frac{1}{2}x-18\cdot\frac{1}{9}=18\cdot\frac{1}{3}$$
$$\overset{9}{\cancel{18}}\cdot\frac{1}{\cancel{2}}x-\overset{2}{\cancel{18}}\cdot\frac{1}{\cancel{9}}=\overset{6}{\cancel{18}}\cdot\frac{1}{\cancel{3}}$$
$$9x-2=6$$
$$9x-2+2=6+2$$
$$9x=8$$
$$\frac{9x}{9}=\frac{8}{9}$$
$$x=\frac{8}{9}$$

Check:
$$\frac{1}{2}x-\frac{1}{9}=\frac{1}{3}$$
$$\frac{1}{2}\left(\frac{8}{9}\right)-\frac{1}{9}\stackrel{?}{=}\frac{1}{3}$$
$$\frac{8}{18}-\frac{2}{18}\stackrel{?}{=}\frac{1}{3}$$
$$\frac{6}{18}\stackrel{?}{=}\frac{1}{3}$$
$$\frac{1}{3}=\frac{1}{3}$$

63.

$$\frac{a}{2}-\frac{a}{12}=\frac{-5a}{4}$$
$$12\cdot\frac{a}{2}-12\cdot\frac{a}{12}=12\cdot\frac{-5a}{4}$$
$$\overset{6}{\cancel{12}}\cdot\frac{a}{\cancel{2}}-\overset{1}{\cancel{12}}\cdot\frac{a}{\cancel{12}}=\overset{3}{\cancel{12}}\cdot\frac{-5a}{\cancel{4}}$$
$$6a-a=-15a$$
$$5a=-15a$$
$$5a+15a=-15a+15a$$
$$20a=0$$
$$a=0$$

Check:
$$\frac{a}{2}-\frac{a}{12}=\frac{-5a}{4}$$
$$\frac{0}{2}-\frac{0}{12}\stackrel{?}{=}\frac{-5\cdot 0}{4}$$
$$0-0\stackrel{?}{=}0$$
$$0=0$$

65.

$$\frac{7}{8}t=-28$$
$$8\left(\frac{7}{8}t\right)=8(-28)$$
$$7t=-224$$
$$\frac{7t}{7}=\frac{-224}{7}$$
$$t=-32$$

Check:
$$\frac{7}{8}t=-28$$
$$\frac{7}{\cancel{8}}\left(-\overset{4}{\cancel{32}}\right)\stackrel{?}{=}-28$$
$$-28=-28$$

67.

$$x+\frac{1}{9}=\frac{4}{9}$$
$$x+\frac{1}{9}-\frac{1}{9}=\frac{4}{9}-\frac{1}{9}$$
$$x=\frac{3}{9}$$
$$x=\frac{1}{3}$$

Check: $x+\frac{1}{9}=\frac{4}{9}$

$$\frac{1}{3}+\frac{1}{9}\stackrel{?}{=}\frac{4}{9}$$
$$\frac{3}{9}+\frac{1}{9}\stackrel{?}{=}\frac{4}{9}$$
$$\frac{4}{9}=\frac{4}{9}$$

69.

$$\frac{5f}{7}=-2$$
$$\frac{7}{5}\cdot\frac{5f}{7}=\frac{7}{5}(-2)$$
$$f=-\frac{14}{5}$$

Check: $\frac{5f}{7}=-2$

$$\frac{\cancel{5}}{\cancel{7}}\left(-\frac{\cancel{14}^{2}}{\cancel{5}}\right)\stackrel{?}{=}-2$$
$$-2=-2$$

71.

$$6x=2x-11$$
$$6x-2x=2x-11-2x$$
$$4x=-11$$
$$\frac{4x}{4}=\frac{-11}{4}$$
$$x=-\frac{11}{4}$$

Check: $6x=2x-11$

$$\cancel{6}^{3}\left(-\frac{11}{\cancel{4}_{2}}\right)\stackrel{?}{=}\cancel{2}\left(-\frac{11}{\cancel{4}_{2}}\right)-11$$
$$-\frac{33}{2}\stackrel{?}{=}-\frac{11}{2}-\frac{22}{2}$$
$$-\frac{33}{2}=\frac{-33}{2}$$

73.

$$2(y-3)=7$$
$$2y-6=7$$
$$2y-6+6=7+6$$
$$2y=13$$
$$\frac{2y}{2}=\frac{13}{2}$$
$$y=\frac{13}{2}$$

Check: $2(y-3)=7$

$$2\left(\frac{13}{2}-3\right)\stackrel{?}{=}7$$
$$2\left(\frac{13}{2}-\frac{6}{2}\right)\stackrel{?}{=}7$$
$$2\left(\frac{7}{2}\right)\stackrel{?}{=}7$$
$$7=7$$

75.

$$\frac{1}{3}=2x-\frac{1}{2}$$
$$(6)\frac{1}{3}=(6)2x-(6)\frac{1}{2}$$
$$\left(\overset{2}{\cancel{6}}\right)\frac{1}{\cancel{3}}=(6)2x-\left(\overset{3}{\cancel{6}}\right)\frac{1}{\cancel{2}}$$
$$2=12x-3$$
$$2+3=12x-3+3$$
$$5=12x$$
$$\frac{5}{12}=\frac{12}{12}x$$
$$\frac{5}{12}=x$$

Check:

$$\frac{1}{3}=2x-\frac{1}{2}$$
$$\frac{1}{3}\overset{?}{=}2\left(\frac{5}{12}\right)-\frac{1}{2}$$
$$\frac{1}{3}\overset{?}{=}\frac{10}{12}-\frac{6}{12}$$
$$\frac{1}{3}\overset{?}{=}\frac{4}{12}$$
$$\frac{1}{3}=\frac{1}{3}$$

77.

$$4=-\frac{3}{5}h$$
$$\left(-\frac{5}{3}\right)4=\left(-\frac{5}{3}\right)\left(-\frac{3}{5}h\right)$$
$$-\frac{20}{3}=h$$

Check: $4=-\frac{3}{5}h$

$$4\overset{?}{=}-\frac{\cancel{3}}{\cancel{5}}\left(-\frac{\overset{4}{\cancel{20}}}{\cancel{3}}\right)$$
$$4\overset{?}{=}-1(-4)$$
$$4=4$$

79.

$$5+\frac{1}{3}x=\frac{1}{2}+\frac{1}{6}x$$
$$6\cdot 5+6\cdot\frac{1}{3}x=6\cdot\frac{1}{2}+6\cdot\frac{1}{6}x$$
$$6\cdot 5+\overset{2}{\cancel{6}}\cdot\frac{1}{\cancel{3}}x=\overset{3}{\cancel{6}}\cdot\frac{1}{\cancel{2}}+\cancel{6}\cdot\frac{1}{\cancel{6}}x$$
$$30+2x=3+x$$
$$30+2x-x=3+x-x$$
$$30+x=3$$
$$30+x-30=3-30$$
$$x=-27$$

Check:

$$5+\frac{1}{3}x=\frac{1}{2}+\frac{1}{6}x$$
$$5+\frac{1}{\cancel{3}}\left(-\overset{9}{\cancel{27}}\right)\overset{?}{=}\frac{1}{2}+\frac{1}{\underset{2}{\cancel{6}}}\left(-\overset{9}{\cancel{27}}\right)$$
$$5+(-9)\overset{?}{=}\frac{1}{2}+\left(-\frac{9}{2}\right)$$
$$-4\overset{?}{=}-\frac{8}{2}$$
$$-4=-4$$

81.

$$\frac{2}{15}x=\frac{4}{5}$$
$$\frac{15}{2}\cdot\frac{2}{15}x=\frac{15}{2}\cdot\frac{4}{5}$$
$$\frac{\cancel{15}}{\cancel{2}}\cdot\frac{\cancel{2}}{\cancel{15}}x=\frac{\overset{3}{\cancel{15}}}{\cancel{2}}\cdot\frac{\overset{2}{\cancel{4}}}{\cancel{5}}$$
$$x=6$$

Check: $\frac{2}{15}x=\frac{4}{5}$

$$\frac{2}{\underset{5}{\cancel{15}}}\left(\overset{2}{\cancel{6}}\right)\overset{?}{=}\frac{4}{15}$$
$$\frac{4}{5}=\frac{4}{5}$$

83.
$$\frac{2}{5}x+1=\frac{1}{3}+x$$
$$15\cdot\frac{2}{5}x+15\cdot 1=15\cdot\frac{1}{3}+15\cdot x$$
$$\cancel{15}^{3}\cdot\frac{2}{\cancel{5}}x+15\cdot 1=\cancel{15}^{5}\cdot\frac{1}{\cancel{3}}+15\cdot x$$
$$6x+15=5+15x$$
$$6x+15-15x=5+15x-15x$$
$$-9x+15=5$$
$$-9x+15-15=5-15$$
$$-9x=-10$$
$$\frac{-9x}{-9}=\frac{-10}{-9}$$
$$x=\frac{10}{9}$$

Check:
$$\frac{2}{5}x+1=\frac{1}{3}+x$$
$$\frac{2}{\cancel{5}}\left(\frac{\cancel{10}^{2}}{9}\right)+1\stackrel{?}{=}\frac{1}{3}+\frac{10}{9}$$
$$\frac{4}{9}+\frac{9}{9}\stackrel{?}{=}\frac{3}{9}+\frac{10}{9}$$
$$\frac{13}{9}=\frac{13}{9}$$

85.
$$\frac{5h}{6}-8=12$$
$$\frac{5h}{6}-8+8=12+8$$
$$\frac{5h}{6}=20$$
$$\frac{6}{5}\cdot\frac{5h}{6}=\frac{6}{5}\cdot 20$$
$$\frac{\cancel{6}}{\cancel{5}}\cdot\frac{\cancel{5}h}{\cancel{6}}=\frac{6}{\cancel{5}}\cdot\cancel{20}^{4}$$
$$h=24$$

Check: $\frac{5h}{6}-8=12$
$$\frac{5}{\cancel{6}}\left(\cancel{24}^{4}\right)-8\stackrel{?}{=}12$$
$$20-8\stackrel{?}{=}12$$
$$12=12$$

87.
$$\frac{1}{2}y-2=\frac{1}{5}y+1$$
$$10\cdot\frac{1}{2}y-10\cdot 2=10\cdot\frac{1}{5}y+10\cdot 1$$
$$\cancel{10}^{5}\cdot\frac{1}{\cancel{2}}y-10\cdot 2=\cancel{10}^{2}\cdot\frac{1}{\cancel{5}}y+10\cdot 1$$
$$5y-20=2y+10$$
$$5y-20+20=2y+10+20$$
$$5y=2y+30$$
$$5y-2y=2y+30-2y$$
$$3y=30$$
$$\frac{3y}{3}=\frac{30}{3}$$
$$y=10$$

Check: $\frac{1}{2}y-2=\frac{1}{5}y+1$
$$\frac{1}{\cancel{2}}\left(\cancel{10}^{5}\right)-2\stackrel{?}{=}\frac{1}{\cancel{5}}\left(\cancel{10}^{2}\right)+1$$
$$5-2\stackrel{?}{=}2+1$$
$$3=3$$

89.

$$\frac{x}{21}+\frac{3x}{7}=1-\frac{5x}{3}$$

$$21\cdot\frac{x}{21}+\overset{3}{\cancel{21}}\cdot\frac{3x}{\underset{1}{\cancel{7}}}=21\cdot1-21\cdot\frac{5x}{3}$$

$$x+9x=21-35x$$

$$10x=21-35x$$

$$10x+35x=21-35x+35x$$

$$45x=21$$

$$\frac{45x}{45}=\frac{21}{45}$$

$$x=\frac{7}{15}$$

Check:

$$\frac{x}{21}+\frac{3x}{7}=1-\frac{5x}{3}$$

$$\frac{1}{\underset{3}{\cancel{21}}}\cdot\frac{\cancel{7}}{15}+\frac{\cancel{3}}{\cancel{7}}\cdot\frac{\cancel{7}}{\underset{5}{\cancel{15}}}\overset{?}{=}1-\frac{\cancel{5}}{3}\cdot\frac{7}{\underset{3}{\cancel{15}}}$$

$$\frac{1}{45}+\frac{1}{5}\overset{?}{=}1-\frac{7}{9}$$

$$\frac{1}{45}+\frac{9}{45}\overset{?}{=}\frac{45}{45}-\frac{35}{45}$$

$$\frac{10}{45}=\frac{10}{45}$$

91.

$$\frac{x}{2}+\frac{x}{3}+\frac{x}{4}=0$$

$$\overset{6}{\cancel{12}}\cdot\frac{x}{\cancel{2}}+\overset{4}{\cancel{12}}\cdot\frac{x}{\cancel{3}}+\overset{3}{\cancel{12}}\cdot\frac{x}{\cancel{4}}=12\cdot0$$

$$6x+4x+3x=0$$

$$13x=0$$

$$x=0$$

Check:

$$\frac{x}{2}+\frac{x}{3}+\frac{x}{4}=0$$

$$\frac{0}{2}+\frac{0}{3}+\frac{0}{4}\overset{?}{=}0$$

$$0=0$$

93.

a.

$$\frac{12}{5}+\frac{n}{2}=\frac{n}{10}$$

$$10\cdot\frac{12}{5}+10\cdot\frac{n}{2}=10\cdot\frac{n}{10}$$

$$\overset{2}{\cancel{10}}\cdot\frac{12}{\cancel{5}}+\overset{5}{\cancel{10}}\cdot\frac{n}{\cancel{2}}=\cancel{10}\cdot\frac{n}{\cancel{10}}$$

$$24+5n=n$$

$$24+5n-5n=n-5n$$

$$24=-4n$$

$$\frac{24}{-4}=\frac{-4n}{-4}$$

$$-6=n$$

b.

$$\frac{12}{5}+\frac{n}{2}=\frac{12}{5}\cdot\frac{2}{2}+\frac{n}{2}\cdot\frac{5}{5}$$

$$=\frac{24}{10}+\frac{5n}{10}$$

$$=\frac{24+5n}{10}$$

95.

a.

$$\frac{x}{4}-\frac{1}{3}=\frac{x}{2}$$

$$12\cdot\frac{x}{4}-12\cdot\frac{1}{3}=12\cdot\frac{x}{2}$$

$$\overset{3}{\cancel{12}}\cdot\frac{x}{\cancel{4}}-\overset{4}{\cancel{12}}\cdot\frac{1}{\cancel{3}}=\overset{6}{\cancel{12}}\cdot\frac{x}{\cancel{2}}$$

$$3x-4=6x$$

$$3x-4-3x=6x-3x$$

$$-4=3x$$

$$\frac{-4}{3}=\frac{3x}{3}$$

$$-\frac{4}{3}=x$$

b.

$$\frac{x}{4}-\frac{1}{3}=\frac{x}{4}\cdot\frac{3}{3}-\frac{1}{3}\cdot\frac{4}{4}$$

$$=\frac{3x}{12}-\frac{4}{12}$$

$$=\frac{3x-4}{12}$$

APPLICATIONS

97. **Analyze**

* $\frac{1}{8}$ of the cars serviced last year had transmission problems.

* The shop repaired <u>32</u> transmission problems last year.

* How many <u>cars</u> did the shop service last year?

Assign

Let x = the number of <u>cars</u> that the repair shop serviced last year.

Form

The key phrase *one-eighth of* suggests <u>multiplication</u>.

$\frac{1}{8}$	of	the number of cars repaired last year	had	transmission problems.

$$\frac{1}{8}\cdot x = 32$$

Solve

$$\frac{1}{8}x = 32$$

$$8\left(\frac{1}{8}x\right) = 8(32)$$

$$x = 256$$

State

The shop repaired <u>256</u> cars last year.

Check

$$\frac{32}{256} = \frac{\cancel{32}}{8\cdot\cancel{32}} = \frac{1}{8}$$

99. t = the total number of teeth

$$\frac{4}{5}t = 16$$

$$\frac{\cancel{5}}{\cancel{4}}\left(\frac{\cancel{4}}{\cancel{5}}t\right) = \frac{5}{\cancel{4}}\left(\overset{4}{\cancel{16}}\right)$$

$$t = 20$$

The child will have 20 baby teeth.

101. p = the number of pages

$$\frac{2}{3}p = 300$$

$$\frac{\cancel{3}}{\cancel{2}}\left(\frac{\cancel{2}}{\cancel{3}}p\right) = \frac{3}{\cancel{2}}\left(\overset{150}{\cancel{300}}\right)$$

$$p = 450$$

The phone book has 450 pages.

103. h = the height of the warehouse

$$\frac{2}{3}h = 18$$

$$\frac{\cancel{3}}{\cancel{2}}\left(\frac{\cancel{2}}{\cancel{3}}h\right) = \frac{3}{\cancel{2}}\left(\overset{9}{\cancel{18}}\right)$$

$$h = 27$$

The warehouse ceiling is 27 feet high.

105. p = the number of square inches

$$\frac{5}{12}p = 100$$

$$\frac{\cancel{12}}{\cancel{5}}\left(\frac{\cancel{5}}{\cancel{12}}p\right) = \frac{12}{\cancel{5}}\left(\overset{20}{\cancel{100}}\right)$$

$$p = 240$$

The page is 240 square inches.

107. c = the class length

$$\frac{1}{4}c + \frac{2}{3}c + 30 = c$$

$$\overset{3}{\cancel{12}}\cdot\frac{1}{\cancel{4}}c + \overset{4}{\cancel{12}}\cdot\frac{2}{\cancel{3}}c + 12\cdot 30 = 12\cdot c$$

$$3c + 8c + 360 = 12c$$

$$11c + 360 = 12c$$

$$11c + 360 - 11c = 12c - 11c$$

$$360 = c$$

The class is 360 minutes long.

109. r = the size of the expanded roster

$$\frac{1}{2}r+\frac{1}{4}r+\frac{3}{20}r+4=r$$

$$\overset{10}{\cancel{20}}\cdot\frac{1}{\cancel{2}}r+\overset{5}{\cancel{20}}\cdot\frac{1}{\cancel{4}}r+\cancel{20}\cdot\frac{3}{\cancel{20}}r+20\cdot 4=20\cdot r$$

$$10r+5r+3r+80=20r$$

$$18r+80=20r$$

$$80=2r$$

$$40=r$$

There are 40 players on the extended roster.

111. h = the number of homes in the subdivision

$$\frac{1}{4}h=9$$

$$4\left(\frac{1}{4}h\right)=4(9)$$

$$h=36$$

There are 36 homes in the subdivision.

WRITING

113. Both sides of the equation should have been multiplied by 30.

115. It means that the fractions are removed by multiplying every term of the equation by a common denominator.

REVIEW

117. a. 13,000,000

b. 12,600,000

c. 12,599,800

Chapter 4 Review

1. numerator 11, denominator 16, proper fraction

2. shaded $\frac{4}{7}$; unshaded $\frac{3}{7}$

3. The figure is not divided into equal parts.

4. $-\frac{2}{3}, \frac{-2}{3}$

5. a. 1 b. 0 c. 18 d. undefined

6. equivalent fractions: $\frac{6}{8}=\frac{3}{4}$

7. $\frac{2}{3}\cdot\frac{6}{6}=\frac{12}{18}$

8. $\frac{3}{8}\cdot\frac{2}{2}=\frac{6}{16}$

9. $\frac{7}{15}=\frac{7}{15}\cdot\frac{3a}{3a}=\frac{21a}{45a}$

10. $\frac{13}{12x}=\frac{13}{12x}\cdot\frac{5}{5}=\frac{65}{60x}$

11. $5=\frac{5}{1}\cdot\frac{9}{9}=\frac{45}{9}$

12. a. No. There is a common factor of 3.

b. Yes. The numerator and denominator have nothing in common except 1.

13. $\frac{15}{45}=\frac{1\cdot\cancel{15}}{3\cdot\cancel{15}}=\frac{1}{3}$

14. $\frac{20x^3}{48x^2}=\frac{\cancel{4}\cdot 5\cdot\cancel{x}\cdot\cancel{x}\cdot x}{\cancel{4}\cdot 12\cdot\cancel{x}\cdot\cancel{x}}=\frac{5x}{12}$

15. $\frac{66}{108}=\frac{\cancel{6}\cdot 11}{\cancel{6}\cdot 18}=\frac{11}{18}$

16. $\frac{117a^2b^6}{208a^5b}=\frac{9\cdot\cancel{13}\cdot\cancel{a^2}\cdot\cancel{b}\cdot b^5}{\cancel{13}\cdot 16\cdot\cancel{a^2}\cdot a^3\cdot\cancel{b}}=\frac{9b^5}{16a^3}$

17. $\frac{81}{64}$ is in lowest terms because 81 and 64 have no common factors other than 1.

18. $\frac{8}{12}=\frac{2\cdot\cancel{4}}{3\cdot\cancel{4}}=\frac{2}{3}$

$\frac{176}{264}=\frac{2\cdot\cancel{88}}{3\cdot\cancel{88}}=\frac{2}{3}$

The fractions are equivalent.

19. Sleeping: $\frac{7}{24}$

Not sleeping: $\frac{17}{24}$

20. a. $\frac{5}{8}$ is being represented as an equivalent fraction by multiplying by a form of 1. To get a denominator of 16, multiply by 1 in the form of $\frac{2}{2}$.

b. $\frac{4}{6}$ is being represented as an equivalent fraction by removing a form of 1. To write $\frac{4}{6}$ in lowest terms, remove a 1 in the form of $\frac{2}{2}$.

21. To multiply two fractions, multiply the <u>numerators</u> and multiply the <u>denominators</u>. Then <u>simplify</u> if possible.

22. $\frac{5}{6}\cdot\frac{2}{3}$

23. $\frac{1}{2}\cdot\frac{1}{3}=\frac{1\cdot 1}{2\cdot 3}=\frac{1}{6}$

24. $\frac{2}{5}\left(-\frac{7}{9}\right)=-\frac{2\cdot 7}{5\cdot 9}=-\frac{14}{45}$

25. $\frac{9c^3}{16d}\cdot\frac{20d}{27c}=\frac{\cancel{9}\cdot\cancel{c}\cdot c^2\cdot\cancel{4}\cdot 5\cdot\cancel{d}}{\cancel{4}\cdot 4\cdot\cancel{d}\cdot 3\cdot\cancel{9}\cdot\cancel{c}}$

$=\frac{c^2\cdot 5}{4\cdot 3}=\frac{5c^2}{12}$

26. $-\frac{5}{6}\left(-\frac{1}{15}\right)\left(-\frac{18}{25}\right)=-\frac{\cancel{5}\cdot\cancel{3}\cdot\cancel{6}}{\cancel{6}\cdot\cancel{3}\cdot\cancel{5}\cdot 5\cdot 5}$

$=-\frac{1}{5\cdot 5}=-\frac{1}{25}$

27. $\frac{3}{5}\cdot 7=\frac{3}{5}\cdot\frac{7}{1}=\frac{21}{5}$

28. $-4\left(-\frac{9m}{16}\right)=\frac{4}{1}\cdot\frac{9m}{16}=\frac{\cancel{4}\cdot 9\cdot m}{\cancel{4}\cdot 4}=\frac{9m}{4}$

29. $-3x\left(\frac{1}{3}\right)=-\frac{3x}{1}\cdot\frac{1}{3}=-\frac{\cancel{3}\cdot x}{\cancel{3}}=-x$

30. $-\frac{6}{7}\left(-\frac{7}{6}\right)=\frac{42}{42}=1$

31. $-\left(\frac{3}{4}\right)^2=-\frac{3}{4}\cdot\frac{3}{4}=-\frac{9}{16}$

32. $\left(-\frac{5a}{2}\right)^3=\left(-\frac{5a}{2}\right)\left(-\frac{5a}{2}\right)\left(-\frac{5a}{2}\right)$

$=-\frac{5a\cdot 5a\cdot 5a}{2\cdot 2\cdot 2}=-\frac{125a^3}{8}$

33. $\left(-\frac{2}{5}\right)^3=\left(-\frac{2}{5}\right)\left(-\frac{2}{5}\right)\left(-\frac{2}{5}\right)=-\frac{8}{125}$

34. $\left(\frac{2}{3}\right)^2=\frac{2}{3}\cdot\frac{2}{3}=\frac{4}{9}$

35. $8\left(\frac{1}{4}\right)=\frac{8}{1}\cdot\frac{1}{4}=\frac{8}{4}=2mi$

36. $\frac{1}{6}(180)=\frac{1}{6}\cdot\frac{180}{1}=\frac{180}{6}=30lb$

37. $A=\frac{1}{2}\cdot b\cdot h$

$A=\frac{1}{2}\cdot 15\cdot 8=\frac{1}{2}\cdot 120$

$A=60in^2$

38. $A=\frac{1}{2}\cdot b\cdot h$

$A=\frac{1}{2}\cdot 22\cdot 15=11\cdot 15$

$A=165ft^2$

39. a. 8 b. $-\frac{12}{11}$ c. $\frac{1}{5}$ d. $\frac{7}{8a}$

40. To divide two fractions, <u>multiply</u> the first fraction by the <u>reciprocal</u> of the second fraction.

41. $\frac{1}{6}\div\frac{11}{25}=\frac{1}{6}\cdot\frac{25}{11}=\frac{25}{66}$

42. $-\frac{7}{32}\div\frac{1}{4}=-\frac{7}{32}\cdot\frac{4}{1}=-\frac{7\cdot\cancel{4}}{\cancel{4}\cdot 8}=-\frac{7}{8}$

43. $-\frac{39m^4}{25n}\div\left(-\frac{13m^3}{10n}\right)=\frac{39m^4}{25n}\cdot\frac{10n}{13m^3}$

$=\frac{3\cdot\cancel{13}\cdot m\cdot\cancel{m^3}\cdot 2\cdot\cancel{5}\cdot\cancel{n}}{\cancel{5}\cdot 5\cdot\cancel{n}\cdot\cancel{13}\cdot\cancel{m^3}}=\frac{3\cdot m\cdot 2}{5}=\frac{6m}{5}$

44. $54d\div\frac{63}{5}=\frac{54d}{1}\cdot\frac{5}{63}=\frac{6\cdot\cancel{9}\cdot d\cdot 5}{7\cdot\cancel{9}}$

$=\frac{6\cdot d\cdot 5}{7}=\frac{30d}{7}$

45. $-\frac{3}{8}\div\frac{1}{4}=-\frac{3}{8}\cdot\frac{4}{1}=-\frac{3\cdot\cancel{4}}{2\cdot\cancel{4}}=-\frac{3}{2}$

46. $\frac{4}{5}\div\frac{1}{2}=\frac{4}{5}\cdot\frac{2}{1}=\frac{8}{5}$

47. $\frac{2}{3}\div(-120)=-\frac{2}{3}\cdot\frac{1}{120}=-\frac{\cancel{2}\cdot 1}{3\cdot\cancel{2}\cdot 60}$

$=-\frac{1}{3\cdot 60}=-\frac{1}{180}$

48. $\frac{7}{15}\div\frac{7}{15}=\frac{7}{15}\cdot\frac{15}{7}=\frac{\cancel{7}}{\cancel{15}}\cdot\frac{\cancel{15}}{\cancel{7}}=1$

49. $\frac{3}{4}\div\frac{1}{16}=\frac{3}{4}\cdot\frac{16}{1}=\frac{3\cdot\cancel{4}\cdot 4}{\cancel{4}\cdot 1}=3\cdot 4=12\text{ pins}$

50. $20\div\frac{2}{3}=\frac{20}{1}\cdot\frac{3}{2}=\frac{\cancel{2}\cdot 10\cdot 3}{1\cdot\cancel{2}}=3\cdot 10=30\text{ cases}$

51. $\frac{2}{7}+\frac{3}{7}=\frac{2+3}{7}=\frac{5}{7}$

52. $\frac{3}{4}-\frac{1}{4}=\frac{3-1}{4}=\frac{2}{4}=\frac{\cancel{2}\cdot 1}{\cancel{2}\cdot 2}=\frac{1}{2}$

53. $\frac{7x}{8}+\frac{3x}{8}=\frac{7x+3x}{8}=\frac{10x}{8}=\frac{\cancel{2}\cdot 5\cdot x}{\cancel{2}\cdot 4}=\frac{5x}{4}$

54. $-\frac{3}{5}-\frac{3}{5}=\frac{-3-3}{5}=-\frac{6}{5}$

55.

a. $\frac{3}{8}+\frac{2}{8}=\frac{3+2}{8}=\frac{5}{8}$

b. $\frac{4}{5}-\frac{3}{5}=\frac{4-3}{5}=\frac{1}{5}$

56. $LCD=2\cdot 3\cdot 3\cdot 5=90$

57. $\frac{1}{6}+\frac{2}{3}=\frac{1}{6}+\frac{2}{3}\cdot\frac{2}{2}=\frac{1}{6}+\frac{4}{6}=\frac{5}{6}$

58. $-\frac{2}{5}-\frac{3}{8}=-\frac{2}{5}\cdot\frac{8}{8}-\frac{3}{8}\cdot\frac{5}{5}=-\frac{16}{40}-\frac{15}{40}=-\frac{31}{40}$

59. $\frac{5}{24}+\frac{3}{16}=\frac{5}{24}\cdot\frac{2}{2}+\frac{3}{16}\cdot\frac{3}{3}=\frac{10}{48}+\frac{9}{48}=\frac{19}{48}$

60. $3-\frac{1}{7}=\frac{3}{1}\cdot\frac{7}{7}-\frac{1}{7}=\frac{21}{7}-\frac{1}{7}=\frac{20}{7}$

61. $-\frac{19}{18}+\frac{5}{12}=-\frac{19}{18}\cdot\frac{2}{2}+\frac{5}{12}\cdot\frac{3}{3}=-\frac{38}{36}+\frac{15}{36}=-\frac{23}{36}$

62. $\frac{17}{20}-\frac{4}{15}=\frac{17}{20}\cdot\frac{3}{3}-\frac{4}{15}\cdot\frac{4}{4}$

$=\frac{51}{60}-\frac{16}{60}=\frac{35}{60}=\frac{\cancel{5}\cdot 7}{\cancel{5}\cdot 12}=\frac{7}{12}$

63. $-6+\frac{13}{6}=-\frac{6}{1}\cdot\frac{6}{6}+\frac{13}{6}=-\frac{36}{6}+\frac{13}{6}=-\frac{23}{6}$

64. $\frac{1}{3}+\frac{1}{4}+\frac{1}{5}=\frac{1}{3}\cdot\frac{20}{20}+\frac{1}{4}\cdot\frac{15}{15}+\frac{1}{5}\cdot\frac{12}{12}$

$=\frac{20}{60}+\frac{15}{60}+\frac{12}{60}=\frac{47}{60}$

65. $\frac{8}{n}+\frac{1}{2}=\frac{8}{n}\cdot\frac{2}{2}+\frac{1}{2}\cdot\frac{n}{n}=\frac{16}{2n}+\frac{n}{2n}=\frac{16+n}{2n}$

66. $\frac{11}{9}-\frac{4}{x}=\frac{11}{9}\cdot\frac{x}{x}-\frac{4}{x}\cdot\frac{9}{9}=\frac{11x}{9x}-\frac{36}{9x}=\frac{11x-36}{9x}$

67. $\frac{a}{3}-\frac{4}{11}=\frac{a}{3}\cdot\frac{11}{11}-\frac{4}{11}\cdot\frac{3}{3}=\frac{11a}{33}-\frac{12}{33}=\frac{11a-12}{33}$

68. $\frac{7}{8}+\frac{r}{7}=\frac{7}{8}\cdot\frac{7}{7}+\frac{r}{7}\cdot\frac{8}{8}=\frac{49}{56}+\frac{8r}{56}=\frac{49+8r}{56}$

69. $\frac{3}{4}-\frac{17}{32}=\frac{3}{4}\cdot\frac{8}{8}-\frac{17}{32}=\frac{24}{32}-\frac{17}{32}=\frac{7}{32}in$

70. $\frac{1}{20}+\frac{2}{5}+\frac{3}{10}=\frac{1}{20}+\frac{2}{5}\cdot\frac{4}{4}+\frac{3}{10}\cdot\frac{2}{2}$

$=\frac{1}{20}+\frac{8}{20}+\frac{6}{20}=\frac{15}{20}=\frac{3\cdot\cancel{5}}{4\cdot\cancel{5}}=\frac{3}{4}$

71. $\frac{2}{9}\square\frac{3}{11}$

$\frac{2}{9}\cdot\frac{11}{11}\square\frac{3}{11}\cdot\frac{9}{9}$

$\frac{22}{99}<\frac{27}{99}$

The second hour was better.

72. $\frac{1}{60}\square\frac{1}{250}$

$\frac{1}{60}\cdot\frac{25}{25}\square\frac{1}{250}\cdot\frac{6}{6}$

$\frac{25}{1500}>\frac{6}{1500}$

The shutter speed of $\frac{1}{250}$ would be better.

73. $4\frac{1}{4}=\frac{4\cdot4+1}{4}=\frac{17}{4}$

74.

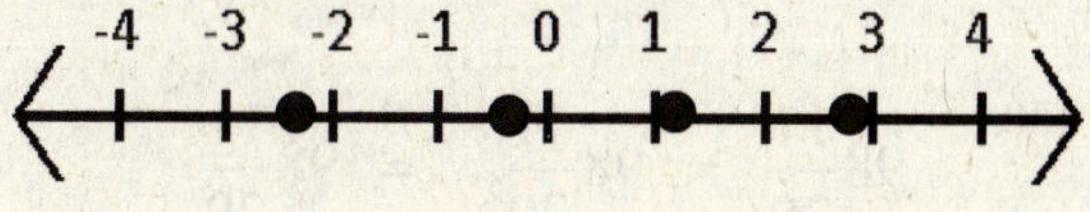

75. $\begin{array}{r} 3 \\ 5\overline{)16} \\ \underline{-15} \\ 1 \end{array}$

$\frac{16}{5}=3\frac{1}{5}$

76. $\begin{array}{r} 3 \\ 12\overline{)47} \\ \underline{-36} \\ 11 \end{array}$

$-\frac{47}{12}=-3\frac{11}{12}$

77. $\begin{array}{r} 17 \\ 3\overline{)51} \\ \underline{-51} \\ 0 \end{array}$

$\frac{51}{3}=17$

78. $\begin{array}{r} 2 \\ 6\overline{)14} \\ \underline{-12} \\ 2 \end{array}$

$\frac{14}{6}=2\frac{2}{6}=2\frac{1}{3}$

79. $9\frac{3}{8}=\frac{8\cdot9+3}{8}=\frac{72+3}{8}=\frac{75}{8}$

80. $-2\frac{1}{5}=-\frac{5\cdot2+1}{5}=-\frac{10+1}{5}=-\frac{11}{5}$

81. $3\frac{11}{14}=\frac{14\cdot3+11}{14}=\frac{42+11}{14}=\frac{53}{14}$

82. $1\frac{99}{100}=\frac{100\cdot1+99}{100}=\frac{100+99}{100}=\frac{199}{100}$

83. $1\frac{2}{5}\cdot1\frac{1}{2}=\frac{7}{5}\cdot\frac{3}{2}=\frac{21}{10}=2\frac{1}{10}$

84. $-3\frac{1}{2}\div3\frac{2}{3}=-\frac{7}{2}\div\frac{11}{3}=-\frac{7}{2}\cdot\frac{3}{11}=-\frac{21}{22}$

85. $-6\left(-6\frac{2}{3}\right)=-\frac{6}{1}\left(-\frac{20}{3}\right)=\frac{2\cdot\cancel{3}\cdot20}{1\cdot\cancel{3}}=2\cdot20=40$

86. $8\div3\frac{1}{5}=\frac{8}{1}\div\frac{16}{5}=\frac{8}{1}\cdot\frac{5}{16}=\frac{\cancel{8}\cdot5}{2\cdot\cancel{8}}=\frac{5}{2}=2\frac{1}{2}$

87.

$$-11\frac{1}{5}\div\left(-\frac{7}{10}\right)=\left(-\frac{56}{5}\right)\cdot\left(-\frac{10}{7}\right)$$

$$=\frac{\cancel{7}\cdot 8\cdot 2\cdot \cancel{5}}{\cancel{5}\cdot\cancel{7}}=8\cdot 2=16$$

88.

$$5\frac{2}{3}\left(-7\frac{1}{5}\right)=\frac{17}{3}\left(-\frac{36}{5}\right)=-\frac{17\cdot\cancel{3}\cdot 12}{\cancel{3}\cdot 5}$$

$$=-\frac{17\cdot 12}{5}=-\frac{204}{5}=-40\frac{4}{5}$$

89.

$$\left(-2\frac{3}{4}\right)^2=\left(-\frac{11}{4}\right)^2=\left(-\frac{11}{4}\right)\left(-\frac{11}{4}\right)$$

$$=\frac{121}{16}=7\frac{9}{16}$$

90.

$$1\frac{5}{16}\cdot 1\frac{7}{9}\cdot 2\frac{2}{3}=\frac{21}{\cancel{16}}\cdot\frac{\cancel{16}}{9}\cdot\frac{8}{3}$$

$$=\frac{\cancel{3}\cdot 7\cdot 8}{9\cdot\cancel{3}}=\frac{56}{9}=6\frac{2}{9}$$

91.

$$5\frac{1}{2}\left(8\frac{3}{4}\right)=\frac{11}{2}\cdot\frac{35}{4}=\frac{385}{8}=48\frac{1}{8}in$$

92.

$$12\left(7\frac{1}{4}\right)=\frac{12}{1}\cdot\frac{29}{4}=\frac{3\cdot\cancel{4}\cdot 29}{1\cdot\cancel{4}}$$

$$=3\cdot 29=87in^2$$

93.

$$90\div 2\frac{1}{4}=90\div\frac{9}{4}=\frac{90}{1}\cdot\frac{4}{9}$$

$$=\frac{\cancel{9}\cdot 10\cdot 4}{1\cdot\cancel{9}}=4\cdot 10=40\text{ posters}$$

94.

$$67\frac{1}{2}\div 7\frac{1}{2}=\frac{135}{2}\div\frac{15}{2}=\frac{135}{2}\cdot\frac{2}{15}$$

$$=\frac{9\cdot\cancel{15}\cdot\cancel{2}}{\cancel{2}\cdot\cancel{15}}=9\text{ loads}$$

95.

$$1\frac{3}{8}+2\frac{1}{5}=\frac{11}{8}+\frac{11}{5}=\frac{11}{8}\cdot\frac{5}{5}+\frac{11}{5}\cdot\frac{8}{8}$$

$$=\frac{55}{40}+\frac{88}{40}=\frac{143}{40}=3\frac{23}{40}$$

96.

$$3\frac{1}{2}+2\frac{2}{3}=\frac{7}{2}+\frac{8}{3}=\frac{7}{2}\cdot\frac{3}{3}+\frac{8}{3}\cdot\frac{2}{2}$$

$$=\frac{21}{6}+\frac{16}{6}=\frac{37}{6}=6\frac{1}{6}$$

97.

$$\begin{array}{rcrcr} 157\frac{11}{30} & = & 157\frac{11}{30}\cdot\frac{2}{2} & = & 157\frac{22}{60} \\ +98\frac{7}{12} & & +98\frac{7}{12}\cdot\frac{5}{5} & & +98\frac{35}{60} \\ \hline & & & & =255\frac{57}{60} \\ & & & & =255\frac{19}{20} \end{array}$$

98.

$$\begin{array}{rcrcr} 6\frac{3}{14} & = & 6\frac{3}{14}\cdot\frac{5}{5} & = & 6\frac{15}{70} \\ +17\frac{7}{10} & & +17\frac{7}{10}\cdot\frac{7}{7} & & +17\frac{49}{70} \\ \hline & & & & =23\frac{64}{70} \\ & & & & =23\frac{32}{35} \end{array}$$

99.

$$\begin{array}{rcrcr} 33\frac{8}{9} & = & 33\frac{8}{9}\cdot\frac{2}{2} & = & 33\frac{16}{18} \\ +49\frac{1}{6} & & +49\frac{1}{6}\cdot\frac{3}{3} & & +49\frac{3}{18} \\ \hline & & & & =82\frac{19}{18} \\ & & & & =82+1\frac{1}{18} \\ & & & & =83\frac{1}{18} \end{array}$$

100.

$$\begin{array}{rcrcr} 98\frac{11}{20} & = & 98\frac{11}{20} & = & 98\frac{11}{20} \\ +14\frac{4}{5} & & +14\frac{4}{5}\cdot\frac{4}{4} & & +14\frac{16}{20} \\ \hline & & & & =112\frac{27}{20} \\ & & & & =112+1\frac{7}{20} \\ & & & & =113\frac{7}{20} \end{array}$$

101.

$$\begin{array}{rcccccc} 23\frac{1}{3} & = & 23\frac{1}{3}\cdot\frac{2}{2} & = & 23\frac{2}{6} & = & 22\frac{8}{6} \\ -\ 2\frac{5}{6} & & -2\frac{5}{6} & & -2\frac{5}{6} & & -2\frac{5}{6} \\ & & & & & & =20\frac{3}{6} \\ & & & & & & =20\frac{1}{2} \end{array}$$

102.

$$\begin{array}{rcl} 39 & = & 38\frac{8}{8} \\ -4\frac{5}{8} & & -4\frac{5}{8} \\ & & 34\frac{3}{8} \end{array}$$

103.

$$10\frac{3}{4}+21\frac{1}{2}+7\frac{2}{3}=10\frac{3}{4}\cdot\frac{3}{3}+21\frac{1}{2}\cdot\frac{6}{6}+7\frac{2}{3}\cdot\frac{4}{4}$$
$$=10\frac{9}{12}+21\frac{6}{12}+7\frac{8}{12}=38\frac{23}{12}$$
$$=38+1\frac{11}{12}=39\frac{11}{12}\,gal$$

104.

$$2-1\frac{3}{8}=1\frac{8}{8}-1\frac{3}{8}=\frac{5}{8}\,in$$

105.

$$\frac{3}{4}+\left(-\frac{1}{3}\right)^2\left(\frac{5}{4}\right)=\frac{3}{4}+\left(\frac{1}{9}\right)\left(\frac{5}{4}\right)$$
$$=\frac{3}{4}+\frac{5}{36}=\frac{3}{4}\cdot\frac{9}{9}+\frac{5}{36}=\frac{27+5}{36}$$
$$=\frac{32}{36}=\frac{8}{9}$$

106.

$$\left(\frac{2}{3}\div\frac{16}{9}\right)-\left(1\frac{2}{3}\cdot\frac{1}{15}\right)=\left(\frac{2}{3}\cdot\frac{9}{16}\right)-\left(\frac{5}{3}\cdot\frac{1}{15}\right)$$
$$=\frac{18}{48}-\frac{5}{45}=\frac{3}{8}-\frac{1}{9}=\frac{3}{8}\cdot\frac{9}{9}-\frac{1}{9}\cdot\frac{8}{8}$$
$$=\frac{27}{72}-\frac{8}{72}=\frac{19}{72}$$

107.

$$\left(\frac{11}{5}-1\frac{2}{3}\right)-\left(-\frac{4}{9}\cdot 18\right)=\left(\frac{11}{5}-\frac{5}{3}\right)+\left(\frac{4}{9}\cdot\frac{18}{1}\right)$$
$$=\left(\frac{33}{15}-\frac{25}{15}\right)+\left(\frac{4}{\not 9}\cdot\frac{2\cdot\not 9}{1}\right)=\frac{8}{15}+8=8\frac{8}{15}$$

108.

$$\left|-\frac{9}{16}\div 2\frac{1}{4}\right|+\left(-3\frac{7}{8}\right)=\left|-\frac{9}{16}\div\frac{9}{4}\right|-\frac{31}{8}$$
$$=\left|-\frac{\not 9}{\not 16_4}\cdot\frac{\not 4}{\not 9}\right|-\frac{31}{8}=\left|-\frac{1}{4}\right|-\frac{31}{8}=\frac{1}{4}-\frac{31}{8}$$
$$=\frac{1}{4}\cdot\frac{2}{2}-\frac{31}{8}=\frac{2}{8}-\frac{31}{8}=-\frac{29}{8}=-3\frac{5}{8}$$

109.

$$\frac{\frac{3}{5}}{-\frac{17}{20}}=\frac{3}{5}\left(-\frac{20}{17}\right)=\frac{3}{\not 5}\left(-\frac{4\cdot\not 5}{17}\right)=-\frac{12}{17}$$

110.

$$\frac{4-\frac{2}{7}}{4\frac{1}{7}}=\frac{\frac{28}{7}-\frac{2}{7}}{\frac{29}{7}}=\frac{\frac{26}{7}}{\frac{29}{7}}=\frac{26}{\not 7}\cdot\frac{\not 7}{29}=\frac{26}{29}$$

111.

$$\frac{\frac{2}{3}-\frac{1}{6}}{-\frac{3}{4}-\frac{1}{2}}=\frac{\frac{2}{3}\cdot\frac{2}{2}-\frac{1}{6}}{-\frac{3}{4}-\frac{1}{2}\cdot\frac{2}{2}}=\frac{\frac{4}{6}-\frac{1}{6}}{-\frac{3}{4}-\frac{2}{4}}$$
$$=\frac{\frac{3}{6}}{-\frac{5}{4}}=\frac{\frac{1}{2}}{-\frac{5}{4}}=\frac{1}{2}\left(-\frac{4}{5}\right)=-\frac{\not 2\cdot 2}{\not 2\cdot 5}=-\frac{2}{5}$$

112.

$$\frac{5\frac{1}{4}}{\frac{7}{4}+\left(-\frac{1}{3}\right)}=\frac{\frac{21}{4}}{\frac{7}{4}\cdot\frac{3}{3}-\frac{1}{3}\cdot\frac{4}{4}}=\frac{\frac{21}{4}}{\frac{21}{12}-\frac{4}{12}}$$
$$=\frac{\frac{21}{4}}{\frac{17}{12}}=\frac{21}{\not 4}\cdot\frac{\not 12^{3}}{17}=\frac{63}{17}=3\frac{12}{17}$$

113.

$$\left(5\frac{1}{5}+1\frac{1}{2}\right)-4\frac{1}{8}=\frac{26}{5}+\frac{3}{2}-\frac{33}{8}$$

$$=\frac{26}{5}\cdot\frac{8}{8}+\frac{3}{2}\cdot\frac{20}{20}-\frac{33}{8}\cdot\frac{5}{5}=\frac{208}{40}+\frac{60}{40}-\frac{165}{40}$$

$$=\frac{103}{40}=2\frac{23}{40}$$

114.

$$\left(4\frac{5}{8}-3\frac{1}{4}\right)+12\frac{11}{16}=\frac{37}{8}-\frac{13}{4}+\frac{203}{16}$$

$$=\frac{37}{8}\cdot\frac{2}{2}-\frac{13}{4}\cdot\frac{4}{4}+\frac{203}{16}=\frac{74}{16}-\frac{52}{16}+\frac{203}{16}$$

$$=\frac{225}{16}=14\frac{1}{16}$$

115.

$$A=\frac{1}{2}h(a+b)=\frac{1}{2}\cdot 2\frac{7}{9}\left(1\frac{1}{8}+4\frac{7}{8}\right)$$

$$=\frac{1}{2}\cdot\frac{25}{9}\cdot 5\frac{8}{8}=\frac{1}{\cancel{2}}\cdot\frac{25}{9}\cdot\overset{3}{\cancel{6}}=\frac{25}{\underset{3}{\cancel{9}}}\cdot\cancel{3}$$

$$=\frac{25}{3}=8\frac{1}{3}$$

116.

$$P=2l+2w=2\left(2\frac{1}{3}\right)+2\left(3\frac{1}{4}\right)$$

$$=2\left(\frac{7}{3}\right)+2\left(\frac{13}{4}\right)=\frac{14}{3}+\frac{26}{4}$$

$$=\frac{14}{3}+\frac{13}{2}=\frac{14}{3}\cdot\frac{2}{2}+\frac{13}{2}\cdot\frac{3}{3}$$

$$=\frac{28}{6}+\frac{39}{6}=\frac{67}{6}=11\frac{1}{6}$$

117.

$$1\frac{1}{2}+2\frac{2}{3}+\frac{3}{4}=\frac{3}{2}+\frac{8}{3}+\frac{3}{4}$$

$$=\frac{3}{2}\cdot\frac{6}{6}+\frac{8}{3}\cdot\frac{4}{4}+\frac{3}{4}\cdot\frac{3}{3}=\frac{18}{12}+\frac{32}{12}+\frac{9}{12}$$

$$=\frac{59}{12}\text{ oz cream}$$

$$\frac{59}{12}\div\frac{5}{6}=\frac{59}{\underset{2}{\cancel{12}}}\cdot\frac{\cancel{6}}{5}=\frac{59}{10}=5\frac{9}{10}$$

He can make 5 full tubes with $\frac{9}{10}$ of a tube left over.

118.

$$18\frac{1}{16}-5\frac{5}{16}-4\frac{3}{4}$$

$$=17\frac{17}{16}-5\frac{5}{16}-4\frac{12}{16}$$

$$=12\frac{12}{16}-4\frac{12}{16}=8in$$

119.

$$x-\frac{1}{40}=\frac{11}{40}$$

$$x-\frac{1}{40}+\frac{1}{40}=\frac{11}{40}+\frac{1}{40}$$

$$x=\frac{12}{40}=\frac{3}{10}$$

Check: $x-\frac{1}{40}=\frac{11}{40}$

$$\frac{3}{10}-\frac{1}{40}\overset{?}{=}\frac{11}{40}$$

$$\frac{12}{40}-\frac{1}{40}\overset{?}{=}\frac{11}{40}$$

$$\frac{11}{40}=\frac{11}{40}$$

120.

$$\frac{5}{4}=n+\frac{2}{3}$$

$$\frac{5}{4}-\frac{2}{3}=n+\frac{2}{3}-\frac{2}{3}$$

$$\frac{15}{12}-\frac{8}{12}=n$$

$$\frac{7}{12}=n$$

Check: $\frac{5}{4}=n+\frac{2}{3}$

$$\frac{5}{4}\overset{?}{=}\frac{7}{12}+\frac{2}{3}$$

$$\frac{15}{12}\overset{?}{=}\frac{7}{12}+\frac{8}{12}$$

$$\frac{15}{12}=\frac{15}{12}$$

121.
$$\frac{2}{9}x = 22$$
$$\frac{9}{2}\cdot\frac{2}{9}x = \frac{9}{2}\cdot 22$$
$$x = \frac{9\cdot \cancel{2}\cdot 11}{\cancel{2}}$$
$$x = 99$$

Check: $\frac{2}{9}x = 22$
$$\frac{2}{\cancel{9}}\cdot \overset{11}{\cancel{99}} \overset{?}{=} 22$$
$$2\cdot 11 \overset{?}{=} 22$$
$$22 = 22$$

122.
$$\frac{4}{25}y = 4$$
$$\frac{25}{4}\cdot\frac{4}{25}y = \frac{25}{\cancel{4}}\cdot\cancel{4}$$
$$y = 25$$

Check: $\frac{4}{25}y = 4$
$$\frac{4}{25}\cdot 25 \overset{?}{=} 4$$
$$4 = 4$$

123.
$$-\frac{7}{3}y = 5$$
$$\left(-\frac{3}{7}\right)\left(-\frac{7}{3}y\right) = \left(-\frac{3}{7}\right)\cdot 5$$
$$y = -\frac{15}{7}$$

Check: $-\frac{7}{3}y = 5$
$$-\frac{7}{3}\cdot\left(-\frac{15}{7}\right) \overset{?}{=} 5$$
$$\frac{\cancel{7}\cdot\cancel{3}\cdot 5}{\cancel{3}\cdot\cancel{7}} \overset{?}{=} 5$$
$$5 = 5$$

124.
$$\frac{1}{8}x = \frac{27}{32}$$
$$8\cdot\frac{1}{8}x = 8\cdot\frac{27}{32}$$
$$x = \frac{\cancel{8}\cdot 27}{4\cdot\cancel{8}}$$
$$x = \frac{27}{4}$$

Check: $\frac{1}{8}x = \frac{27}{32}$
$$\frac{1}{8}\cdot\frac{27}{4} \overset{?}{=} \frac{27}{32}$$
$$\frac{27}{32} = \frac{27}{32}$$

125.
$$37x = -\frac{19}{3}$$
$$\frac{1}{37}\cdot 37x = \frac{1}{37}\left(-\frac{19}{3}\right)$$
$$x = -\frac{19}{111}$$

Check: $37x = -\frac{19}{3}$
$$37\cdot\left(-\frac{19}{111}\right) \overset{?}{=} -\frac{19}{3}$$
$$-\frac{\cancel{37}\cdot 19}{3\cdot\cancel{37}} \overset{?}{=} -\frac{19}{3}$$
$$-\frac{19}{3} = -\frac{19}{3}$$

126.

$$\frac{7}{8}x+21=7$$
$$\frac{7}{8}x+21-21=7-21$$
$$\frac{7}{8}x=-14$$
$$\frac{8}{7}\cdot\frac{7}{8}x=\left(\frac{8}{7}\right)\left(-14\right)$$
$$x=8(-2)$$
$$x=-16$$

Check: $\frac{7}{8}x+21=7$

$$\frac{7}{8}(-16)+21\overset{?}{=}7$$
$$-14+21\overset{?}{=}7$$
$$7=7$$

127.

$$\frac{d}{4}=\frac{d}{7}-6$$
$$28\cdot\frac{d}{4}=28\cdot\frac{d}{7}-28\cdot 6$$
$$7d=4d-168$$
$$7d-4d=4d-168-4d$$
$$3d=-168$$
$$\frac{3d}{3}=\frac{-168}{3}$$
$$d=-56$$

Check: $\frac{d}{4}=\frac{d}{7}-6$

$$\frac{-56}{4}\overset{?}{=}\frac{-56}{7}-6$$
$$-14\overset{?}{=}-8-6$$
$$-14=-14$$

128.

$$\frac{1}{6}x-\frac{1}{2}=\frac{1}{3}x$$
$$6\cdot\frac{1}{6}x-6\cdot\frac{1}{2}=6\cdot\frac{1}{3}x$$
$$x-3=2x$$
$$x-3-x=2x-x$$
$$-3=x$$

Check: $\frac{1}{6}x-\frac{1}{2}=\frac{1}{3}x$

$$\frac{1}{6}(-3)-\frac{1}{2}\overset{?}{=}\frac{1}{3}(-3)$$
$$-\frac{1}{2}-\frac{1}{2}\overset{?}{=}-1$$
$$-1=-1$$

129. x = the number of pages in the textbook

$$\frac{2}{3}x=220$$
$$\frac{3}{2}\cdot\frac{2}{3}x=\frac{3}{2}\cdot 220$$
$$x=3\cdot 110$$
$$x=330$$

The book has 330 pages.

130. x = the length of the seminar

$$\frac{1}{4}x+35+\frac{2}{5}x=x$$
$$20\cdot\frac{1}{4}x+20\cdot 35+20\cdot\frac{2}{5}x=20\cdot x$$
$$5x+700+8x=20x$$
$$13x+700=20x$$
$$13x+700-13x=20x-13x$$
$$700=7x$$
$$\frac{700}{7}=\frac{7x}{7}$$
$$100=x$$

The seminar is 100 minutes.

Chapter 4 Test

1.

a. For the fraction $\frac{6}{7}$, the <u>numerator</u> is 6 and the <u>denominator</u> is 7.

b. Two fractions are <u>equivalent</u> if they represent the same number.

c. A fraction is in <u>simplest</u> form when the numerator and denominator have no common factors other than 1.

d. To <u>simplify</u> a fraction, we remove common factors of the numerator and denominator.

e. The <u>reciprocal</u> of $\frac{4}{5}$ is $\frac{5}{4}$.

f. A <u>mixed</u> number, such as $1\frac{9}{16}$, is the sum of a whole number and a proper fraction.

g. $\frac{\frac{1}{8}}{\frac{7}{12}}$ and $\frac{\frac{3}{4}+\frac{1}{3}}{\frac{5}{12}-\frac{1}{4}}$ are examples of <u>complex</u> fractions.

2. a. $\frac{4}{5}$ b. $\frac{1}{5}$

3. $2\frac{1}{6}=\frac{2\cdot 6+1}{6}=\frac{13}{6}$

4.

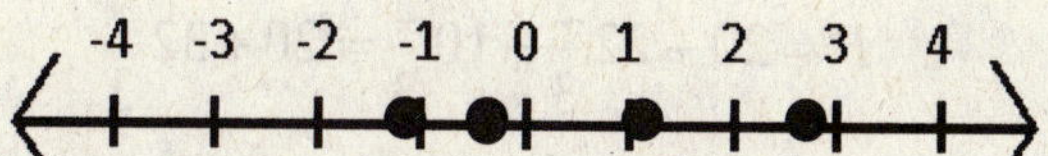

5. Yes: $\frac{5}{15}=\frac{1\cdot \cancel{5}}{3\cdot \cancel{5}}=\frac{1}{3}$

6.

a. $\frac{4}{5}=\frac{4}{5}\cdot\frac{9}{9}=\frac{36}{45}$

b. $\frac{7}{8}=\frac{7}{8}\cdot\frac{3x}{3x}=\frac{21x}{24x}$

7. a. 0 b. undefined

8.

a. $\frac{27}{36}=\frac{3\cdot \cancel{9}}{4\cdot \cancel{9}}=\frac{3}{4}$

b. $\frac{72n^3}{180n}=\frac{2\cdot \cancel{36}\cdot \cancel{n}\cdot n\cdot n}{5\cdot \cancel{36}\cdot \cancel{n}}=\frac{2\cdot n\cdot n}{5}=\frac{2n^2}{5}$

9. $\frac{3}{16}+\frac{7}{16}=\frac{10}{16}=\frac{\cancel{2}\cdot 5}{\cancel{2}\cdot 8}=\frac{5}{8}$

10. $-\frac{3}{4}\left(\frac{1}{5}\right)=-\frac{3}{20}$

11. $\frac{4a}{3b}\div\frac{a^2}{9b^5}=\frac{4a}{3b}\cdot\frac{9b^5}{a^2}=\frac{4\cdot \cancel{a}\cdot \cancel{3}\cdot 3\cdot \cancel{b}\cdot b^4}{\cancel{3}\cdot \cancel{b}\cdot \cancel{a}\cdot a}$

$=\frac{4\cdot 3\cdot b^4}{a}=\frac{12b^4}{a}$

12. $\frac{11}{12}-\frac{11}{30}=\frac{11}{12}\cdot\frac{5}{5}-\frac{11}{30}\cdot\frac{2}{2}$

$=\frac{55}{60}-\frac{22}{60}=\frac{33}{60}=\frac{\cancel{3}\cdot 11}{\cancel{3}\cdot 20}=\frac{11}{20}$

13. $-\frac{3}{7}+2=-\frac{3}{7}+\frac{2}{1}\cdot\frac{7}{7}=-\frac{3}{7}+\frac{14}{7}=\frac{11}{7}$

14. $\frac{9}{10}\left(-\frac{4}{15}\right)\left(-\frac{25}{18}\right)=\frac{\cancel{3}\cdot \cancel{3}\cdot \cancel{2}\cdot \cancel{2}\cdot \cancel{5}\cdot \cancel{5}}{\cancel{2}\cdot \cancel{5}\cdot \cancel{3}\cdot \cancel{5}\cdot \cancel{2}\cdot \cancel{3}\cdot 3}=\frac{1}{3}$

15. $\frac{8}{9}\square\frac{9}{10}$

$\frac{8}{9}\cdot\frac{10}{10}\square\frac{9}{10}\cdot\frac{9}{9}$

$\frac{80}{90}<\frac{81}{90}$

The larger fraction is $\frac{9}{10}$.

16. a. $-\frac{53}{17}$ b. $\frac{3a^2}{7}$

17. $\frac{x}{6}-\frac{4}{5}=\frac{x}{6}\cdot\frac{5}{5}-\frac{4}{5}\cdot\frac{6}{6}=\frac{5x}{30}-\frac{24}{30}=\frac{5x-24}{30}$

18. $\frac{2}{5}\cdot 100=\frac{2}{5}\cdot\frac{100}{1}=\frac{2\cdot \cancel{5}\cdot 20}{\cancel{5}\cdot 1}=2\cdot 20=40$

19.
$$\frac{2}{3}+\frac{1}{8}+\frac{1}{6}=\frac{2}{3}\cdot\frac{8}{8}+\frac{1}{8}\cdot\frac{3}{3}+\frac{1}{6}\cdot\frac{4}{4}$$
$$=\frac{16}{24}+\frac{3}{24}+\frac{4}{24}=\frac{23}{24}$$

20.
$$\begin{array}{r} 9 \\ 6\overline{)55} \\ -54 \\ \hline 1 \end{array}$$

a. $\dfrac{55}{6}=9\dfrac{1}{6}$

b. $1\dfrac{18}{21}=\dfrac{1\cdot 21+18}{21}=\dfrac{39}{21}$

21.
$$\begin{array}{rcrcr} 157\frac{3}{10} & = & 157\frac{3}{10}\cdot\frac{3}{3} & = & 157\frac{9}{30} \\ +103\frac{13}{15} & & +103\frac{13}{15}\cdot\frac{2}{2} & & +103\frac{26}{30} \\ \hline & & & & =260\frac{35}{30} \\ & & & & =260+1\frac{5}{30} \\ & & & & =261\frac{1}{6} \end{array}$$

22.
$$\begin{array}{rcrcrcr} 67\frac{1}{4} & = & 67\frac{1}{4}\cdot\frac{3}{3} & = & 67\frac{3}{12} & = & 66\frac{15}{12} \\ -29\frac{5}{6} & & -29\frac{5}{6}\cdot\frac{2}{2} & & -29\frac{10}{12} & & -29\frac{10}{12} \\ \hline & & & & & & 37\frac{5}{12} \end{array}$$

23.
$$6\frac{1}{4}\div 3\frac{3}{4}=\frac{25}{4}\div\frac{15}{4}=\frac{25}{4}\cdot\frac{4}{15}$$
$$=\frac{\cancel{5}\cdot 5\cdot\cancel{4}}{\cancel{4}\cdot 3\cdot\cancel{5}}=\frac{5}{3}=1\frac{2}{3}$$

24. a. Foreman:
$$250-210\frac{1}{2}=249\frac{2}{2}-210\frac{1}{2}=39\frac{1}{2}\,lb$$

b. Foreman: $39\frac{1}{2}-34=5\frac{1}{2}\,in$

c. Ali: $15-14\frac{3}{4}=14\frac{4}{4}-14\frac{3}{4}=\frac{1}{4}\,in$

25.
$$P=2l+2w=2\cdot\frac{1}{3}+2\cdot\frac{1}{9}=\frac{2}{3}+\frac{2}{9}$$
$$=\frac{2}{3}\cdot\frac{3}{3}+\frac{2}{9}=\frac{6}{9}+\frac{2}{9}=\frac{8}{9}$$

26.
$$13\frac{1}{2}\div 9=\frac{27}{2}\div\frac{9}{1}=\frac{\overset{3}{\cancel{27}}}{2}\cdot\frac{1}{\cancel{9}}$$
$$=\frac{3}{2}=\$1\frac{1}{2}\text{ million}$$

27.
$$10\frac{1}{2}+2\left(\frac{5}{8}\right)=\frac{21}{2}+\frac{10}{8}=\frac{21}{2}\cdot\frac{4}{4}+\frac{10}{8}$$
$$=\frac{84}{8}+\frac{10}{8}=\frac{94}{8}=11\frac{6}{8}=11\frac{3}{4}\,in$$

28.
$$A=\frac{1}{2}b\cdot h=\frac{1}{\cancel{2}}\left(10\frac{2}{3}\right)\overset{10}{\cancel{20}}$$
$$=\frac{32}{3}\cdot\frac{10}{1}=\frac{320}{3}=106\frac{2}{3}\,in^2$$

$$P=20+22\frac{2}{3}+10\frac{2}{3}=20+32\frac{4}{3}$$
$$=20+32+1\frac{1}{3}=53\frac{1}{3}\,in$$

29.
$$40\left(1\frac{1}{2}\right)=40\left(\frac{3}{2}\right)=\frac{120}{2}=60\text{ calories}$$

30.
$$8\div\frac{2}{3}=\frac{8}{1}\cdot\frac{3}{2}=\frac{24}{2}=12\,\text{servings}$$

31. $$\left(\frac{2}{3}\cdot\frac{5}{16}\right)-\left(-1\frac{3}{5}\div 4\frac{4}{5}\right)$$
$$=\frac{10}{48}-\left(-\frac{8}{5}\div\frac{24}{5}\right)$$
$$=\frac{5}{24}+\left(\frac{8}{5}\cdot\frac{5}{24}\right)=\frac{5}{24}+\frac{40}{120}=\frac{5}{24}+\frac{1}{3}$$
$$=\frac{5}{24}+\frac{1}{3}\cdot\frac{8}{8}=\frac{5}{24}+\frac{8}{24}=\frac{13}{24}$$

32. $$\left(\frac{1}{2}\right)^3\div\left(\frac{3}{4}-\frac{1}{3}\right)$$
$$=\left(\frac{1}{2}\right)\left(\frac{1}{2}\right)\left(\frac{1}{2}\right)\div\left(\frac{3}{4}\cdot\frac{3}{3}-\frac{1}{3}\cdot\frac{4}{4}\right)$$
$$=\frac{1}{8}\div\left(\frac{9}{12}-\frac{4}{12}\right)=\frac{1}{8}\div\frac{5}{12}$$
$$=\frac{1}{8}\cdot\frac{12}{5}=\frac{3\cdot\cancel{4}}{2\cdot\cancel{4}\cdot 5}=\frac{3}{2\cdot 5}=\frac{3}{10}$$

33. $$\frac{\frac{5}{6}}{\frac{7}{8}}=\frac{5}{6}\cdot\frac{8}{7}=\frac{5\cdot\cancel{2}\cdot 4}{\cancel{2}\cdot 3\cdot 7}=\frac{20}{21}$$

34. $$\frac{\frac{1}{2}+\frac{1}{3}}{-\frac{1}{6}-\frac{1}{3}}=\frac{\frac{1}{2}\cdot\frac{3}{3}+\frac{1}{3}\cdot\frac{2}{2}}{-\frac{1}{6}-\frac{1}{3}\cdot\frac{2}{2}}=\frac{\frac{3}{6}+\frac{2}{6}}{-\frac{1}{6}-\frac{2}{6}}$$
$$=\frac{\frac{5}{6}}{-\frac{3}{6}}=\frac{5}{6}\left(-\frac{6}{3}\right)=-\frac{5}{3}$$

35. $$\frac{5}{16}y=\frac{5}{6}$$
$$\frac{16}{5}\cdot\frac{5}{16}y=\frac{16}{5}\cdot\frac{5}{6}$$
$$y=\frac{\cancel{2}\cdot 8\cdot\cancel{5}}{\cancel{5}\cdot\cancel{2}\cdot 3}$$
$$y=\frac{8}{3}$$

Check: $$\frac{5}{16}y=\frac{5}{6}$$
$$\frac{5}{16}\cdot\frac{8}{3}\stackrel{?}{=}\frac{5}{6}$$
$$\frac{5\cdot 8}{2\cdot 8\cdot 3}\stackrel{?}{=}\frac{5}{6}$$
$$\frac{5}{6}=\frac{5}{6}$$

36. $$-\frac{3}{8}=a+\frac{1}{2}$$
$$\cancel{8}\left(-\frac{3}{\cancel{8}}\right)=8\cdot a+\overset{4}{\cancel{8}}\cdot\frac{1}{\cancel{2}}$$
$$-3=8a+4$$
$$-3-4=8a+4-4$$
$$-7=8a$$
$$\frac{-7}{8}=\frac{8a}{8}$$
$$-\frac{7}{8}=a$$

Check: $$-\frac{3}{8}=a+\frac{1}{2}$$
$$-\frac{3}{8}\stackrel{?}{=}-\frac{7}{8}+\frac{1}{2}$$
$$-\frac{3}{8}\stackrel{?}{=}-\frac{7}{8}+\frac{4}{8}$$
$$-\frac{3}{8}=-\frac{3}{8}$$

37.

$$\frac{6}{11}n+35=-7$$

$$\cancel{11}\cdot\frac{6}{\cancel{11}}n+11\cdot 35=11(-7)$$

$$6n+385=-77$$

$$6n+385-385=-77-385$$

$$6n=-462$$

$$\frac{6n}{6}=-\frac{462}{6}$$

$$n=-77$$

Check: $\frac{6}{11}n+35=-7$

$$\frac{6}{\cancel{11}}\left(-\overset{1}{\cancel{77}}\right)+35\overset{?}{=}-7$$

$$6(-7)+35\overset{?}{=}-7$$

$$-42+35\overset{?}{=}-7$$

$$-7=-7$$

38.

$$\frac{7}{8}x-\frac{5}{16}=\frac{1}{4}x$$

$$\overset{2}{\cancel{16}}\cdot\frac{7}{\cancel{8}}x-\cancel{16}\cdot\frac{5}{\cancel{16}}=\overset{4}{\cancel{16}}\cdot\frac{1}{\cancel{4}}x$$

$$14x-5=4x$$

$$14x-5-14x=4x-14x$$

$$-5=-10x$$

$$\frac{-5}{-10}=\frac{-10x}{-10}$$

$$\frac{1}{2}=x$$

Check: $\frac{7}{8}x-\frac{5}{16}=\frac{1}{4}x$

$$\frac{7}{8}\cdot\frac{1}{2}-\frac{5}{16}\overset{?}{=}\frac{1}{4}\cdot\frac{1}{2}$$

$$\frac{7}{16}-\frac{5}{16}\overset{?}{=}\frac{1}{8}$$

$$\frac{2}{16}\overset{?}{=}\frac{1}{8}$$

$$\frac{1}{8}=\frac{1}{8}$$

39. x = the length of the class

$$10+\frac{1}{2}x+\frac{1}{3}x+25=x$$

$$35+\frac{3}{6}x+\frac{2}{6}x=x$$

$$35+\frac{5}{6}x=x$$

$$35+\frac{5}{6}x-\frac{5}{6}x=x-\frac{5}{6}x$$

$$35=\frac{1}{6}x$$

$$6\cdot 35=\cancel{6}\cdot\frac{1}{\cancel{6}}x$$

$$210=x$$

The class is 210 minutes long.

40. a. This demonstrates removing a form of 1 by taking the same factor from the numerator and denominator.

b. This demonstrates the idea of equivalent fractions.

c. This demonstrates building an equivalent fraction by multiplying by a form of 1.

Chapters 1 – 4 Cumulative Review

1. a. 5

b. 8 hundred thousands

c. 5,896,600

d. 5,900,000

2. trillions

3. Miami-Dade, Kings, Dallas, Queens, Riverside, San Bernardino

4. a. $75+150+75+150=225+225=450\,ft$

b.

$$\begin{array}{r} 150 \\ \underline{\times 75} \\ 750 \\ \underline{+10,500} \\ 11,250\,ft^2 \end{array}$$

5.
$$\begin{array}{r} {}^{2\,2\ 1\,2}\\ 7{,}897\\ 6{,}909\\ 1{,}812\\ +14{,}378\\ \hline 30{,}996 \end{array}$$

6.
$$\begin{array}{r} 20{,}000\\ -3{,}456\\ \hline 16{,}544 \end{array}$$

Check: $16{,}544+3{,}456=20{,}000$

7.
$$\begin{array}{r} 20\\ \times 12\\ \hline 40\\ +200\\ \hline 240 \end{array}$$

$$\begin{array}{r} 240\\ \times 10\\ \hline 2{,}400 \end{array}$$

2,400 stickers per packet

8.
$$\begin{array}{r} 5{,}345\\ \times 56\\ \hline 32{,}070\\ +267{,}250\\ \hline 299{,}320 \end{array}$$

9.
$$\begin{array}{r} 991\\ 35\overline{)34685}\\ -315\\ \hline 318\\ -315\\ \hline 35\\ -35\\ \hline 0 \end{array}$$

$35\cdot 991=34{,}685$

10. a. 1, 2, 3, 4, 6, 8, 12, 24

b. $450=45\cdot 10=5\cdot 9\cdot 2\cdot 5$
$=5\cdot 3\cdot 3\cdot 2\cdot 5=2\cdot 3^2\cdot 5^2$

11.
$16=2\cdot 2\cdot 2\cdot 2$
$20=2\cdot 2\cdot 5$
$LCM(16,20)=2\cdot 2\cdot 2\cdot 2\cdot 5=80$

12.
$63=3\cdot 3\cdot 7$
$84=2\cdot 2\cdot 3\cdot 7$
$GCF(63,84)=3\cdot 7=21$

13.
$15+5\left[12-\left(2^2+4\right)\right]$
$=15+5\left[12-(4+4)\right]$
$=15+5[12-8]=15+5[4]$
$=15+20=35$

14.
$$\frac{158{,}000+163{,}000+147{,}000}{3}$$
$$=\frac{468{,}000}{3}=\$156{,}000$$

15.
$18=n-47$
$18+47=n-47+47$
$65=n$
Check: $18=n-47$
$18\overset{?}{=}65-47$
$18=18$

16.
$23x=483$
$\dfrac{23x}{23}=\dfrac{483}{23}$
$x=21$
Check: $23x=483$
$23\cdot 21\overset{?}{=}483$
$483=483$

17. a. $\{\ldots,-2,-1,0,1,2,\ldots\}$

b. true

18. $-20+6+(-1)=-14+(-1)=-15$

19. Subtract the absolute value of the smaller from the larger – since the larger number is being subtracted, the result will be negative. $453-129=324$ so $129-453=-324$.

20. $-50-(-60)=-50+60=10$

21. $-25(8)=-200\,ft$

22. $-55\div 5=-11°F/hr$

23. $7(8)+(-3)(-6)(-2)=56+(-36)=20$

24. $(-2)^3-3^3=(-2)(-2)(-2)-3\cdot 3\cdot 3$
$=-8-27=-35$

25. $-5+3|-4-(-6)|=-5+3|-4+6|$
$=-5+3|2|=-5+6=1$

26. $\dfrac{2(3^2-4^2)}{-2(3)-1}=\dfrac{2(9-16)}{-6-1}=\dfrac{2(-7)}{-7}=2$

27.
$$-12y-9=39$$
$$-12y-9+9=39+9$$
$$-12y=48$$
$$\frac{-12y}{-12}=\frac{48}{-12}$$
$$y=-4$$

Check: $-12y-9=39$
$$-12(-4)-9\stackrel{?}{=}39$$
$$48-9\stackrel{?}{=}39$$
$$39=39$$

28.
$$-9=4+\frac{m}{-4}$$
$$(-4)(-9)=-4\cdot 4+(-4)\left(\frac{m}{-4}\right)$$
$$36=-16+m$$
$$36+16=-16+m+16$$
$$52=m$$

Check: $-9=4+\dfrac{m}{-4}$
$$-9\stackrel{?}{=}4+\frac{52}{-4}$$
$$-9\stackrel{?}{=}4-13$$
$$-9=-9$$

29. x = the amount raised
$$-132+x=-64$$
$$-132+x+132=-64+132$$
$$x=68$$
The cage was raised 68 feet.

30. x = the profit the second year
$$-4,028+x=33,611$$
$$-4,028+x+4,028=33,611+4,028$$
$$x=37,639$$
The store made $37,639 the second year.

31. a. $x+15$ b. $x-8$

c. $4x$ d. $\dfrac{x}{10}$

32. $3x-x^3=3(-4)-(-4)^3$
$=-12-(-64)=-12+64=52$

33. a. $-8k(-6)=48k$

b. $(-4x)(-7y)=28xy$

34. a. $2(3x-4)=2\cdot 3x-2\cdot 4=6x-8$

b. $-5(3x-2y+4)$
$=-5\cdot 3x-(-5)\cdot 2y+(-5)4$
$=-15x+10y-20$

35. $-3x+8x=5x$

36. $4a^2-\left(-3a^2\right)=4a^2+3a^2=7a^2$

37. $4x-3y-5x+2y=4x-5x-3y+2y$
$=-x-y$

38. $-2(3x-4)+2x=-6x+8+2x=-4x+8$

39.
$$6x-12=2x+4$$
$$6x-12-2x=2x+4-2x$$
$$4x-12=4$$
$$4x-12+12=4+12$$
$$4x=16$$
$$\frac{4x}{4}=\frac{16}{4}$$
$$x=4$$
Check: $6x-12=2x+4$
$$6\cdot4-12\overset{?}{=}2\cdot4+4$$
$$24-12\overset{?}{=}8+4$$
$$12=12$$

40.
$$3(2y-8)=-2(y-4)$$
$$6y-24=-2y+8$$
$$6y-24+2y=-2y+8+2y$$
$$8y-24=8$$
$$8y-24+24=8+24$$
$$8y=32$$
$$\frac{8y}{8}=\frac{32}{8}$$
$$y=4$$
Check: $3(2y-8)=-2(y-4)$
$$3(2\cdot4-8)\overset{?}{=}-2(4-4)$$
$$3(8-8)\overset{?}{=}-2(0)$$
$$3\cdot0\overset{?}{=}0$$
$$0=0$$

41.
$$-(19-3s)-(8s+1)=35$$
$$-19+3s-8s-1=35$$
$$-20-5s=35$$
$$-20+20-5s=35+20$$
$$-5s=55$$
$$\frac{-5s}{-5}=\frac{55}{-5}$$
$$s=-11$$
Check: $-(19-3s)-(8s+1)=35$
$$-[19-3(-11)]-[8(-11)+1]\overset{?}{=}35$$
$$-(19+33)-(-88+1)\overset{?}{=}35$$
$$-52-(-87)\overset{?}{=}35$$
$$35=35$$

42.
$$14m=m+5m+3m-40$$
$$14m=9m-40$$
$$14m-9m=9m-40-9m$$
$$5m=-40$$
$$\frac{5m}{5}=\frac{-40}{5}$$
$$m=-8$$
Check: $14m=m+5m+3m-40$
$$14(-8)\overset{?}{=}-8+5(-8)+3(-8)-40$$
$$-112\overset{?}{=}-8-40-24-40$$
$$-112=-112$$

43. x = the number of three-hour shifts remaining
$$3x+37=100$$
$$3x+37-37=100-37$$
$$3x=63$$
$$\frac{3x}{3}=\frac{63}{3}$$
$$x=21$$
She must observe another 21 three-hour shifts.

44.
$$P = 2l + 2w$$
$$210 = 2(4w) + 2w$$
$$210 = 8w + 2w$$
$$210 = 10w$$
$$\frac{210}{10} = \frac{10w}{10}$$
$$21 = w$$
$$4 \cdot 21 = 84 = l$$
The rectangle is 21 feet by 84 feet.

45. $$\frac{21}{28} = \frac{3 \cdot \cancel{7}}{4 \cdot \cancel{7}} = \frac{3}{4}$$

46. $$\frac{40x^6y^4}{16x^3y^5} = \frac{5 \cdot \cancel{8} \cdot \cancel{x^3} \cdot x^3 \cdot \cancel{y^4}}{2 \cdot \cancel{8} \cdot \cancel{x^3} \cdot y \cdot \cancel{y^4}} = \frac{5 \cdot x^3}{2 \cdot y} = \frac{5x^3}{2y}$$

47. $$\frac{6}{5}\left(-\frac{2}{3}\right) = -\frac{2 \cdot \cancel{3} \cdot 2}{5 \cdot \cancel{3}} = -\frac{2 \cdot 2}{5} = -\frac{4}{5}$$

48. $$\frac{14p^2}{8} \div \frac{7p^3}{2} = \frac{14p^2}{8} \cdot \frac{2}{7p^3}$$
$$= \frac{\cancel{2} \cdot \cancel{7} \cdot \cancel{p} \cdot \cancel{p} \cdot \cancel{2}}{\cancel{2} \cdot \cancel{2} \cdot 2 \cdot \cancel{7} \cdot \cancel{p} \cdot \cancel{p} \cdot p} = \frac{1}{2p}$$

49. $$\frac{2}{3} + \frac{3}{4} = \frac{2}{3} \cdot \frac{4}{4} + \frac{3}{4} \cdot \frac{3}{3} = \frac{8}{12} + \frac{9}{12} = \frac{17}{12} = 1\frac{5}{12}$$

50. $$\frac{4}{m} - \frac{3}{5} = \frac{4}{m} \cdot \frac{5}{5} - \frac{3}{5} \cdot \frac{m}{m} = \frac{20}{5m} - \frac{3m}{5m} = \frac{20 - 3m}{5m}$$

51. $$\frac{3}{4} - \frac{1}{16} = \frac{3}{4} \cdot \frac{4}{4} - \frac{1}{16} = \frac{12}{16} - \frac{1}{16} = \frac{11}{16}\,in$$

52.

a.
$$\begin{array}{r} 10 \\ 7\overline{)75} \\ \underline{-70} \\ 5 \end{array}$$
$$\frac{75}{7} = 10\frac{5}{7}$$

b. $$-6\frac{5}{8} = -\frac{8 \cdot 6 + 5}{8} = -\frac{53}{8}$$

53. $$2\frac{2}{5}\left(3\frac{1}{12}\right) = \frac{\cancel{12}}{5} \cdot \frac{37}{\cancel{12}} = \frac{37}{5} = 7\frac{2}{5}$$

54. $$15\frac{1}{3} \div 2\frac{2}{9} = \frac{46}{3} \div \frac{20}{9} = \frac{46}{3} \cdot \frac{9}{20}$$
$$= \frac{\cancel{2} \cdot 23 \cdot \cancel{3} \cdot 3}{\cancel{3} \cdot \cancel{2} \cdot 10} = \frac{69}{10} = 6\frac{9}{10}$$

55. $$4\frac{2}{3} + 5\frac{1}{4} = \left(4 + \frac{2}{3} \cdot \frac{4}{4}\right) + \left(5 + \frac{1}{4} \cdot \frac{3}{3}\right)$$
$$= 4\frac{8}{12} + 5\frac{3}{12} = 9\frac{11}{12}$$

56. $$14\frac{2}{5} - 8\frac{2}{3} = \left(14 + \frac{2}{5} \cdot \frac{3}{3}\right) - \left(8 + \frac{2}{3} \cdot \frac{5}{5}\right)$$
$$= 14\frac{6}{15} - 8\frac{10}{15} = 13\frac{21}{15} - 8\frac{10}{15} = 5\frac{11}{15}$$

57. $$4\left(1\frac{1}{2}\right) \text{ by } 8\left(3\frac{1}{2}\right) = \left(\overset{2}{\cancel{4}} \cdot \frac{3}{\cancel{2}}\right) \text{ by } \left(\overset{4}{\cancel{8}} \cdot \frac{7}{\cancel{2}}\right)$$
$$= 6in \text{ by } 28in$$

58. $$500 - 225\frac{3}{4} = 499\frac{4}{4} - 225\frac{3}{4} = 274\frac{1}{4}\,gal$$

59. $$1\frac{1}{3} + 1\frac{1}{3} + \frac{3}{4} = 2\frac{2}{3} + \frac{3}{4} = 2\frac{2}{3} \cdot \frac{4}{4} + \frac{3}{4} \cdot \frac{3}{3}$$
$$= 2\frac{8}{12} + \frac{9}{12} = 2\frac{17}{12} = 2 + 1\frac{5}{12} = 3\frac{5}{12}\,ft$$

60. $$\left(-\frac{3}{4} \cdot \frac{9}{16}\right) + \left(\frac{1}{2} - \frac{1}{8}\right) = -\frac{27}{64} + \left(\frac{4}{8} - \frac{1}{8}\right)$$
$$= -\frac{27}{64} + \frac{3}{8} = -\frac{27}{64} + \frac{24}{64} = -\frac{3}{64}$$

61. $$\frac{\frac{2}{3}}{\frac{4}{5}} = \frac{2}{3} \cdot \frac{5}{4} = \frac{10}{12} = \frac{5}{6}$$

62. $$\frac{\frac{3}{7}+\left(-\frac{1}{2}\right)}{1\frac{3}{4}}=\frac{\frac{6}{14}-\frac{7}{14}}{\frac{7}{4}}=\frac{-\frac{1}{14}}{\frac{7}{4}}$$
$$=-\frac{1}{14}\cdot\frac{4}{7}=-\frac{\not{2}\cdot 2}{\not{2}\cdot 7\cdot 7}=-\frac{2}{49}$$

63. $$x+\frac{1}{5}=-\frac{11}{15}$$
$$x+\frac{1}{5}-\frac{1}{5}=-\frac{11}{15}-\frac{1}{5}$$
$$x=-\frac{11}{15}-\frac{3}{15}$$
$$x=-\frac{14}{15}$$
Check: $$x+\frac{1}{5}=-\frac{11}{15}$$
$$-\frac{14}{15}+\frac{1}{5}\stackrel{?}{=}-\frac{11}{15}$$
$$-\frac{14}{15}+\frac{3}{15}\stackrel{?}{=}-\frac{11}{15}$$
$$-\frac{11}{15}=-\frac{11}{15}$$

64. $$\frac{3}{4}x=\frac{5}{8}x+\frac{1}{2}$$
$$\overset{2}{\not{8}}\cdot\frac{3}{\not{4}}x=\not{8}\cdot\frac{5}{\not{8}}x+\overset{4}{\not{8}}\cdot\frac{1}{\not{2}}$$
$$6x=5x+4$$
$$6x-5x=5x+4-5x$$
$$x=4$$
Check: $$\frac{3}{4}x=\frac{5}{8}x+\frac{1}{2}$$
$$\frac{3}{\not{4}}\cdot\not{4}\stackrel{?}{=}\frac{5}{\underset{2}{\not{8}}}\cdot\not{4}+\frac{1}{2}$$
$$3\stackrel{?}{=}\frac{5}{2}+\frac{1}{2}$$
$$3\stackrel{?}{=}\frac{6}{2}$$
$$3=3$$

65. $$\frac{2}{3}x=-10$$
$$\frac{\not{3}}{\not{2}}\cdot\frac{\not{2}}{\not{3}}x=\frac{3}{\not{2}}\left(-\overset{-5}{\not{10}}\right)$$
$$x=3(-5)$$
$$x=-15$$
Check: $$\frac{2}{3}x=-10$$
$$\frac{2}{\not{3}}\left(-\overset{-5}{\not{15}}\right)\stackrel{?}{=}-10$$
$$2(-5)\stackrel{?}{=}-10$$
$$-10=-10$$

66. $$3y-8=0$$
$$3y-8+8=0+8$$
$$3y=8$$
$$\frac{3y}{3}=\frac{8}{3}$$
$$y=\frac{8}{3}$$
Check: $$3y-8=0$$
$$3\left(\frac{8}{3}\right)-8\stackrel{?}{=}0$$
$$8-8\stackrel{?}{=}0$$
$$0=0$$

67. x = the old shaving time
$$\frac{2}{3}x=90$$
$$\frac{3}{2}\cdot\frac{2}{3}x=\frac{3}{2}\cdot 90$$
$$x=3(45)$$
$$x=135$$
It would have taken 135 minutes with the old model.

68. x = the max points earnable

$$\frac{3}{4}x+\frac{1}{5}x+50=x$$

$$\overset{5}{\cancel{20}}\cdot\frac{3}{\cancel{4}}x+\overset{4}{\cancel{20}}\cdot\frac{1}{\cancel{5}}x+20\cdot 50=20\cdot x$$

$$15x+4x+1{,}000=20x$$

$$19x+1{,}000=20x$$

$$19x+1{,}000-19x=20x-19x$$

$$1{,}000=x$$

The most points possible is 1,000.

CHAPTER 5 Decimals

Section 5.1: An Introduction to Decimals

VOCABULARY

1. Decimals are written by entering the digits 0, 1, 2, 3, 4, 5, 6, 7, 8, and 9 into place-value columns that are separated by a decimal <u>point</u>.

3. We can show the value represented by each digit of the decimal 98.6213 by using <u>expanded</u> form:

$$90+8+\frac{6}{10}+\frac{2}{100}+\frac{1}{1,000}+\frac{3}{10,000}$$

CONCEPTS

5. *From left to right*: Thousands, Hundreds, Tens, Ones, Tenths, Hundredths, Thousandths, Ten-thousandths.

7. a. The value of each place in the whole-number part of decimal number is <u>10</u> times greater than the column directly to its right.

   b. The value of each place in the fractional part of a decimal number is $\underline{\frac{1}{10}}$ of the value of the place directly to its left.

9. a. $\frac{7}{10}$; 0.7　　b. $\frac{47}{100}$; 0.47

11. Whole-number part ; Fractional part

NOTATION

13. The columns to the right of the decimal point in a decimal number form its fractional part. Their place value names are similar to those in the whole-number part, but they end in the letters "<u>ths</u>."

15. 79,816.0245

GUIDED PRACTICE

17. a. 9 tenths　　b. 6
    c. 4　　d. 5 ones

19. a. 8 millionths　　b. 0
    c. 5　　d. 6 ones

21. $37.89=30+7+\frac{8}{10}+\frac{9}{100}$

23. $124.575=100+20+4+\frac{5}{10}+\frac{7}{100}+\frac{5}{1000}$

25. $7,498.6468=7000+400+90+8$
$+\frac{6}{10}+\frac{4}{100}+\frac{6}{1,000}+\frac{8}{10,000}$

27. $6.40941=6+\frac{4}{10}+\frac{0}{100}+\frac{9}{1,000}$
$+\frac{4}{10,000}+\frac{1}{100,000}$

29. 0.3 = three tenths = $\frac{3}{10}$

31. 50.41 = fifty and forty-one hundredths = $50\frac{41}{100}$

33. 19.529 = nineteen and five hundred twenty-nine thousandths = $19\frac{529}{1,000}$

35. 304.0003 = three hundred four and three ten-thousandths = $304\frac{3}{10,000}$

37. -0.00137 = negative one hundred thirty-seven hundred-thousandths = $-\frac{137}{100,000}$

39. -1,072.499 = negative one thousand seventy-two and four hundred ninety-nine thousandths = $-1,072\frac{499}{1,000}$

41. 6.187

43. 10.0056

45. -16.39

47. 104.000004

49. $2.59>2.55$

51. $45.103<45.108$

53. $3.28724>3.2871$

55. $379.67 > 379.6088$

57. $-23.45 < -23.1$

59. $-0.065 > -0.066$

61. 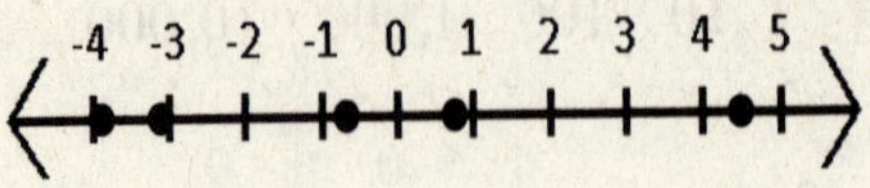

63. 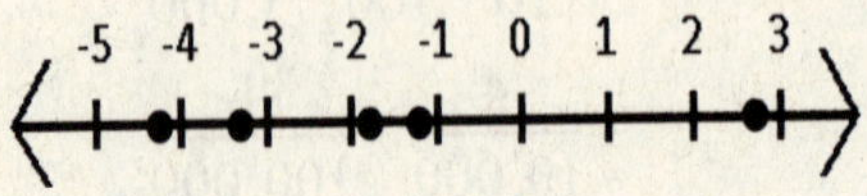

65. 506.2

67. 33.08

69. 4.234

71. 0.3656

73. -0.14

75. -2.7

77. 3.150

79. 1.414213

81. 16.100

83. 290.30350

85. \$0.28

87. \$27,842

APPLICATIONS

89. -0.7

91. \$1,025.78

93.
```
---------|---------|---------|-------*-|---------|
        .1        .2        .3        .4        .5
```

95. two-thousandths, $\frac{2}{1,000} = \frac{1}{500}$

97. \$0.16 , \$1.02 , \$1.20 , \$0.00 , \$0.10

99. Candlemaking, Crafts, Hobbies, Folk Dolls, Modern Art

101. Cylinders 2 and 4 should be replaced.

103. bacterium, plant cell, animal cell, asbestos fiber

105. June 3, 4, 5, 6, 7

WRITING

107. Ten is ten times larger than one, while one-tenth is ten times smaller than one.

109. When reading a number, the only time it is proper to use the word "and" is at the place of the decimal point.

111. Decade : 10 years

Decathlon : 10 events

Decimal : 10 parts

REVIEW

113. a. Perimeter

$$3\frac{1}{2}+2\frac{3}{4}+3\frac{1}{2}+2\frac{3}{4}$$
$$=\frac{7}{2}+\frac{11}{4}+\frac{7}{2}+\frac{11}{4}$$
$$=\frac{14}{4}+\frac{11}{4}+\frac{14}{4}+\frac{11}{4}$$
$$=\frac{50}{4}$$
$$=12\frac{1}{2}\ ft.$$

b. Area

$$3\frac{1}{2}\cdot 2\frac{3}{4}=\frac{7}{2}\cdot\frac{11}{4}=\frac{77}{8}=9\frac{5}{8}\ ft^2$$

Section 5.2: Adding and Subtracting Decimals

VOCABULARY

1.
$$\begin{array}{r} 1.72 \leftarrow \text{addend} \\ 4.68 \leftarrow \text{addend} \\ \underline{+2.02} \leftarrow \text{addend} \\ 8.42 \leftarrow \text{sum} \end{array}$$

3.
$$\begin{array}{r} 12.9 \leftarrow \text{minuend} \\ \underline{-4.3} \leftarrow \text{subtrahend} \\ 8.6 \leftarrow \text{difference} \end{array}$$

5. To see whether the result of an addition is reasonable, we can round the addends and estimate the sum or difference.

CONCEPTS

7. The subtraction is incorrect: $15.2+12.5 \neq 28.7$

9. To subtract signed decimals, add the opposite of the decimal that is being subtracted.

11. a. $6.8-1.2=6.8+(-1.2)$

b. $29.03-(-13.55)=29.03+13.55$

c. $-5.1-7.4=-5.1+(-7.4)$

NOTATION

13.
$$\begin{array}{r} 46.600 \\ 11.000 \\ \underline{+15.702} \end{array}$$

GUIDED PRACTICE

15.
$$\begin{array}{r} 32.5 \\ \underline{+7.4} \\ 39.9 \end{array}$$

17.
$$\begin{array}{r} 3.04 \\ 4.12 \\ \underline{+1.43} \\ 8.59 \end{array}$$

19.
$$\begin{array}{r} {}^{2\,1} \\ 36.821 \\ 7.300 \\ 42.000 \\ \underline{+15.440} \\ 101.561 \end{array}$$

21.
$$\begin{array}{r} {}^{1\,2} \\ 27.471 \\ 6.400 \\ 157.000 \\ \underline{+\ \ 12.120} \\ 202.991 \end{array}$$

23.
$$\begin{array}{r} 6.83 \\ \underline{-3.52} \\ 3.31 \end{array}$$

25.
$$\begin{array}{r} 8.97 \\ \underline{-6.22} \\ 2.75 \end{array}$$

27.
$$\begin{array}{r} {}^{4\ 14} \\ 495.4 \\ \underline{-153.7} \\ 341.7 \end{array}$$

29.
$$\begin{array}{r} {}^{7\ 11} \\ 878.1 \\ \underline{-174.6} \\ 703.5 \end{array}$$

31.
$$\begin{array}{r} {}^{9} \\ {}^{2\,10\,10} \\ 18.300 \\ \underline{-11.065} \\ 7.235 \end{array}$$

33.
$$\begin{array}{r} {}^{9} \\ {}^{8\,10\,10} \\ 66.900 \\ \underline{-23.037} \\ 43.863 \end{array}$$

35. $-6.3+(-8.4)=-14.7$

37. $-5.6+(-5.6)=-11.2$

39. $4.12+(-18.8)=-14.68$

41. $6.45+(-12.6)=-6.15$

43. $-1.9-0.2=-1.9+(-0.2)=-2.1$

45. $-42.5-2.8=-42.5+(-2.8)=-45.3$

47. $-4.49-(-11.3)=-4.49+11.3=6.81$

49. $-6.78-(-24.6)=-6.78+24.6=17.82$

51. $-11.1-(-14.4+7.8)=-11.1-(-6.6)$
$=-11.1+6.6=-4.5$

53. $-16.4-(-18.9+5.9)=-16.4-(-13)$
$=-16.4+13=-3.4$

55. $510.65+279.19\approx 510+280=790$

57. $671.01-88.35\approx 700-90=610$

TRY IT YOURSELF

59. $-45.6+34.7=-10.9$

61. $-9.5-7.1=-9.5+(-7.1)=-16.6$

63. $15.95+(-15.95)=0$

65.
$$\begin{array}{r} 21.88 \\ +33.12 \\ \hline 55.00 \end{array}$$

67. $2.55-(-1.1)=2.55+1.1=3.65$

69. $645+9.90005+0.12+3.02002=658.04007$

71. $0-47.5=0+(-47.5)=-47.5$

73. $(3.4-6.6)+7.3=-3.2+7.3=4.1$

75. $247.9+40+0.56=288.46$

77.
$$\begin{array}{r} 78.10 \\ -7.81 \\ \hline 70.29 \end{array}$$

79. $-7.8+(-6.5)=-14.3$

81. $19.93+(-20.21)+1.37=-0.28+1.37$
$=1.09$

83. $2.43+5.6=8.03$

85. $|-14.1+6.9|+8=|-7.2|+8$
$=7.2+8=15.2$

87. $5-0.023=4.977$

89. $-2.002-(-4.6)=-2.002+4.6=2.598$

APPLICATIONS

91. Refrigerator: \$610.80 + \$205.00 = \$815.80

Washing Machine: \$389.50 + 155.50 = \$545.00

Dryer: \$363.99 + \$167.50 = \$531.49

93. Design 1: 1.74 mi, 2.32 mi, 1.47 + 2.32 = 4.06 mi

Design 2: 2.90 mi, 0 mi, 2.90 + 0 = 2.90 mi

95. $0.218+0.218+1.939=2.375in$

97. $52.07-10.49=41.58$ sec

99. Subtotal: \$523.19

Total Deposit: \$523.19 - \$25.00 = \$498.19

101. Monday: 99.7° - 98.6° = 1.1° above normal

Tuesday: 98.6° + 2.5° = 101.1° in the A.M.

Wednesday: 98.6° - 98.6° = 0° above normal

Thursday: 100.0° - 98.6° = 1.4° above normal

Friday: 98.6° + 0.9° = 99.5° in the A.M.

103. $14.57+9.65+16.18=40.4mi$

$40.4-20.39=20.01mi$

105. a. $\$149.79-\$47.85=\$101.94$

b. $\$47.85+\$7.95=\$55.80$

WRITING

107. When we line up the decimal point, we are also lining up like place values and we can only add like place values.

109. The 37 should be written as 37.00 so the decimals line up.

111. To subtract the 6 we need to borrow from one place to the left. Since it is empty we must go one further left and borrow a 1 from the hundredths place. This gives us a 10 in the thousandths place from which we can now borrow.

REVIEW

113.

a. $\frac{4}{5}+\frac{5}{12}=\frac{4}{5}\cdot\frac{12}{12}+\frac{5}{12}\cdot\frac{5}{5}$
$=\frac{48}{60}+\frac{25}{60}=\frac{73}{60}=1\frac{13}{60}$

b. $\frac{4}{5}-\frac{5}{12}=\frac{4}{5}\cdot\frac{12}{12}-\frac{5}{12}\cdot\frac{5}{5}$
$=\frac{48}{60}-\frac{25}{60}=\frac{23}{60}$

c. $\frac{\cancel{4}^{1}}{\cancel{5}_{1}}\cdot\frac{\cancel{5}^{1}}{\cancel{12}_{3}}=\frac{1}{3}$

d. $\frac{4}{5}\div\frac{5}{12}=\frac{4}{5}\cdot\frac{12}{5}=\frac{48}{25}=1\frac{23}{25}$

Section 5.3: Multiplying Decimals

VOCABULARY

1.
$$\begin{array}{rl} 3.4 & \leftarrow \textit{factor} \\ \underline{\times 2.6} & \leftarrow \textit{factor} \\ 204 & \leftarrow \textit{partial product} \\ \underline{680} & \leftarrow \textit{partial procuct} \\ 8.84 & \leftarrow \textit{product} \end{array}$$

CONCEPTS

3.

a. $\begin{array}{r} 3.8 \\ \underline{\times 0.6} \\ 2.28 \end{array}$

b. $\begin{array}{r} 1.79 \\ \underline{\times 8.1} \\ 179 \\ \underline{14320} \\ 14.499 \end{array}$

c. $\begin{array}{r} 2.0 \\ \underline{\times 7} \\ 14.0 \end{array}$

d. $\begin{array}{r} 0.013 \\ \underline{\times 0.02} \\ 0.00026 \end{array}$

5. a. Negative × Negative = Positive

b. Negative × Positive = Negative

NOTATION

7. a. 10, 100, 1,000, 10,000, 100,000

b. 0.1, 0.01, 0.001, 0.0001, 0.00001

GUIDED PRACTICE

9. $4.8\cdot 6.2=29.76$
$$\begin{array}{r} 4.8 \\ \underline{\times 6.2} \\ 96 \\ \underline{288} \\ 29.76 \end{array}$$

11. $5.6(8.9)=49.84$
$$\begin{array}{r} 5.6 \\ \underline{\times 8.9} \\ 504 \\ \underline{448} \\ 49.84 \end{array}$$

13. $0.003(2.7) = 0.081$

$$\begin{array}{r} 2.7 \\ \underline{\times 0.003} \\ 0.0081 \end{array}$$

15. $5.8 \times 0.009 = 0.0522$

$$\begin{array}{r} 5.8 \\ \underline{\times 0.009} \\ 0.0522 \end{array}$$

17. $179(6.3) = 1,127.7$

$$\begin{array}{r} 179 \\ \underline{\times 6.3} \\ 537 \\ \underline{1074} \\ 1127.7 \end{array}$$

19. $316 \times 7.4 = 2,338.4$

$$\begin{array}{r} 316 \\ \underline{\times 7.4} \\ 1264 \\ \underline{2212} \\ 2338.4 \end{array}$$

21. $6.84 \cdot 100$ 2 zeros: move 2 right : 684

23. $0.041(10,000)$ 4 zeros: move 4 right: 410

25. $647.59 \cdot 0.01$ 2 places: move 2 left : 6.4759

27. $1.15(0.001)$ 3 places: move 3 left : 0.00115

29. 14.2 million 6 zeros: move 6 right : 14,200,000

31. 98.2 billion 9 zeros: move 9 right : 98,200,000,000

33. 1.421 trillion
12 zeros: move 12 right : 1,421,000,000,000

35. 657.1 billion 9 zeros: move 9 right : 657,100,000,000

37. $-1.9(7.2) = -13.68$

$$\begin{array}{r} 1.9 \\ \underline{\times 7.2} \\ 38 \\ \underline{133} \\ 13.68 \end{array}$$

39. $-3.3(-1.6) = 5.28$

$$\begin{array}{r} 3.3 \\ \underline{\times 1.6} \\ 198 \\ \underline{33} \\ 5.28 \end{array}$$

41. $(-10,000)(-44.83)$

$(-)(-) = +, \text{so } 10,000(44.83)$

$4 \text{ right}: 448,300$

43. $678.231(-1,000)$

$(+)(-) = -, \text{so}$

$3 \text{ right} = 678,231$

$-678,231$

45. $(3.4)^2 = (3.4)(3.4) = 11.56$

47. $(-0.03)^2 = (-0.03)(-0.03) = 0.0009$

49.

$$\begin{aligned} & -(-0.2)^2 + 4|-2.3 + 1.5| \\ &= -(-0.2)^2 + 4|-0.8| \\ &= -0.04 + 4(0.8) \\ &= -0.04 + 3.2 \\ &= 3.16 \end{aligned}$$

51.

$$\begin{aligned} & -(-0.8)^2 + 7|-5.1 - 4.8| \\ &= -(-0.8)^2 + 7|-9.9| \\ &= -0.64 + 7(9.9) \\ &= -0.64 + 69.3 \\ &= 68.66 \end{aligned}$$

53. $A = 85.50 + 85.50 \cdot 0.08 \cdot 5$
$= 85.50 + 34.2$
$= 119.70$

55. $A = (5.3)(7.2) = 38.16$

57. $P = 2(3.7) + 2(3.6) = 7.4 + 7.2 = 14.6$

59. $C = 2(3.14)(2.5) = 15.7$

61. $46 \cdot 5.3 \approx 50 \cdot 5 = 250$

63. $17.11 \cdot 3.85 \approx 17.1 \cdot 3.9 = 66.69$

TRY IT YOURSELF

65. $-0.56(0.33) = -0.1848$

$$\begin{array}{r} 0.56 \\ \underline{\times 0.33} \\ 168 \\ \underline{168} \\ 0.1848 \end{array}$$

67. $(-0.1)^2 = 0.01$

69. $(-0.7 - 0.5)(2.4 - 3.1)$
$= (-1.2)(-0.7)$
$= 0.84$

71. $0.008 \times 0.09 = 0.00072$

$$\begin{array}{r} 0.008 \\ \underline{\times 0.09} \\ 0.00072 \end{array}$$

73. $-0.2 \cdot 1{,}000{,}000$
6 zeros: move 6 right: $-200{,}000$

75. $(-5.6)(-2.2)$
$(5.6)(2.2) = 12.32$
$(-)(-) = (+)$, so
12.32

77. $-4.6(23.4 - 19.6)$
$= -4.6(3.8)$
$= -17.48$

79. $(-4.9)(-0.001)$: 3 left : 0.0049

81. $(-0.2)^2 + 2(7.1)$
$= 0.04 + 14.2$
$= 14.24$

83. $2.13 \times 4.05 = 8.6265$

$$\begin{array}{r} 2.13 \\ \underline{\times 4.05} \\ 1065 \\ \underline{8520} \\ 8.6265 \end{array}$$

85. $(-7)(8.1781)$
$(7)(8.1781) = 57.2467$
$(-)(+) = (-)$, so
-57.2467

87. $-1{,}000(0.02239)$ 3 zeros: move 3 right: -22.39

89. $(0.5 + 0.6)^2(-3.2)$
$= (1.1)^2(-3.2)$
$= (1.21)(-3.2)$
$= -3.872$

91. $-0.2(306)(-0.4)$
$= (-61.2)(-0.4)$
$= 24.28$

93. $-0.01(|-2.6 - 6.7|)^2$
$= -0.01(|-9.3|)^2$
$= -0.01(9.3)^2$
$= -0.01(86.49)$
$= -0.8649$

95.

Decimal	Its square
0.1	0.01
0.2	0.04
0.3	0.09
0.4	0.16
0.5	0.25
0.6	0.36
0.7	0.49
0.8	0.64
0.9	0.81

APPLICATIONS

97. $500\cdot(0.0038)=1.9in.$

99. $\$37.35\cdot 38\cdot 52.2=74087.46\approx\$74{,}100$

101.

Type of nut	Price per pound	Pounds	Cost
Almonds	\$5.95	16	\$95.20
Walnuts	\$4.95	25	\$123.75

103.

$$34\,\mathring{A}=34(0.000000004)$$
$$=0.000000136in.$$

$$3.4\,\mathring{A}=3.4(0.000000004)$$
$$=0.0000000136in.$$

$$10\,\mathring{A}=10(0.000000004)$$
$$=0.00000004in.$$

105. a. $6\cdot 0.35=2.1mi.$

b. $10\cdot 0.35=3.5mi.$

c. $2.1mi.+3.5mi.=5.6mi.$

107. $719\cdot 0.14277\approx\$102.65$

109. a. 19,600,000 acres

b. 7,061,000,000 people

c. 2,962,900,000,000 miles

111. a. $8\times 24=192\,ft.^2$

b. $25.5\times 8.75=223.125\,ft^2$

c. $223.125-192=31.125\,ft^2$

113.

Ticket type	Price	Number sold	Receipts
Floor	\$12.50	1,000	\$12,500
Balcony	\$15.75	100	\$1,575

b. \$14,075

115. $2(45.5+20.5+2.2)=2(68.2)=136.4lbs.$

117. $0.57+3(.09)=0.57+0.27=0.84in.$

WRITING

119. Count the total number of places to the right of each decimal point and have that many places to the right of the decimal point in the answer.

121. To multiply a decimal by a power of 10 that is greater than 1, move the decimal to the right the number of zeros that appear in the power of 10. To multiply by a power of 10 that is less than 1, move the decimal to the left the same number of decimal places as there are in the power of 10.

123. We line the decimal points up in addition so that the like place values line up and like pieces are being added. In multiplication, we determine the position of the decimal point in the answer by counting the total number of digits to the right of the decimal regardless of position.

REVIEW

125. $220=2\cdot 110=2\cdot 10\cdot 11$
$=2\cdot 2\cdot 5\cdot 11=2^2\cdot 5\cdot 11$

127. $162=2\cdot 81=2\cdot 9\cdot 9$
$=2\cdot 3\cdot 3\cdot 3\cdot 3=2\cdot 3^4$

Section 5.4: Dividing Decimals

VOCABULARY

1.

$$\begin{array}{c} \textit{quotient} \\ \downarrow \\ \textit{divisor} \rightarrow 5\overline{)15.85} \leftarrow \textit{dividend} \end{array}$$

(quotient: 3.17)

CONCEPTS

3. a. $4\overline{)21.04}$ with quotient 5.26 b. $3\overline{)0.024}$ with quotient 0.008

5. a. 1 right: $13\overline{)106.6}$

b. 2 right: $371\overline{)1669.5}$

7. $\frac{10}{10}$

9. thousandths

11. a. To find the quotient of a decimal and 10, 100, 1,000, and so on, move the decimal point to the <u>left</u> the same number of places as there are zeros in the power of 10.

b. To find the quotient of a decimal and 0.1, 0.01, 0.001, and so on, move the decimal point to the <u>right</u> the same number of decimal places as there are in the power of 10.

NOTATION

13. The red arrows indicate that the decimal points of both the divisor and dividend were moved 2 units to the right.

GUIDED PRACTICE

15.

$$\begin{array}{r} 2.1 \\ 6\overline{)12.6} \\ -\underline{12}\downarrow \\ 06 \\ -\underline{6} \\ 0 \end{array}$$

Check: $\begin{array}{r} 2.1 \\ \underline{\times 6} \\ 12.6 \end{array}$

17.

$$\begin{array}{r} 9.2 \\ 3\overline{)27.6} \\ -\underline{27}\downarrow \\ 06 \\ -\underline{6} \\ 0 \end{array}$$

Check: $\begin{array}{r} 9.2 \\ \underline{\times 3} \\ 27.6 \end{array}$

19.

$$\begin{array}{r} 4.27 \\ 23\overline{)98.21} \\ -\underline{92}\downarrow \\ 62 \\ -\underline{46}\downarrow \\ 161 \\ -\underline{161} \\ 0 \end{array}$$

Check: $4.27 \cdot 23 = 98.21$

21.

$$\begin{array}{r} 8.65 \\ 37\overline{)320.05} \\ -\underline{296}\downarrow \\ 240 \\ -\underline{222}\downarrow \\ 185 \\ -\underline{185} \\ 0 \end{array}$$

Check: $8.65 \cdot 37 = 320.05$

23.

$$\begin{array}{r} 3.35 \\ 4\overline{)13.40} \\ -\underline{12}\downarrow \\ 14 \\ -\underline{12}\downarrow \\ 20 \\ -\underline{20} \\ 0 \end{array}$$

Check: $3.35 \cdot 4 = 13.40$

25.

$$\begin{array}{r} 4.56 \\ 5\overline{)22.80} \\ -\underline{20}\downarrow \\ 28 \\ -\underline{25}\downarrow \\ 30 \\ -\underline{30} \\ 0 \end{array}$$

Check: $4.56 \cdot 5 = 22.8$

27. $\frac{0.1932}{0.42} = \frac{19.32}{42}$

$$\begin{array}{r} 0.46 \\ 42\overline{)19.32} \\ -16\ 8 \\ \hline 2\ 52 \\ -2\ 52 \\ \hline 0 \end{array}$$

Check: $0.42 \cdot 0.46 = 0.1932$

29.

$$0.29\overline{)0.1131} = 29\overline{)11.31} \quad \begin{array}{r} 0.39 \\ -8\ 7 \\ \hline 2\ 61 \\ -2\ 61 \\ \hline 0 \end{array}$$

Check: $0.39 \cdot 0.29 = 0.1131$

31. $\frac{11.83}{0.6} = 6\overline{)118.300}$

$$\begin{array}{r} 19.716.... \\ 6\overline{)118.300} \\ \underline{-6} \\ 58 \\ \underline{-54} \\ 43 \\ \underline{-42} \\ 10 \\ \underline{-6} \\ 40 \\ \underline{-36} \\ 4 \end{array}$$

≈ 19.72

Check: $19.72 \cdot 0.6 \approx 11.83$

33. $\frac{17.09}{0.7} = 7\overline{)170.900}$

$$\begin{array}{r} 24.414.... .. \\ 7\overline{)170.900} \\ \underline{-14} \\ 30 \\ \underline{-28} \\ 29 \\ \underline{-28} \\ 10 \\ \underline{-7} \\ 30 \\ \underline{-28} \\ 2 \end{array}$$

≈ 24.41

Check: $24.41 \cdot 0.7 \approx 17.09$

35. $289.842 \div 72.1 \approx 280 \div 70 = 4$

37. $383.76 \div 7.8 \approx 400 \div 8 = 50$

39. $3{,}883.284 \div 48.12 \approx 4{,}000 \div 50 = 80$

41. $6.1\overline{)15{,}819.74} \approx 15{,}000 \div 5 = 3{,}000$

43. $451.78 \div 100$ 2 zeros: move 2 left : 4.5178

45. $\frac{30.09}{10,000}$ 4 zeros : move 4 left : 0.003009

47. $1.25 \div 0.1$ 1 place: move 1 right : 12.5

49. $\frac{545.2}{0.001}$ 3 places: move 3 right : $545,200$

51. $-110.336 \div 12.8$:

$$128\overline{)\ 1103.36} = 8.62$$

$$\begin{array}{r} 1103.36 \\ -1024 \\ \hline 793 \\ -768 \\ \hline 256 \\ -256 \\ \hline 0 \end{array}$$

$(-)\div(+)=(-)$

−8.62

53. $-91.304 \div (-22.6)$

$$226\overline{)913.04} = 4.04$$

$$\begin{array}{r} 913.04 \\ -904 \\ \hline 90 \\ -0 \\ \hline 904 \\ -904 \\ \hline 0 \end{array}$$

$(-)\div(-)=(+)$

4.04

55. $\frac{-20.3257}{-0.001}$

3 places: move 3 right

$(-)\div(-)=(+)$

$20,325.7$

57. $0.003 \div (-100)$

2 zeros: move 2 left

$(+)\div(-)=(-)$

−0.00003

59. $\frac{2(0.614)+2.3854}{0.2-0.9}$

$=\frac{1.228+2.3854}{-0.7}$

$=\frac{3.6134}{-0.7}$

$=-5.162$

61. $\frac{5.409-3(1.8)}{(0.3)^2}$

$=\frac{5.409-5.4}{0.09}$

$=\frac{0.009}{0.09}$

$=0.1$

63. $t=\frac{d}{r}=\frac{211.75}{60.5}=3.5$

65. $r=\frac{d}{t}=\frac{219.375}{3.75}=58.5$

TRY IT YOURSELF

67. $4.5\overline{)11.97}$

$$45\overline{)119.70} = 2.66$$

$$\begin{array}{r} 119.70 \\ -90 \\ \hline 297 \\ -270 \\ \hline 270 \\ -270 \\ \hline 0 \end{array}$$

69. $\frac{75.04}{10}$ =7.504

71.

$$\begin{array}{r} 0.0045 \\ 8\overline{)0.0360} \\ \underline{-32} \\ 40 \\ \underline{-40} \\ 0 \end{array}$$

73.

$$\begin{array}{r} 0.321 \\ 9\overline{)2.889} \\ \underline{-27} \\ 18 \\ \underline{-18} \\ 09 \\ \underline{-9} \\ 0 \end{array}$$

75.

$$\frac{-3(0.2)-2(3.3)}{30(0.4)^2}$$
$$=\frac{-.6-6.6}{30(0.16)}$$
$$=\frac{-7.2}{4.8}$$
$$=-1.5$$

77. $1.2202 \div (-0.01)$

2 places: move 2 right

$(+)\div(-)=(-)$

-122.02

79. $-5.714 \div 2.4$

$$\begin{array}{r} 2.38 \\ 24\overline{)571.40} \\ \underline{-48} \\ 91 \\ \underline{-72} \\ 194 \\ \underline{-192} \\ 2 \end{array}$$

$-2.38 \approx -2.4$

81. $-39 \div (-4)$

$$\begin{array}{r} 9.75 \\ 4\overline{)39.00} \\ \underline{-36} \\ 30 \\ \underline{-28} \\ 20 \\ \underline{-20} \\ 0 \end{array}$$

83. $7.8915 \div .00001$

5 places: move 5 right

789,150

85. $\frac{-30.6}{5.1} = -6$

$$\begin{array}{r} 6 \\ 51.\overline{)306.} \\ \underline{-306} \\ 0 \end{array}$$

87. $12.243 \div 0.9 = 122.43 \div 9 \approx 13.60$

$$\begin{array}{r} 13.603 \\ 9\overline{)122.430} \\ \underline{-9} \\ 32 \\ \underline{-27} \\ 54 \\ \underline{-54} \\ 030 \\ \underline{-27} \\ 3 \end{array}$$

89. $1000\overline{)34.8}$

3 zeros: move 3left

0.0348

91. $\dfrac{40.7(3-8.3)}{0.4-0.61}$
$=\dfrac{40.7(-5.3)}{-0.21}$
$=\dfrac{-215.71}{-0.21}$
$\approx 1{,}027.19$

93. $\dfrac{0.25}{1.6}=\dfrac{2.5}{16}=0.15625$

APPLICATIONS

95. $14 \div 0.05 = 280$ slices

97. $60 \div 0.00003 = 2{,}000{,}000$ calculations

99. $8.5 \div 0.017 = 500$ squeezes

101. $27.5 \div 2.5 = 11$ hours: 6:00 PM

103. $1{,}510.4+1{,}503.8+1{,}462.4+1{,}514.2$
$+1{,}544.6+1{,}575.0=9{,}110.4$
$9{,}110.4 \div 6 = 1{,}518.4$ million

105. $0.219+0.233+0.204$
$+0.297+0.202=1.155$
$1.155 \div 5 = 0.231$ sec

WRITING

107. Begin by moving the decimal point of both the divisor and dividend the same number of places so the divisor is a whole number, attaching zeros if necessary. Then divide as if it were an integer problem, placing the decimal point in the quotient directly above the decimal point in the dividend.

109. In this case equivalent means they have the same answer.

111. Round to the nearest 10th to get 2.4 and 0.8 which gives 3 as an approximation.

REVIEW

113. $10 = 2\cdot 5$
$25 = 5\cdot 5$
a. $\text{GCF} = 5$
b. $\text{LCM} = 2\cdot 5\cdot 5 = 50$

Section 5.5: Fractions and Decimals

VOCABULARY

1. A fraction and a decimal are said to be <u>equivalent</u> if they name the same number.

3. 0.75, 0.625, and 3.5 are examples of <u>terminating</u> decimals.

CONCEPTS

5. $\dfrac{7}{8}$ means $7 \div 8$.

7. A decimal point and two additional <u>zeros</u> were written to the right of 3.

9. Sometimes, when finding the decimal equivalent of a fraction, the division process never gives a remainder of 0. We call the resulting decimal a <u>repeating</u> decimal.

11. a. 0.38

b. 0.212

NOTATION

13. a. $0.7=\dfrac{7}{10}$

b. $0.77=\dfrac{77}{100}$

GUIDED PRACTICE

15.
$$\begin{array}{r} 0.5 \\ 2\overline{)1.0} \\ -10 \\ \hline 0 \end{array}$$

17.
$$\begin{array}{r} 0.875 \\ 8\overline{)7.000} \\ -64\downarrow \\ \hline 60 \\ -56\downarrow \\ \hline 40 \\ -40 \\ \hline 0 \end{array}$$

19.

$$\begin{array}{r} 0.55 \\ 20\overline{)11.00} \\ \underline{-100}\downarrow \\ 100 \\ \underline{-100} \\ 0 \end{array}$$

21.

$$\begin{array}{r} 2.6 \\ 5\overline{)13.0} \\ \underline{-10}\downarrow \\ 30 \\ \underline{-30} \\ 0 \end{array}$$

23.

$$\begin{array}{r} 0.5625 \\ 16\overline{)\ 9.0000} \\ \underline{-80} \\ 100 \\ \underline{-96} \\ 40 \\ \underline{-32} \\ 80 \\ \underline{-80} \\ 0 \end{array}$$

25.

$$\begin{array}{r} 0.53125 \\ 32\overline{)17.0000} \\ \underline{-160}\downarrow \\ 100 \\ \underline{-96}\downarrow \\ 40 \\ \underline{-32}\downarrow \\ 80 \\ \underline{-64}\downarrow \\ 160 \\ \underline{-160} \\ 0 \end{array}$$

Since the original problem was negative,

−0.53125

27. $\frac{3}{5}=\frac{3}{5}\cdot\frac{2}{2}=\frac{6}{10}=0.6$

29. $\frac{9}{40}=\frac{9}{40}\cdot\frac{25}{25}=\frac{225}{1000}=0.225$

31. $\frac{19}{25}=\frac{19}{25}\cdot\frac{4}{4}=\frac{76}{100}=0.76$

33. $\frac{1}{500}=\frac{1}{500}\cdot\frac{2}{2}=\frac{2}{1000}=0.002$

35.

$$\begin{array}{r} 0.75 \\ 4\overline{)3.00} \\ \underline{-28} \\ 20 \\ \underline{-20} \\ 0 \end{array}$$

so $3\frac{3}{4}=3.75$

37.

$$\begin{array}{r} 0.6875 \\ 16\overline{)11.0000} \\ \underline{-96} \\ 140 \\ \underline{-128} \\ 120 \\ \underline{-112} \\ 80 \\ \underline{-80} \\ 0 \end{array}$$

so $12\frac{11}{16}=12.6875$

39.

$$\begin{array}{r} 0.11\cdots \\ 9\overline{)1.00} \\ \underline{-9} \\ 10 \\ \underline{-9} \\ 1 \end{array}$$

so $\frac{1}{9}=0.\overline{1}$

41.

$$\begin{array}{r} 0.5833\cdots \\ 12\overline{)7.0000} \\ \underline{-60} \\ 100 \\ \underline{-96} \\ 40 \\ \underline{-36} \\ 40 \end{array}$$

$$\frac{7}{12} = .58333... = 0.58\overline{3}$$

43.

$$\begin{array}{r} 0.077\cdots \\ 90\overline{)7.000} \\ \underline{-630} \\ 700 \\ \underline{-630} \\ 700 \end{array}$$

$$\frac{7}{90} = .0777... = 0.0\overline{7}$$

45.

$$\begin{array}{r} 0.0166\cdots \\ 60\overline{)1.0000} \\ \underline{-60} \\ 400 \\ \underline{-360} \\ 400 \\ \underline{-360} \\ 400 \end{array}$$

$$\frac{1}{60} = .016666... = 0.01\overline{6}$$

47.

$$\begin{array}{r} 0.454\cdots \\ 11\overline{)5.000} \\ \underline{-44} \\ 60 \\ \underline{-55} \\ 50 \\ \underline{-44} \\ 60 \end{array}$$

$$-\frac{5}{11} = -.454545... = -0.\overline{45}$$

49.

$$\begin{array}{r} 0.606\cdots \\ 33\overline{)20.000} \\ \underline{-198} \\ 20 \\ \underline{-0} \\ 200 \\ \underline{-198} \\ 20 \end{array}$$

$$-\frac{20}{33} = -.606060... = -0.\overline{60}$$

51.

$$\begin{array}{r} 0.233 \\ 30\overline{)7.000} \\ \underline{-60} \\ 100 \\ \underline{-90} \\ 100 \\ \underline{-90} \\ 10 \end{array}$$

$$\frac{7}{30} = 0.2\overline{3} \approx 0.23$$

53.

$$\begin{array}{r} 0.488 \\ 45\overline{)22.000} \\ \underline{-180} \\ 400 \\ \underline{-360} \\ 400 \\ \underline{-360} \\ 40 \end{array}$$

$$\frac{22}{45} = 0.4\overline{8} \approx 0.49$$

55.

$$\begin{array}{r} 1.846 \\ 13\overline{)24.000} \\ \underline{-13} \\ 110 \\ \underline{-104} \\ 60 \\ \underline{-52} \\ 80 \\ \underline{-78} \\ 2 \end{array}$$

$$\frac{24}{13} = 1.\overline{846153} \approx 1.85$$

57.

$$\begin{array}{r} 1.083 \\ 12\overline{)13.000} \\ \underline{-12} \\ 10 \\ \underline{-0} \\ 100 \\ \underline{-96} \\ 40 \\ \underline{-36} \\ 4 \end{array}$$

$$-\frac{13}{12} = -1.08\overline{3} \approx -1.08$$

59.

$$\begin{array}{r} 0.1515 \\ 33\overline{)5.0000} \\ \underline{-33} \\ 170 \\ \underline{-165} \\ 50 \\ \underline{-33} \\ 17 \end{array}$$

$$\frac{5}{33} = 0.151515... \approx 0.152$$

61.

$$\begin{array}{r} 0.3703 \\ 27\overline{)10.0000} \\ \underline{-81} \\ 190 \\ \underline{-189} \\ 10 \\ \underline{-0} \\ 100 \\ \underline{-81} \\ 19 \end{array}$$

$$\frac{10}{27} = 0.370370370... \approx 0.370$$

63.

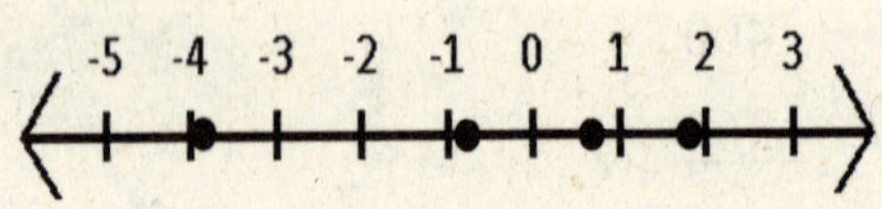

65.

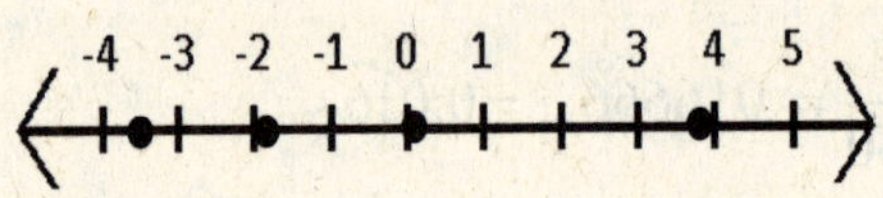

67.

$\frac{7}{8} \square 0.895$

$0.875 \square 0.895$

$0.875 < 0.895$

69.

$0.\overline{7} \square \frac{17}{22}$

$0.777... \square 0.7727...$

$0.777... > 0.7727...$

71. $\frac{52}{25}\square 2.08$

$2.08\square 2.08$

$2.08 = 2.08$

73. $-\frac{11}{20}\square -0.\overline{48}$

$-.55\square -0.4848...$

$-.55 < -0.4848...$

75. $6\frac{1}{2}, 6.25, \frac{19}{3}$

$= 6.5, 6.25, 6.\overline{3}$

so: $6.25, \frac{19}{3}, 6\frac{1}{2}$

77. $-0.\overline{81}, -\frac{8}{9}, -\frac{6}{7}$

$= -0.\overline{81}, -0.888..., -0.857...$

so: $-\frac{8}{9}, -\frac{6}{7}, -0.\overline{81}$

79. $\frac{1}{9}+0.3 = \frac{1}{9}+\frac{3}{10} = \frac{10}{90}+\frac{27}{90} = \frac{37}{90}$

81. $0.9-\frac{7}{12} = \frac{9}{10}-\frac{7}{12} = \frac{54}{60}-\frac{35}{60} = \frac{19}{60}$

83. $\frac{5}{11}(0.3) = \frac{5}{11}\cdot\frac{3}{10} = \frac{15}{110} = \frac{3}{22}$

85. $\frac{1}{4}(0.25)+\frac{15}{16} = \frac{1}{4}\cdot\frac{25}{100}+\frac{15}{16}$

$= \frac{25}{400}+\frac{375}{400} = \frac{400}{400} = 1$

87. $0.24+\frac{1}{3} \approx 0.24+0.33 = 0.57$

89. $5.69-\frac{5}{12} \approx 5.69-0.42 = 5.27$

91. $0.43-\frac{1}{12} \approx 0.43-0.08 = 0.35$

93. $\frac{1}{15}-0.55 \approx 0.07-0.55 = -0.48$

95. $(3.5+6.7)\left(-\frac{1}{4}\right) = (10.2)(-0.25) = -2.55$

97. $\left(\frac{1}{5}\right)^2(1.7) = (0.2)^2(1.7)$

$= (0.04)(1.7) = 0.068$

99. $7.5-(0.78)\left(\frac{1}{2}\right)^2 = 7.5-(0.78)(0.5)^2$

$= 7.5-(0.78)(0.25) = 7.5-0.195 = 7.305$

101. $\frac{3}{8}(3.2)+\left(4\frac{1}{2}\right)\left(-\frac{1}{4}\right)$

$= (0.375)(3.2)+(4.5)(-0.25)$

$= 1.2+(-1.125) = 0.075$

APPLICATIONS

103. Each line represents 0.0625 of an inch.
$0.0625, 0.375, 0.5625, 0.9375$

105. $3 \div 40 = 0.075$ This is the thicker line.

107. 23.4 sec, 23.8 sec, 24.2 sec, 32.6 sec

109. $6\cdot\left(\frac{1}{2}\cdot 6\cdot 5.2\right) = 6(15.6) = 93.6 in^2$

111. $\frac{2}{3}\cdot 4.14+\frac{3}{4}\cdot 5.68 = \$7.02$

113. To write a fraction as a decimal, use long division to divide the numerator by the denominator.

115. Since only the threes are repeated, the number should be written as $0.1\overline{3}$.

117. By writing 0.76 as $\frac{76}{100}$ and then reducing.

119. a. $\{0,1,2,3,4,5,6,7,8,9\}$

b. $\{2,3,5,7,11,13,17,19,23,29\}$

c. $\{..., -3, -2, -1, 0, 1, 2, 3, ...\}$

Section 5.6: Square Roots

VOCABULARY

1. When we raise a number to the second power, we are squaring it, or finding its square.

3. The symbol $\sqrt{\ }$ is called a radical symbol.

5. Whole numbers such as 36 and 49, that are squares of whole numbers, are called perfect squares.

CONCEPTS

7. a. The square of 5 is 25, because $5^2 = 25$.

   b. The square of $\frac{1}{4}$ is $\frac{1}{16}$, because $\left(\frac{1}{4}\right)^2 = \frac{1}{16}$.

9. a. $\sqrt{49} = 7$ because $7^2 = 49$.

   b. $\sqrt{4} = 2$ because $2^2 = 4$.

11. a. $\sqrt{1} = 1$

    b. $\sqrt{0} = 0$

13. Step 2: Evaluate all exponential expressions and any square roots.

15. 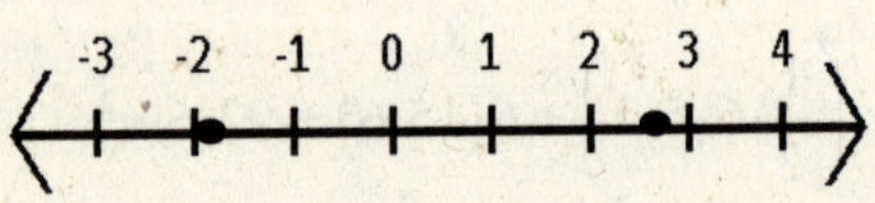

NOTATION

17. a. The symbol $\sqrt{\ }$ is used to indicate a positive square root.

    b. The symbol $-\sqrt{\ }$ is used to indicate the negative square root of a positive number.

19. $-\sqrt{49} + \sqrt{64} = -7 + 8$
    $= 1$

GUIDED PRACTICE

21. Since $5^2 = (-5^2) = 25,$ the square roots are $5, -5$.

23. Since $4^2 = (-4)^2 = 16,$ the square roots are $4, -4$.

25. $\sqrt{16} = 4$ since $4^2 = 16$.

27. $\sqrt{9} = 3$ since $3^2 = 9$.

29. $-\sqrt{144} = -12$

31. $-\sqrt{49} = -7$

33. $\sqrt{961} = 31$

35. $\sqrt{3,969} = 63$

37. $\sqrt{\frac{4}{25}} = \frac{2}{5}$ since $\left(\frac{2}{5}\right)^2 = \frac{4}{25}$.

39. $-\sqrt{\frac{16}{9}} = -\frac{4}{3}$

41. $-\sqrt{\frac{1}{81}} = -\frac{1}{9}$

43. $\sqrt{0.64} = 0.8$ since $(0.8)^2 = 0.64$.

45. $-\sqrt{0.81} = -0.9$

47. $\sqrt{0.09} = 0.3$

49. $\sqrt{36} + \sqrt{1} = 6 + 1 = 7$

51. $\sqrt{81} + \sqrt{49} = 9 + 7 = 16$

53. $-\sqrt{144} - \sqrt{16} = -12 - 4 = -16$

55. $-\sqrt{225} + \sqrt{144} = -15 + 12 = -3$

57. $4\sqrt{25} = 4 \cdot 5 = 20$

59. $-10\sqrt{196} = -10 \cdot 14 = -140$

61. $-4\sqrt{169} + 2\sqrt{4} = -4 \cdot 13 + 2 \cdot 2$
    $= -52 + 4 = -48$

63. $-8\sqrt{16} + 5\sqrt{225} = -8 \cdot 4 + 5 \cdot 15$
    $= -32 + 75 = 43$

65.
$$15+4\left[5^2-(6-1)\sqrt{4}\right]$$
$$=15+4\left[25-(5)2\right]$$
$$=15+4[25-10]$$
$$=15+4[15]$$
$$=15+60$$
$$=75$$

67.
$$50-\left[\left(6^2-24\right)+9\sqrt{25}\right]$$
$$=50-\left[(36-24)+9\cdot5\right]$$
$$=50-[12+45]$$
$$=50-57$$
$$=-7$$

69.
$$\sqrt{196}+3\left(5^2-2\sqrt{225}\right)$$
$$=14+3(25-2\cdot15)$$
$$=14+3(25-30)$$
$$=14+3(-5)$$
$$=14+(-15)$$
$$=-1$$

71.
$$\frac{\sqrt{16}-6(2^2)}{\sqrt{4}}=\frac{4-6(4)}{2}=\frac{4-24}{2}$$
$$=\frac{-20}{2}=-10$$

73.
$$\sqrt{\frac{1}{16}}-\sqrt{\frac{9}{25}}=\frac{1}{4}-\frac{3}{5}=\frac{5}{20}-\frac{12}{20}=-\frac{7}{20}$$

75.
$$5\left(-\sqrt{49}\right)(-2)^2=5(-7)(4)=-140$$

77.
$$\left(6^2\right)\sqrt{0.04}+2.36=36\cdot0.2+2.36$$
$$=7.2+2.36=9.56$$

79.
$$-\left(-3\sqrt{1.44}+5\right)$$
$$=-(-3\cdot1.2+5)$$
$$=-(-3.6+5)$$
$$=-1.4$$

81.
$$c=\sqrt{9^2+12^2}=\sqrt{81+144}=\sqrt{225}=15$$

83.
$$a=\sqrt{25^2-24^2}=\sqrt{625-576}=\sqrt{49}=7$$

85.

Number	Square Root
1	1
2	1.414
3	1.732
4	2
5	2.236
6	2.449
7	2.646
8	2.828
9	3
10	3.162

87. $\sqrt{5}\approx2.24$

89. $\sqrt{66}\approx8.12$

91. $\sqrt{24.05}\approx4.904$

93. $-\sqrt{11.1}\approx-3.332$

APPLICATIONS

95. a. $\sqrt{25}\,ft.=5\,ft.$

b. $\sqrt{100}\,ft.=10\,ft.$

97. $\sqrt{16,200}\approx127.3\,ft.$

99. $\sqrt{1,764in.}=42in.$: 42" screen.

WRITING

101. He is dividing the number by 2, rather than finding the number that will give 16 when raised to the second power.

103. A non-terminating decimal is one where division will never produce a remainder of 0. The decimal obtained from dividing 1 by 6 is an example of this.

105. It does not represent a real number because the square of any real number will always be a real number so it is not possible to have a real square root of a negative number.

107. Because 2.449 is an approximation since the actual square root is a non-terminating decimal.

REVIEW

109. $5(-2q) = -10q$

111. $-7y + 2 = 44$

$-7y + 2 - 2 = 44 - 2$

$-7y = 42$

$\frac{-7y}{-7} = \frac{42}{-7}$

$y = -6$

Section 5.7: Solving Equations that Involve Decimals

VOCABULARY

1. In algebra, we often replace one algebraic expression with another that is equivalent and simpler in form. That process is called simplifying an expression.

3. When we write $6.2x + 2.7x$ as $8.9x$ we say we have combined like terms.

CONCEPTS

5. a. $4(3.2t) = (4 \cdot 3.2)t = 12.8t$

b. Associative Property of Multiplication

c. $-6.1y \cdot 2 = -6.1 \cdot 2 \cdot y = -12.2y$

d. Commutative Property of Multiplication

7. a. $4.2m + 6.3m = (4.2 + 6.3)m = 10.5m$

b. $3.6n^2 - 5.8n^2 = (3.6 - 5.8)n^2 = -2.2n^2$

c. $1.2 + 3.2d + 1.5 = 1.2 + 1.5 + 3.2d$

$= 2.7 + 3.2d$

d. Like terms can be combined by adding or subtracting the coefficients of the terms and keeping the same variables with the same exponents.

9. * To isolate the variable term $2.1m$, we undo the subtraction of 9.1 by adding 9.1 to both sides.

* To isolate the variable m, we undo the multiplication of 2.1 by dividing both sides by 2.1.

NOTATION

11. $0.6a - 2.3 = -1.82$

$0.6a - 2.3 + 2.3 = -1.82 + 2.3$

$0.6a = 0.48$

$\frac{0.6a}{0.6} = \frac{0.48}{0.6}$

$a = 0.8$

Check: $0.6a - 2.3 = -1.82$

$0.6(0.8) - 2.3 \stackrel{?}{=} -1.82$

$0.48 - 2.3 \stackrel{?}{=} -1.82$

$-1.82 = -1.82$

0.8 is the solution.

GUIDED PRACTICE

13. $3.2 \cdot 4t = (3.2 \cdot 4)t = 12.8t$

15. $8.06m(-7) = [8.06(-7)]m = -56.42m$

17. $-5(-6.7t) = [-5(-6.7)]t = 33.5t$

19. $2(4.4c)(3) = (2 \cdot 4.4 \cdot 3)c = 26.4c$

21. $3.7(4x + 3) = 3.7 \cdot 4x + 3.7 \cdot 3 = 14.8x + 11.1$

23. $-6(1.9m - 2.8) = (-6)1.9m - (-6)2.8$

$= -11.4m + 16.8$

25. $(y - 9.4)0.06 = y(0.06) - 9.4(0.06)$

$= 0.06y - 0.564$

27. $-2.5(-1.2t + 11)$

$= (-2.5)(-1.2t) + (-2.5)(11)$

$= 3t - 27.5$

29. $3.2x + 4.7x = (3.2 + 4.7)x = 7.9x$

31. $-2.4v+(-9.8v)+3.4v$
$=(-2.4-9.8+3.4)v=-8.8v$

33. $0.32b^2-0.59b^2=(0.32-0.59)b^2$
$=-0.27b^2$

35. $7.55a+3+1.12a-1.56$
$=(7.55+1.12)a+(3-1.56)$
$=8.67a+1.44$

37. $5.2(m+3)-1.2-(3.7m-4.1)$
$=5.2m+15.6-1.2-3.7m+4.1$
$=1.5m+18.5$

39. $3.8(d+2)-4.5-(1.8d-3.7)$
$=3.8d+7.6-4.5-1.8d+3.7$
$=2d+6.8$

41. $7(3.4y-1.1)-(6.7y+4.9)+y$
$=23.8y-7.7-6.7y-4.9+y$
$=18.1y-12.6$

43. $9(2.7b-4.2)-(14.2b+37.8)-b$
$=24.3b-37.8-14.2b-37.8-b$
$=9.1b-75.6$

45. $x+8.1=9.8$
$x+8.1-8.1=9.8-8.1$
$x=1.7$
Check: $x+8.1=9.8$
$1.7+8.1\stackrel{?}{=}9.8$
$9.8=9.8$

47. $6.75+y=8.99$
$6.75+y-6.75=8.99-6.75$
$y=2.24$
Check: $6.75+y=8.99$
$6.75+2.24\stackrel{?}{=}8.99$
$8.99=8.99$

49. $m-1.4=-5.8$
$m-1.4+1.4=-5.8+1.4$
$m=-4.4$
Check: $m-1.4=-5.8$
$-4.4-1.4\stackrel{?}{=}-5.8$
$-5.8=-5.8$

51. $7.08=c-0.03$
$7.08+0.03=c-0.03+0.03$
$7.11=c$
Check: $7.08=c-0.03$
$7.08\stackrel{?}{=}7.11-0.03$
$7.08=7.08$

53. $\frac{x}{6}=-4.7$
$6\cdot\frac{x}{6}=6(-4.7)$
$x=-28.2$
Check: $\frac{x}{6}=-4.7$
$\frac{-28.2}{6}\stackrel{?}{=}-4.7$
$-4.7=-4.7$

55. $\frac{y}{7}=0.06$
$7\cdot\frac{y}{7}=7(0.06)$
$y=0.42$
Check: $\frac{y}{7}=0.06$
$\frac{0.42}{7}\stackrel{?}{=}0.06$
$0.06=0.06$

57. $-3.51=-2.7x$

$$\frac{-3.51}{-2.7}=\frac{-2.7x}{-2.7}$$

$1.3=x$

Check: $-3.51=-2.7x$

$$-3.51\overset{?}{=}-2.7(1.3)$$

$$-3.51=-3.51$$

59. $-1.95=0.5f$

$$\frac{-1.95}{0.5}=\frac{0.5f}{0.5}$$

$-3.9=f$

Check: $-1.95=0.5f$

$$-1.95\overset{?}{=}0.5(-3.9)$$

$$-1.95=-1.95$$

61. $3.2x-3.01=-5.25$

$3.2x-3.01+3.01=-5.25+3.01$

$3.2x=-2.24$

$$\frac{3.2x}{3.2}=\frac{-2.24}{3.2}$$

$x=-0.7$

Check $3.2x-3.01=-5.25$

$$3.2(-0.7)-3.01\overset{?}{=}-5.25$$

$$-2.24-3.01\overset{?}{=}-5.25$$

$$-5.25=-5.25$$

63. $-1.5b+2.7=1.2$

$-1.5b+2.7-2.7=1.2-2.7$

$-1.5b=-1.5$

$$\frac{-1.5b}{-1.5}=\frac{-1.5}{-1.5}$$

$b=1$

Check: $-1.5b+2.7=1.2$

$$-1.5(1)+2.7\overset{?}{=}1.2$$

$$-1.5+2.7\overset{?}{=}1.2$$

$$1.2=1.2$$

65. $0.4y-1=0.8y+3.4$

$0.4y-1+1=0.8y+3.4+1$

$0.4y=0.8y+4.4$

$0.4y-0.8y=0.8y+4.4-0.8y$

$-0.4y=4.4$

$$\frac{-0.4y}{-0.4}=\frac{4.4}{-0.4}$$

$y=-11$

Check: $0.4y-1=0.8y+3.4$

$$0.4(-11)-1\overset{?}{=}0.8(-11)+3.4$$

$$-4.4-1\overset{?}{=}-8.8+3.4$$

$$-5.4=-5.4$$

67. $53.7r+2.6=46.3r+17.4$

$53.7r+2.6-2.6=46.3r+17.4-2.6$

$53.7r=46.3r+14.8$

$53.7r-46.3r=46.3r+14.8-46.3r$

$7.4r=14.8$

$$\frac{7.4r}{7.4}=\frac{14.8}{7.4}$$

$r=2$

Check: $53.7r+2.6=46.3r+17.4$

$$53.7(2)+2.6\overset{?}{=}46.3(2)+17.4$$

$$107.4+2.6\overset{?}{=}92.6+17.4$$

$$110=110$$

69. $1.2m-1.2=1.1m+1.4$

$1.2m-1.2+1.2=1.1m+1.4+1.2$

$1.2m=1.1m+2.6$

$1.2m-1.1m=1.1m+2.6-1.1m$

$0.1m=2.6$

$$\frac{0.1m}{0.1}=\frac{2.6}{0.1}$$

$m=26$

Check: $1.2m-1.2=1.1m+1.4$

$$1.2(26)-1.2\overset{?}{=}1.1(26)+1.4$$

$$31.2-1.2\overset{?}{=}28.6+1.4$$

$$30=30$$

71. $2.36x + 0.09 = 2.47x - 0.35$

$2.36x + 0.09 - 0.09 = 2.47x - 0.35 - 0.09$

$2.36x = 2.47x - 0.44$

$2.36x - 2.47x = 2.47x - 0.44 - 2.47x$

$-0.11x = -0.44$

$\frac{-0.11x}{-0.11} = \frac{-0.44}{-0.11}$

$x = 4$

Check: $2.36x + 0.09 = 2.47x - 0.35$

$2.36(4) + 0.09 \stackrel{?}{=} 2.47(4) - 0.35$

$9.44 + 0.09 \stackrel{?}{=} 9.88 - 0.35$

$9.53 = 9.53$

73. $3(y - 1.1) + 0.5 = -0.4$

$3y - 3.3 + 0.5 = -0.4$

$3y - 2.8 = -0.4$

$3y - 2.8 + 2.8 = -0.4 + 2.8$

$3y = 2.4$

$\frac{3y}{3} = \frac{2.4}{3}$

$y = 0.8$

Check: $3(y - 1.1) + 0.5 = -0.4$

$3(0.8 - 1.1) + 0.5 \stackrel{?}{=} -0.4$

$3(-0.3) + 0.5 \stackrel{?}{=} -0.4$

$-0.9 + 0.5 \stackrel{?}{=} -0.4$

$-0.4 = -0.4$

75. $8(m - 1.9) + 17.1 = -14.5$

$8m - 15.2 + 17.1 = -14.5$

$8m + 1.9 = -14.5$

$8m + 1.9 - 1.9 = -14.5 - 1.9$

$8m = -16.4$

$\frac{8m}{8} = \frac{-16.4}{8}$

$m = -2.05$

Check: $8(m - 1.9) + 17.1 = -14.5$

$8(-2.05 - 1.9) + 17.1 \stackrel{?}{=} -14.5$

$8(-3.95) + 17.1 \stackrel{?}{=} -14.5$

$-31.6 + 17.1 \stackrel{?}{=} -14.5$

$-14.5 = -14.5$

TRY IT YOURSELF

77. $2x = -8.72$

$\frac{2x}{2} = \frac{-8.72}{2}$

$x = -4.36$

79. $2.3 = 3(a - 1.1) + 3.2$

$2.3 = 3a - 3.3 + 3.2$

$2.3 = 3a - 0.1$

$2.3 + 0.1 = 3a - 0.1 + 0.1$

$2.4 = 3a$

$\frac{2.4}{3} = \frac{3a}{3}$

$0.8 = a$

81. $5.6x - 8.3 - 6.1x + 12.2$

$= (5.6 - 6.1)x + (-8.3 + 12.2) = -0.5x + 3.9$

83. $1.5x - 0.3 = 1.3x + 2.7$

$1.5x - 0.3 + 0.3 = 1.3x + 2.7 + 0.3$

$1.5x = 1.3x + 3$

$1.5x - 1.3x = 1.3x + 3 - 1.3x$

$0.2x = 3$

$\frac{0.2x}{0.2} = \frac{3}{0.2}$

$x = 15$

85. $5.3m - 1.2 = 4.9m - 1.6$

$5.3m - 1.2 + 1.2 = 4.9m - 1.6 + 1.2$

$5.3m = 4.9m - 0.4$

$5.3m - 4.9m = 4.9m - 0.4 - 4.9m$

$0.4m = -0.4$

$\frac{0.4m}{0.4} = \frac{-0.4}{0.4}$

$m = -1$

87. $1.2x - 1.3 = 2.4x + 0.02$

$1.2x - 1.3 + 1.3 = 2.4x + 0.02 + 1.3$

$1.2x = 2.4x + 1.32$

$1.2x - 2.4x = 2.4x + 1.32 - 2.4x$

$-1.2x = 1.32$

$\frac{-1.2x}{-1.2} = \frac{1.32}{-1.2}$

$x = -1.1$

89. $-6.7 = \frac{x}{2.04} - 2.7$

$-6.7 + 2.7 = \frac{x}{2.04} - 2.7 + 2.7$

$-4 = \frac{x}{2.04}$

$-4(2.04) = \frac{x}{2.04}(2.04)$

$-8.16 = x$

91. $0.06x + 0.09(100 - x) = 8.85$

$0.06x + 9 - 0.09x = 8.85$

$0.06x + 9 - 0.09x - 9 = 8.85 - 9$

$-0.03x = -0.15$

$\frac{-0.03x}{-0.03} = \frac{-0.15}{-0.03}$

$x = 5$

93. $3.1r - 5.5r - 1.3r$

$= (3.1 - 5.5 - 1.3)r = -3.7r$

95. $\frac{1}{3}x = -7.06$

$(3)\frac{1}{3}x = (3)(-7.06)$

$x = -21.18$

97. $2(x + 3.9) = 3.4$

$2x + 7.8 = 3.4$

$2x + 7.8 - 7.8 = 3.4 - 7.8$

$2x = -4.4$

$\frac{2x}{2} = \frac{-4.4}{2}$

$x = -2.2$

99. $7.75 = n - (-7.85)$

$7.75 = n + 7.85$

$7.75 - 7.85 = n + 7.85 - 7.85$

$-0.1 = n$

101. $-5.6 - h = -17.1$

$-5.6 - h + 5.6 = -17.1 + 5.6$

$-h = -11.5$

$\frac{-h}{-1} = \frac{-11.5}{-1}$

$h = 11.5$

103. $\frac{x}{100} = 0.004$

$100 \cdot \frac{x}{100} = 100 \cdot 0.004$

$x = 0.4$

APPLICATIONS

105. Analyze

Her base pay is \$48 a day.
She makes 95¢ for each signature.
She wants to make \$200 a day.
How many signatures does she need?

Assign

Let x = the number of signatures she needs

Form

Her base pay	+	the amount per signature	x	the number of signatures	=\$200

Solve

$48+0.95x=200$

$48+0.95x-48=200-48$

$0.95x=152$

$\frac{0.95x}{0.95}=\frac{152}{0.95}$

$x=160$

State

She needs to collect 160 signatures to make \$200 a day.

Check

$0.95\cdot160+48=152+48=200$

107. x = the amount of federal funding requested

$6.8+12.5+x=27.9$

$19.3+x=27.9$

$19.3+x-19.3=27.9-19.3$

$x=8.6$

They should ask for \$8.6 million.

109. x = the original GPA

$x-0.18=3.09$

$x-0.18+0.18=3.09+0.18$

$x=3.27$

She previously had a 3.27 GPA.

111. x = the maximum number of words printed

$20+0.15x=50$

$20+0.15x-20=50-20$

$0.15x=30$

$\frac{0.15x}{0.15}=\frac{30}{0.15}$

$x=200$

200 words is the maximum.

113. x = the number of balloons used

$100+0.08x=300$

$100+0.08x-100=300-100$

$0.08x=200$

$\frac{0.08x}{0.08}=\frac{200}{0.08}$

$x=2500$

They will use 2,500 balloons.

115. x = the number of hours of tutoring affordable

$60+9.75x=450$

$60+9.75x-60=450-60$

$9.75x=390$

$\frac{9.75x}{9.75}=\frac{390}{9.75}$

$x=40$

They can get 40 hours of tutoring.

117. x = the number of cassettes transferred

$$14.50x + 31 = 350$$
$$14.50x + 31 - 31 = 350 - 31$$
$$14.50x = 319$$
$$\frac{14.50x}{14.50} = \frac{319}{14.50}$$
$$x = 22$$

They can convert 22 VHS tapes.

WRITING

119. Because multiplying by 100 just involves moving the decimal point 2 units to the right.

REVIEW

121. a. $\frac{3}{4} + \frac{2}{3} = \frac{9}{12} + \frac{8}{12} = \frac{17}{12} = 1\frac{5}{12}$

b. $\frac{1}{x} + \frac{2}{3} = \frac{3}{3x} + \frac{2x}{3x} = \frac{3+2x}{3x}$

123. a. $\frac{4}{5} - \frac{1}{7} = \frac{28}{35} - \frac{5}{35} = \frac{23}{35}$

b. $\frac{4}{5} - \frac{n}{7} = \frac{28}{35} - \frac{5n}{35} = \frac{28-5n}{35}$

Chapter 5 Review

1. a. $0.67; \frac{67}{100}$

b.

2. a. 7 hundredths b. 3

c. 8 d. 5 ten-thousandths

3. $10 + 6 + \frac{4}{10} + \frac{5}{100} + \frac{2}{1,000} + \frac{3}{10,000}$

4. two and three tenths; $2\frac{3}{10}$

5. negative six hundred fifteen and fifty-nine hundredths

$-615\frac{59}{100}$

6. six hundred one ten-thousandths; $\frac{601}{10,000}$

7. one hundred-thousandth; $\frac{1}{100,000}$

8. 100.61

9. 11.997

10. 301.000016

11. $5.68 < 5.75$

12. $106.8199 < 106.82$

13. $-78.23 > -78.303$

14. $-555.098 > -555.0991$

15.

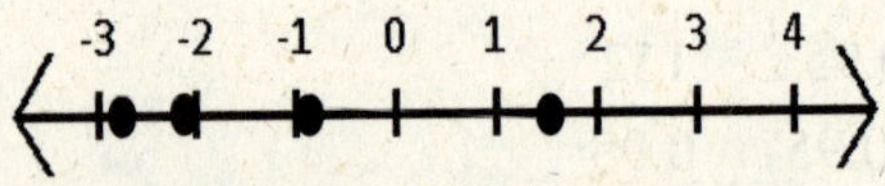

16. a. true b. false c. true d. true

17. 16.100

18. -0.1

19. 11.3150

20. 0.222228

21. \$0.67

22. \$13

23. Washington, Diaz, Chou, Singh, Gerbac

24. Sunday: 1.8

Monday: 0.6

Tuesday: 2.4

Wednesday: 3.8

25.

$$\begin{array}{r} {}^{1\,1}\\ 19.5 \\ 34.4 \\ +\underline{12.8} \\ 66.7 \end{array}$$

26.

$$\begin{array}{r} \not{4}\,\not{5}.\not{8}\,\not{0}\,\not{0} \\ -17.372 \\ \hline 28.428 \end{array}$$

(borrows: 3 15 7 9/10 10)

27.

$$\begin{array}{r} 3\not{4}\,\not{5}.\not{2}\,\not{1}\,4 \\ -27.390 \\ \hline 317.824 \end{array}$$

(borrows: 3 14 11 11)

28.

$$\begin{array}{r} {}^{2\,2\,1}\\ 8.61 \\ 5.97 \\ +\ 9.72 \\ \hline 24.30 \end{array}$$

29. $-16.1+8.4=-7.7$

30. $-4.8-(-7.9)=-4.8+7.9=3.1$

31. $-3.55+(-1.25)=-4.8$

32. $-15.1-13.99=-29.09$

33. $-8.8+(-7.3-9.5)=-8.8+(-16.8)=-25.6$

34. $(5-0.096)-(-0.035)=4.904+0.035$
$=4.939$

35. a. $610+150=760$

b. $300-20=280$

36.

$$\begin{array}{r} {}^{1\,2\,2}\\ 1.55 \\ 1.95 \\ 1.35 \\ 1.75 \\ 2.15 \\ +\ \underline{2.00} \\ 10.75 \text{ mm} \end{array}$$

37.

$$\begin{array}{r} \not{5}\,\not{2}.\not{2}\,\not{0} \\ -\ \ 3.99 \\ \hline \$48.21 \end{array}$$

(borrows: 4 11/12 11/12 10)

38. $13.4-(2.75+2.5)=13.4-5.25=8.15in$

39.

$$\begin{array}{r} 2.3 \\ \times\underline{6.9} \\ 207 \\ +\underline{1380} \\ 15.87 \end{array}$$

40.

$$\begin{array}{r} 1.7 \\ \times\underline{0.004} \\ 0.0068 \end{array}$$

41. $1.5(-0.9)=-1.35$

42. $(-0.003)(-0.02)=0.00006$

43. $1{,}000(90.1452)=90{,}145.2$

44. $0.001(2.897)=0.002897$

45. $(0.2)^2=(0.2)(0.2)=0.04$

46. $(0.6+0.7)^2-(-3)(-4.1)=(1.3)^2-(12.3)$
$=1.69-12.3=-10.61$

47. $(-3.3)^2(0.00001)=10.89(0.00001)$
$=0.0001089$

48. $(0.1)^3 + 2|-45.63 - 12.24|$
$= 0.001 + 2|-57.87| = 0.001 + 2 \cdot 57.87$
$= 0.001 + 115.74 = 115.741$

49. a. $9,600,000 km^2$

b. 61,400,000,000 trees

50. a. $200 \cdot 8 = 1,600$

b. $12.4 \cdot 7.4 = 91.76$

51. $A = 70.05 + 70.05 \cdot 0.08 \cdot 5$
$= 70.05 + 5.604 \cdot 5 = 70.05 + 28.02$
$= 98.07$

52.
$$\begin{array}{r} 12.95 \\ \times \underline{1.5} \\ 6475 \\ +\underline{12950} \\ 19.425 \end{array}$$
Approximately $19.43

53. $0.03 + 3(0.015) - 0.005$
$= 0.03 + 0.045 - 0.005$
$= 0.075 - 0.005 = 0.07 in$

54. Length = $11 - 1 - 0.6 = 10 - 0.6 = 9.4 in$

Width = $8.5 - 0.7 - 0.5 = 7.8 - 0.5 = 7.3 in$

Area = $9.4(7.3) = 68.62 in^2$

55. $10.516 \div 4.4 = 2.39$
$$\begin{array}{r} 2.39 \\ 44\overline{)105.16} \\ \underline{88} \\ 171 \\ \underline{132} \\ 396 \\ \underline{396} \\ 0 \end{array}$$
Check: $4.4(2.39) = 10.516$

56.
$$\begin{array}{r} 1.29 \\ 23\overline{)29.67} \\ -\underline{23} \\ 66 \\ -\underline{46} \\ 207 \\ -\underline{207} \\ 0 \end{array}$$
Check: $23(1.29) = 29.67$

57. $-80.625 \div 12.9 = -806.25 \div 129$
$$\begin{array}{r} -6.25 \\ 129\overline{)-806.25} \\ -\underline{774} \\ 322 \\ -\underline{258} \\ 645 \\ -\underline{645} \\ 0 \end{array}$$
Check: $-6.25(12.9) = -80.625$

58. $\dfrac{0.0742}{1.4} = \dfrac{0.742}{14}$
$$\begin{array}{r} 0.053 \\ 14\overline{)0.742} \\ -\underline{70} \\ 42 \\ -\underline{42} \\ 0 \end{array}$$
Check: $0.053(1.4) = 0.0742$

59. $\dfrac{15.75}{0.25}=\dfrac{1575}{25}$

$$\begin{array}{r} 63 \\ 25\overline{)1575} \\ \underline{-150} \\ 75 \\ \underline{-75} \\ 0 \end{array}$$

Check: $0.25(63)=15.75$

60. $\dfrac{-0.003726}{-0.0046}=\dfrac{37.26}{46}$

$$\begin{array}{r} 0.81 \\ 46\overline{)37.26} \\ \underline{-368} \\ 46 \\ \underline{-46} \\ 0 \end{array}$$

Check: $0.81(-0.0046)=-0.003726$

61. $89.76\div 1{,}000=0.08976$

Check: $1{,}000(0.08976)=89.76$

62. $\dfrac{0.0112}{-10}=-0.00112$

Check: $-0.00112(-10)=0.0112$

63. $-0.8765\div(-0.001)=876.5$

Check: $876.5(-0.001)=-0.8765$

64. $77.021\div 0.0001=770{,}210$

Check: $770{,}210(0.0001)=77.021$

65. $4{,}800\div 40=480\div 4=120$

66. $27{,}000\div 9=3{,}000$

67.

$$\begin{array}{r} 12.94 \\ 61\overline{)789.80} \\ \underline{-61} \\ 179 \\ \underline{-122} \\ 578 \\ \underline{-549} \\ 290 \end{array}$$

To the nearest tenth, 12.9

68.

$$\begin{array}{r} 776.857 \\ 7\overline{)5438.000} \\ \underline{-49} \\ 53 \\ \underline{-49} \\ 48 \\ \underline{-42} \\ 60 \\ \underline{-56} \\ 40 \\ \underline{-35} \\ 50 \end{array}$$

To the nearest hundredth, -776.86

69. $\dfrac{(1.4)^2-2(-4.6)}{0.5+0.3}=\dfrac{1.96+9.2}{0.8}$

$=\dfrac{11.16}{0.8}=\dfrac{111.6}{8}=13.95$

70. $C=\dfrac{5}{9}(68.9-32)$

$C=\dfrac{5}{\cancel{9}_{1}}\cdot\dfrac{\overset{4.1}{\cancel{36.9}}}{1}=5\cdot 4.1=20.5$

71.

$$\begin{array}{r} 8.34 \\ 5\overline{)41.70} \\ \underline{-40} \\ 17 \\ \underline{-15} \\ 20 \\ \underline{-20} \\ 0 \end{array}$$

\$8.34 per person

72.

$$\frac{0.44+0.50+0.46+0.52+0.63}{5}$$

$$=\frac{2.55}{5}=0.51ppb$$

73.

$$\frac{15.4}{1.1}=\frac{154}{11}$$

$$\begin{array}{r} 14 \\ 11\overline{)154} \\ \underline{-14} \\ 14 \\ \underline{-14} \\ 0 \end{array}$$

14 servings per container

74.

$$\frac{0.2375}{0.025}=\frac{237.5}{25}$$

$$\begin{array}{r} 9.5 \\ 25\overline{)237.5} \\ \underline{-225} \\ 125 \\ \underline{-125} \\ 0 \end{array}$$

9.5 revolutions needed

75.

$$\begin{array}{r} 0.875 \\ 8\overline{)7.000} \\ \underline{-64} \\ 60 \\ \underline{-56} \\ 40 \\ \underline{-40} \\ 0 \end{array}$$

$$\frac{7}{8}=0.875$$

76.

$$\begin{array}{r} 0.4 \\ 5\overline{)2.0} \\ \underline{-20} \\ 0 \end{array}$$

$$-\frac{2}{5}=-0.4$$

77.

$$\begin{array}{r} 0.5625 \\ 16\overline{)9.0000} \\ \underline{-80} \\ 100 \\ \underline{-96} \\ 40 \\ \underline{-32} \\ 80 \\ \underline{-80} \\ 0 \end{array}$$

$$\frac{9}{16}=0.5625$$

78.

$$\begin{array}{r} 0.06 \\ 50\overline{)3.00} \\ \underline{-300} \\ 0 \end{array}$$

$$\frac{3}{50}=0.06$$

79.

$$\begin{array}{r} 0.5454 \\ 11\overline{)6.0000} \\ -\underline{55} \\ 50 \\ -\underline{44} \\ 60 \\ -\underline{55} \\ 50 \\ -\underline{44} \\ 6 \end{array}$$

$$\frac{6}{11} = 0.\overline{54}$$

80.

$$\begin{array}{r} 1.33 \\ 3\overline{)4.00} \\ -\underline{3} \\ 10 \\ -\underline{9} \\ 10 \\ -\underline{9} \\ 1 \end{array}$$

$$-\frac{4}{3} = -1.\overline{3}$$

81.

$$\begin{array}{r} 0.056 \\ 125\overline{)7.000} \\ -\underline{625} \\ 750 \\ -\underline{750} \\ 0 \end{array}$$

$$3\frac{7}{125} = 3.056$$

82.

$$\begin{array}{r} 0.577 \\ 45\overline{)26.000} \\ -\underline{225} \\ 350 \\ -\underline{315} \\ 350 \\ -\underline{315} \\ 350 \end{array}$$

$$\frac{26}{45} = 0.5\overline{7}$$

83.

$$\begin{array}{r} 0.575 \\ 33\overline{)19.000} \\ -\underline{165} \\ 250 \\ -\underline{231} \\ 190 \end{array}$$

$$\frac{19}{33} \approx 0.58$$

84.

$$\begin{array}{r} 0.033 \\ 30\overline{)1.000} \\ -\underline{90} \\ 100 \\ -\underline{90} \\ 100 \end{array}$$

$$\frac{31}{30} = 1\frac{1}{30} \approx 1.03$$

85.

$$\frac{13}{25}\square 0.499$$

$$0.52 > 0.499$$

86.

$$-\frac{4}{15}\square -0.2\overline{6}$$

$$-0.2\overline{6} = -0.2\overline{6}$$

87.

$$\frac{10}{33} = .\overline{30}$$

$$0.3, \frac{10}{33}, 0.\overline{3}$$

88. 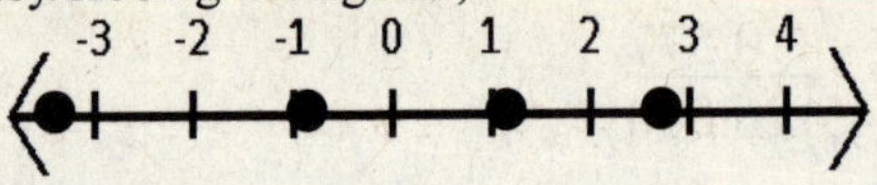

89. $\frac{1}{3}+0.4=\frac{1}{3}+\frac{4}{10}=\frac{1}{3}+\frac{2}{5}$

$=\frac{5}{15}+\frac{6}{15}=\frac{11}{15}$

90. $\frac{5}{6}+0.19=\frac{5}{6}+\frac{19}{100}=\frac{500}{600}+\frac{114}{600}$

$=\frac{614}{600}=1\frac{14}{600}=1\frac{7}{300}$

91. $\frac{1}{2}(9.7+8.9)(10)=0.5(18.6)(10)$

$=9.3(10)=93$

92. $7.5-(0.78)\left(\frac{1}{2}\right)^2=7.5-(0.78)(.5)^2$

$=7.5-0.78(0.25)=7.5-0.195$

$=7.305$

93. $A=\frac{1}{2}bh=0.5(6.4)(10.9)$

$=3.2(10.9)=34.88in^2$

94. $\frac{3}{4}(13.80)+\frac{1}{3}(35.70)$

$=10.35+11.90=\$22.25$

95. -5 , 5

96. $\sqrt{49}=7$ because $7^2=49$.

97. $\sqrt{49}=7$

98. $-\sqrt{16}=-4$

99. $\sqrt{\frac{64}{169}}=\frac{8}{13}$

100. $\sqrt{0.81}=0.9$

101. 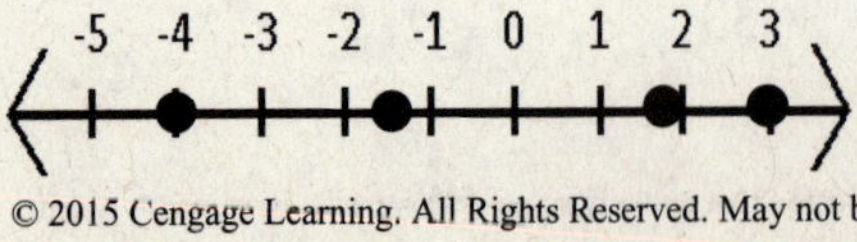

102. a. 4.36 b. 24.45 c. 3.57

103. $-3\sqrt{49}-6=-3\cdot 7-6=-21-6=-27$

104. $\sqrt{\frac{100}{9}}+\sqrt{225}=\frac{10}{3}+15=3\frac{1}{3}+15=18\frac{1}{3}$

105. $40+6\left[5^2-(7-2)\sqrt{16}\right]$

$=40+6[25-5\cdot 4]=40+6[25-20]$

$=40+6[5]=40+30=70$

106. $1-7\left[6^2+(1+2)\sqrt{81}\right]$

$=1-7[36+3\cdot 9]=1-7[36+27]$

$=1-7[63]=1-441=-440$

107. $b=\sqrt{c^2-a^2}=\sqrt{17^2-15^2}$

$=\sqrt{289-225}=\sqrt{64}=8$

108. $\sqrt{1,089}=33in$

109. $\sqrt{81}<\sqrt{83}<\sqrt{100}$

$9<\sqrt{83}<10$

110. $(2.646)^2=7.001316\neq 7$

It is close, but not an exact result.

111. $4.6(7w)=(4.6\cdot 7)w=32.2w$

112. $3.8(-2t)(4)=(-3.8\cdot 2\cdot 4)t=-30.4t$

113. $2.7(2q-1)=2.7(2q)-2.7(1)=5.4q-2.7$

114. $-7(2.9x-8.4)=-7\cdot 2.9x-(-7)(8.4)$

$=-20.3x+58.8$

115. $8.2p+5.9p=(8.2+5.9)p=14.1p$

116. $0.57m^2-0.69m^2=(0.57-0.69)m^2=-0.12m^2$

117. $5.7a-12.4-2.9a=5.7a-2.9a-12.4$

$=2.8a-12.4$

118. $2(0.3t-0.4)+3(0.8t-0.2)$
$=0.6t-0.8+2.4t-0.6=3t-1.4$

119. $y+12.4=-6.01$
$y+12.4-12.4=-6.01-12.4$
$y=-18.41$
Check: $y+12.4=-6.01$
$-18.41+12.4\stackrel{?}{=}-6.01$
$-6.01=-6.01$

120. $x-0.23=5$
$x-0.23+0.23=5+0.23$
$x=5.23$
Check: $x-0.23=5$
$5.23-0.23\stackrel{?}{=}5$
$5=5$

121. $\frac{x}{1.78}=-3$
$1.78\left(\frac{x}{1.78}\right)=1.78(-3)$
$x=-5.34$
Check: $\frac{x}{1.78}=-3$
$\frac{-5.34}{1.78}\stackrel{?}{=}-3$
$-3=-3$

122. $-1.61b=-27.37$
$\frac{-1.61b}{-1.61}=\frac{-27.37}{-1.61}$
$b=17$
Check: $-1.61b=-27.37$
$-1.61(17)=-27.37$
$-27.37=-27.37$

123. $1.7y+1.24=-1.4y-0.62$
$1.7y+1.24+1.4y=-1.4y-0.62+1.4y$
$3.1y+1.24=-0.62$
$3.1y+1.24-1.24=-0.62-1.24$
$3.1y=-1.86$
$\frac{3.1y}{3.1}=\frac{-1.86}{3.1}$
$y=-0.6$
Check:
$1.7y+1.24=-1.4y-0.62$
$1.7(-0.6)+1.24\stackrel{?}{=}-1.4(-0.6)-0.62$
$-1.02+1.24\stackrel{?}{=}0.84-0.62$
$0.22=0.22$

124. $0.08(x-100)=32.5-0.07x$
$0.08x-8=32.5-0.07x$
$0.08x-8+8=32.5-0.07x+8$
$0.08x=40.5-0.07x$
$0.08x+0.07x=40.5-0.07x+0.07x$
$0.15x=40.5$
$\frac{0.15x}{0.15}=\frac{40.5}{0.15}$
$x=270$
Check:
$0.08(x-100)=32.5-0.07x$
$0.08(270-100)\stackrel{?}{=}32.5-0.07(270)$
$0.08(170)\stackrel{?}{=}32.5-18.9$
$13.6=13.6$

125. x = the cost in 2007
$x+0.82=3.63$
$x+0.82-0.82=3.63-0.82$
$x=2.81$
In 2007 gas cost \$2.81 / gallon.

126. g = the number of games

$4+4.50g=40$

$4+4.5g-4=40-4$

$4.5g=36$

$\frac{4.5g}{4.5}=\frac{36}{4.5}$

$x=8$

8 games can be bowled for $40.

Chapter 5 Test

1. a. addend

addend

sum

b. minuend

subtrahend

difference

c. factor

factor

product

d. quotient

divisor dividend

e. 0.6666.... and 0.83333... are examples of repeating decimals.

f. The $\sqrt{}$ symbol is called a radical symbol.

g. When we write $4.1x+2.7x$ as $6.8x$, we say we have combined like terms.

h. To solve an equation means to find all the values of the variable that make the equations true.

2. $\frac{79}{100}$; 0.79

3. a. 1 thousandth b. 4

c. 6 d. 2 tens

4. Selway, Monroe, Paston, Covington, Cadia

5. 4,519.0027

6. a. $80+\frac{7}{10}+\frac{4}{100}$; eighty and seventy-four hundredths; $80\frac{74}{100}$

b. $\frac{8}{100}+\frac{1}{10{,}000}+\frac{3}{100{,}000}$; eight thousand thirteen hundred-thousandths; $\frac{8{,}013}{100{,}000}$

7. a. 461.7

b. 2,733.050

c. -1.983373

8. $0.65

9.

$$\begin{array}{r} {}^{1\,1\;\;1} \\ 4.560 \\ 2.000 \\ 0.896 \\ +\ 3.300 \\ \hline 10.756 \end{array}$$

10.

$$\begin{array}{r} 45.200 \\ -39.079 \\ \hline 6.121 \end{array}$$

11. $(0.32)^2=(0.32)(0.32)$

$$\begin{array}{r} 0.32 \\ \times 0.32 \\ \hline 64 \\ +\ 960 \\ \hline 0.1024 \end{array}$$

12. $\dfrac{0.1368}{0.24}=\dfrac{13.68}{24}$

$$\begin{array}{r} 0.57 \\ 24\overline{)13.68} \\ -\underline{120} \\ 168 \\ -\underline{168} \\ 0 \end{array}$$

13. $-3.2(0.6)=-1.92$

14.
$$\begin{array}{r} 8.7 \\ \times\ \underline{0.004} \\ 0.0348 \end{array}$$

15.
$$\begin{array}{r} 1.181 \\ 11\overline{)13.000} \\ -\underline{11} \\ 20 \\ -\underline{11} \\ 90 \\ -\underline{88} \\ 20 \end{array}$$

$\dfrac{13}{11}=1.\overline{18}$

16. $-2.4-(-1.6)=-2.4+1.6=-0.8$

17. $12.146\div5.3=121.46\div53$

$$\begin{array}{r} 2.290 \\ 53\overline{)121.460} \\ -\underline{106} \\ 154 \\ -\underline{106} \\ 486 \\ -\underline{477} \\ 90 \end{array}$$

$\dfrac{121.46}{-5.3}\approx-2.29$

18. a. $30\cdot7=210$

b. $4000\div20=200$

19. a. Move the decimal 3 left: 0.567909

b. Move the decimal 2 right: 0.458

20. 61,400,000,000

21. $A=l\cdot w$

$A=2.5\cdot0.5$

$A=1.25mi^2$

22. $2.26\div565=0.004in$

23. Friday: $\$130.25+\$162.25=\$292.50$

Saturday: $\$140.50+\$175.75=\$316.25$

The Saturday sales were higher.

$\$316.25-\$292.50=\$23.75$ higher.

24. $8.33\left(2\dfrac{1}{2}\right)=8.33(2.5)=20.825lb$

25. $C=2\pi r$

$C=2\cdot3.14(1.7)$

$C=6.28(1.7)$

$C=10.676$

26. a. $\dfrac{17}{50}=\dfrac{17}{50}\cdot\dfrac{2}{2}=\dfrac{34}{100}=0.34$

b.
$$\begin{array}{r} 0.4166 \\ 12\overline{)5.0000} \\ -\underline{48} \\ 20 \\ -\underline{12} \\ 80 \\ -\underline{72} \\ 80 \\ -\underline{72} \end{array}$$

$\dfrac{5}{12}=0.41\overline{6}$

27. $4.1-(3.2)(0.4)^2=4.1-(3.2)(0.16)$
$=4.1-0.512=3.588$

28. $\left(\frac{2}{5}\right)^2+6\left|-6.2-3\frac{1}{4}\right|$
$=(0.4)^2+6|-6.2-3.25|$
$=0.16+6|-9.45|=0.16+6\cdot 9.45$
$=0.16+56.7=56.86$

29. $8-2\left(2^4-60+6\sqrt{81}\right)$
$=8-2(16-60+6\cdot 9)$
$=8-2(-44+54)=8-2(10)$
$=8-20=-12$

30. $\frac{2}{3}+0.7=\frac{2}{3}+\frac{7}{10}=\frac{20}{30}+\frac{21}{30}=\frac{41}{30}$

31. a.

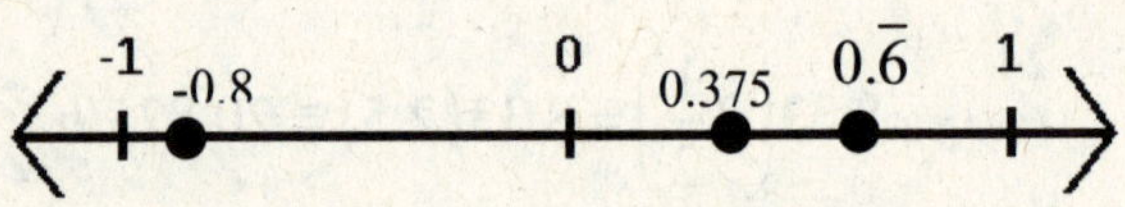

b.

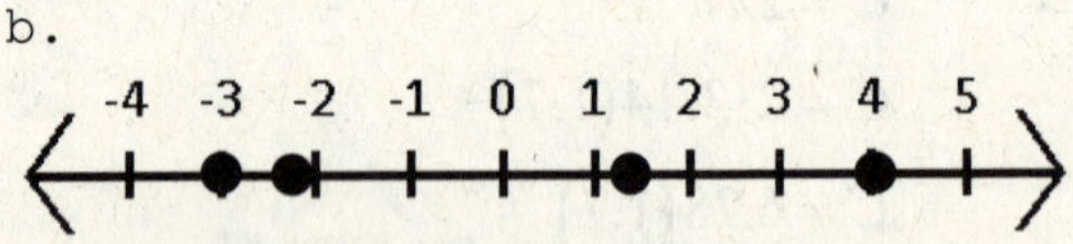

32. $\frac{3}{4}(\$5.60)+\frac{1}{3}(\$4.35)=\$4.20+\$1.45$
$=\$5.65$

33. $\sqrt{a^2+b^2}=\sqrt{12^2+35^2}$
$=\sqrt{144+1225}=\sqrt{1369}=37$

34. a. $-\frac{27}{25}=-\frac{27}{25}\cdot\frac{4}{4}=-\frac{108}{100}=-1.08$

b. $\frac{9}{16}=\frac{9}{16}\cdot\frac{625}{625}=\frac{5625}{10{,}000}=0.5625$

$2\frac{9}{16}=2.5624$

35. $\sqrt{144}=12$ because $12^2=144$.

36. a. $-6.78>-6.79$

b. $0.3<\frac{3}{8}=0.375$

c. $\sqrt{\frac{16}{81}}=\frac{4}{9}=0.\overline{4}$

d. $0.45<0.\overline{45}$

37. $-2\sqrt{25}+3\sqrt{49}=-2\cdot 5+3\cdot 7$
$=-10+21=11$

38. $\sqrt{\frac{1}{36}}-\sqrt{\frac{1}{25}}=\frac{1}{6}-\frac{1}{5}=\frac{1}{6}\cdot\frac{5}{5}-\frac{1}{5}\cdot\frac{6}{6}$
$=\frac{5}{30}-\frac{6}{30}=-\frac{1}{30}$

39. a. $-\sqrt{0.04}=-0.2$

b. $\sqrt{1.69}=1.3$

c. $\sqrt{225}=15$

d. $-\sqrt{121}=-11$

40. It is smaller because it is farther to the left on the number line.

41. a. $9(1.6t)=(9\cdot 1.6)t=14.4t$

b. $-3(4.7a)(2)=(-3\cdot 4.7\cdot 2)a=-28.2a$

42. a. $-5.7c+4.5c=(-5.7+4.5)c=-1.2c$

b. $2.1(x-3)+3.1(x-4)$
$=2.1x-6.3+3.1x-12.4=5.2x-18.7$

43. $-2.4d=16.8$
$\frac{-2.4d}{-2.4}=\frac{16.8}{-2.4}$
$d=-7$
Check: $-2.4d=16.8$
$-2.4(-7)\stackrel{?}{=}16.8$
$16.8=16.8$

44.
$$\frac{x}{2.04}+2.2=6.7$$
$$\frac{x}{2.04}+2.2-2.2=6.7-2.2$$
$$\frac{x}{2.04}=4.5$$
$$2.04\left(\frac{x}{2.04}\right)=2.04(4.5)$$
$$x=9.18$$
Check: $\frac{x}{2.04}+2.2=6.7$
$$\frac{9.18}{2.04}+2.2\stackrel{?}{=}6.7$$
$$4.5+2.2\stackrel{?}{=}6.7$$
$$6.7=6.7$$

45.
$$2(y+4.3)-0.3=7.1$$
$$2y+8.6-0.3=7.1$$
$$2y+8.3=7.1$$
$$2y+8.3-8.3=7.1-8.3$$
$$2y=-1.2$$
$$\frac{2y}{2}=\frac{-1.2}{2}$$
$$y=-0.6$$
Check: $2(y+4.3)-0.3=7.1$
$$2[-0.6+4.3]-0.3\stackrel{?}{=}7.1$$
$$2(3.7)-0.3\stackrel{?}{=}7.1$$
$$7.4-0.3\stackrel{?}{=}7.1$$
$$7.1=7.1$$

46.
$$0.3x+6.9=1.6x-8.7$$
$$0.3x+6.9-0.3x=1.6x-8.7-0.3x$$
$$6.9=1.3x-8.7$$
$$6.9+8.7=1.3x-8.7+8.7$$
$$15.6=1.3x$$
$$\frac{15.6}{1.3}=\frac{1.3x}{1.3}$$
$$12=x$$
Check: $0.3x+6.9=1.6x-8.7$
$$0.3(12)+6.9\stackrel{?}{=}1.6(12)-8.7$$
$$3.6+6.9\stackrel{?}{=}19.2-8.7$$
$$10.5=10.5$$

47.
$$4.37-1.86-2.09=C$$
$$2.51-2.09=C$$
$$0.42g=C$$

48. a = the number of announcements they can print
$$24+0.95a=100$$
$$24+0.95a-24=100-24$$
$$0.95a=76$$
$$\frac{0.95a}{0.95}=\frac{76}{0.95}$$
$$a=80$$
They can print 80 announcements.

Chapters 1 – 5 Cumulative Review

1. a. one hundred fifty-four thousand, three hundred two

b. 100,000 + 50,000 + 4,000 + 300 + 2

2. $(3+4)+5=3+(4+5)$

3.
$$\begin{array}{r} {}^{1\,1\;\,1\,1}\\ 9{,}339 \\ 471 \\ +\ 6{,}883 \\ \hline 16{,}693 \end{array}$$

4.

$$\begin{array}{r} \overset{2}{\cancel{3}}\,\overset{\overset{9}{\cancel{10}}}{\cancel{0}}\,\overset{11}{\cancel{1}} \\ -1\ 9\ 9 \\ \hline 1\ 0\ 2 \end{array}$$

5.

$$\begin{array}{r} 275 \\ \times 275 \\ \hline 1,375 \\ 19,250 \\ +55,000 \\ \hline 75,625\,ft^2 \end{array}$$

6.

$$\begin{array}{r} 27 \\ 43\overline{)1203} \\ -86 \\ \hline 343 \\ -301 \\ \hline 42 \end{array}$$

$27R42$

7. 1, 2, 4, 5, 10, 20

8.

$220 = 2\cdot 110 = 2\cdot 2\cdot 55$

$= 2\cdot 2\cdot 5\cdot 11 = 2^2\cdot 5\cdot 11$

9.

$100 = 2\cdot 2\cdot 5\cdot 5$

$120 = 2\cdot 2\cdot 2\cdot 3\cdot 5$

$LCM(100,120) = 2\cdot 2\cdot 2\cdot 3\cdot 5\cdot 5 = 600$

$GCF(100,120) = 2\cdot 2\cdot 5 = 20$

10.

$$\frac{7+1+8+2+2}{5} = \frac{20}{5} = 4$$

11.

a. $x+19=285$

$x+19-19=285-19$

$x=266$

Check: $x+19=285$

$266+19\overset{?}{=}285$

$285=285$

b. $19x=285$

$\frac{19x}{19}=\frac{285}{19}$

$x=15$

Check: $19x=285$

$19(15)\overset{?}{=}285$

$285=285$

12.

$|-50|\,\square\,-(-40)$

$50>40$

13. $-8+(-5)=-13$

14. Subtraction is the same as <u>adding</u> the opposite.

15. $72-(-11)=72+11=83°F$ increase

16.

$-3(-5)(2)(-9)=15(2)(-9)$

$=30(-9)=-270$

17.

$(-1)^5=(-1)(-1)(-1)(-1)(-1)$

$=1\cdot 1(-1)=-1$

18. $-350(6)=-2,100\,ft$

19. $\frac{-15}{-5}=3 \Leftrightarrow 3(-5)=-15$

20. Begin by dividing out -1,000 from each term.

$420\div 7=60$

21.

$(-6)^2-2(5-4\cdot 2)=36-2(5-8)$

$=36-2(-3)=36+6=42$

22.
$$2y+8=-6$$
$$2y+8-8=-6-8$$
$$2y=-14$$
$$\frac{2y}{2}=\frac{-14}{2}$$
$$y=-7$$
Check: $2y+8=-6$
$$2(-7)+8\stackrel{?}{=}-6$$
$$-14+8\stackrel{?}{=}-6$$
$$-6=-6$$

23. x = the amount to raise the heat
$$-38+x=674$$
$$-38+x+38=674+38$$
$$x=712$$
The temperature must be raised $712°F$.

24. $\frac{c}{5}$

25. $60h$ min

26. $\frac{233+218+206}{3}=\frac{657}{3}=219$

27. $-5(-6x)=[(-5)(-6)]x=30x$

28. $5x+5x+5x=(5+5+5)x=15x$

29.
$$9a+3(a-7)=18-a$$
$$9a+3a-21=18-a$$
$$12a-21=18-a$$
$$12a-21+a=18-a+a$$
$$13a-21=18$$
$$13a-21+21=18+21$$
$$13a=39$$
$$\frac{13a}{13}=\frac{39}{13}$$
$$a=3$$
Check: $9a+3(a-7)=18-a$
$$9(3)+3(3-7)\stackrel{?}{=}18-(3)$$
$$27+3(-4)\stackrel{?}{=}15$$
$$27-12\stackrel{?}{=}15$$
$$15=15$$

30. f = the number of first class seats
$$f+10f=132$$
$$11f=132$$
$$\frac{11f}{11}=\frac{132}{11}$$
$$f=12$$
There are 12 first class seats.

31. $\frac{6}{13}$

32. $\frac{90}{126}=\frac{5\cdot \cancel{18}}{7\cdot \cancel{18}}=\frac{5}{7}$

33. $\frac{3}{8}\cdot\frac{7}{16}=\frac{21}{128}$

34.
$$-\frac{15}{8a^4}\div\frac{10}{a^5}=-\frac{15}{8a^4}\cdot\frac{a^5}{10}$$
$$=-\frac{3\cdot\cancel{5}\cdot\cancel{a}\cdot\cancel{a}\cdot\cancel{a}\cdot\cancel{a}\cdot a}{8\cdot\cancel{a}\cdot\cancel{a}\cdot\cancel{a}\cdot\cancel{a}\cdot 2\cdot\cancel{5}}=-\frac{3a}{16}$$

35. $\frac{a}{9}+\frac{5}{7}=\frac{a}{9}\cdot\frac{7}{7}+\frac{5}{7}\cdot\frac{9}{9}=\frac{7a}{63}+\frac{45}{63}=\frac{7a+45}{63}$

36. $$-4\frac{1}{4}\left(-4\frac{1}{2}\right) = -\frac{17}{4}\left(-\frac{9}{2}\right) = \frac{153}{8} = 19\frac{1}{8}$$

37.
$$\begin{array}{rrr} 76\frac{1}{6} & = 76\frac{4}{24} & = 75\frac{28}{24} \\ \underline{-49\frac{7}{8}} & \underline{-49\frac{21}{24}} & \underline{-49\frac{21}{24}} \\ & & 26\frac{7}{24} \end{array}$$

38. $$\frac{\frac{5}{27}}{\frac{-5}{9}} = \frac{5}{27}\left(-\frac{9}{5}\right) = -\frac{\cancel{5}\cdot\cancel{9}}{3\cdot\cancel{9}\cdot\cancel{5}} = -\frac{1}{3}$$

39. $$\frac{1}{4}\cdot\frac{7}{16} = \frac{7}{64}$$

40. $$3\frac{3}{8} + 7\frac{3}{4} = 3\frac{3}{8} + 7\frac{6}{8} = 10\frac{9}{8}$$
$$= 10 + 1\frac{1}{8} = 11\frac{1}{8}\,in$$

41.
$$\left(\frac{9}{20} \div 2\frac{2}{5}\right) + \left(\frac{3}{4}\right)^2$$
$$= \left(\frac{9}{20} \div \frac{12}{5}\right) + \left(\frac{3}{4}\right)\left(\frac{3}{4}\right)$$
$$= \left(\frac{9}{20}\cdot\frac{5}{12}\right) + \left(\frac{9}{16}\right)$$
$$= \left(\frac{\overset{3}{\cancel{9}}}{\underset{4}{\cancel{20}}}\cdot\frac{\cancel{5}}{\underset{4}{\cancel{12}}}\right) + \left(\frac{9}{16}\right)$$
$$= \frac{3}{16} + \frac{9}{16} = \frac{12}{16} = \frac{3}{4}$$

42.
$$\frac{3}{4}m = -27$$
$$\frac{4}{3}\cdot\frac{3}{4}m = \frac{4}{3}(-27)$$
$$m = 4(-9)$$
$$m = -36$$
Check: $\frac{3}{4}m = -27$
$$\frac{3}{4}(-36) = -27$$
$$3(-9) = -27$$
$$-27 = -27$$

43. $0.001\,in$

44.

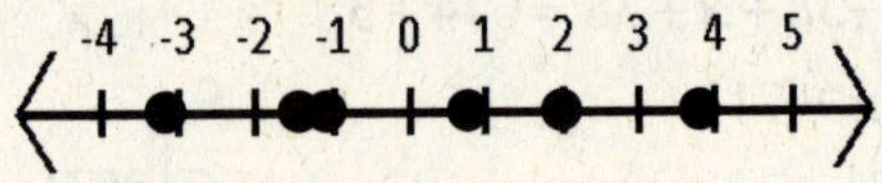

45.
$$\begin{array}{r} \overset{6}{\cancel{7}}.\overset{10}{\cancel{0}}01 \\ \underline{-5.900} \\ 1.101 \end{array}$$

46.
$$-1.8(4.52)$$
$$\begin{array}{r} 4.52^2 \\ \underline{\times 1.8^1} \\ 3616 \\ \underline{+4520} \\ 8.136^3 \end{array}$$
$$-1.8(4.52) = -8.136$$

47. $$56.012(0.001) = 0.056012$$

48.
$$\frac{-21.28}{-3.8} = \frac{212.8}{38}$$
$$\begin{array}{r} 5.6 \\ 38\overline{)212.8} \\ \underline{-190} \\ 228 \\ \underline{-228} \\ 0 \end{array}$$

49.
$$A = 2\cdot\frac{1}{2}\cdot 7.5\cdot 21 = 7.5\cdot 21$$
$$A = 157.5in^2$$

50.
$$C = \frac{5}{9}(451-32) = \frac{5}{9}(419)$$
$$C = 232.\overline{7} \approx 232.8°F$$

51.
$$\begin{array}{r} 0.4166 \\ 12\overline{)5.0000} \\ \underline{-48} \\ 20 \\ \underline{-12} \\ 80 \\ \underline{-72} \\ 80 \\ \underline{-72} \end{array}$$
$$\frac{5}{12} = 0.41\overline{6}$$

52.
$$\frac{3}{8}(-3.2)+\left(4\frac{1}{2}\right)\left(-\frac{1}{4}\right)$$
$$= 0.375(-3.2)+(4.5)(-0.25)$$
$$= -1.2-1.125 = -2.325$$

53.
$$-4\sqrt{36}+2\sqrt{81} = -4\cdot 6+2\cdot 9$$
$$= -24+18 = -6$$

54.
$$-5.6-2h = -17.1-h$$
$$-5.6-2h+2h = -17.1-h+2h$$
$$-5.6 = -17.1+h$$
$$-5.6+17.1 = -17.1+h+17.1$$
$$11.5 = h$$
Check: $-5.6-2h = -17.1-h$
$$-5.6-2(11.5)\stackrel{?}{=}-17.1-(11.5)$$
$$-5.6-23\stackrel{?}{=}-17.1-11.5$$
$$-28.6 = -28.6$$

CHAPTER 6 Ratio, Proportion, and Measurement

Section 6.1: Ratios

VOCABULARY

1. A ratio is the quotient of two numbers or the quotient of two quantities that have the same units.

3. A unit rate is a rate in which the denominator is 1.

CONCEPTS

5. 3 is a factor of both 15 and 24.

7. Multiplying numerator and denominator by 10 will give $\frac{5}{6}$.

9. $\frac{11 \text{ minutes}}{60 \text{ minutes}} = \frac{11}{60}$

NOTATION

11. $\frac{13}{9}$, 13 to 9, 13:9

GUIDED PRACTICE

13. $\frac{5}{8}$

15. $\frac{11}{16}$

17. $\frac{25}{15} = \frac{5 \cdot \cancel{5}}{3 \cdot \cancel{5}} = \frac{5}{3}$

19. $\frac{63}{36} = \frac{7 \cdot \cancel{9}}{4 \cdot \cancel{9}} = \frac{7}{4}$

21. $\frac{22}{33} = \frac{2 \cdot \cancel{11}}{3 \cdot \cancel{11}} = \frac{2}{3}$

23. $\frac{17}{34} = \frac{\cancel{17}}{2 \cdot \cancel{17}} = \frac{1}{2}$

25. $\frac{4oz}{12oz} = \frac{\cancel{4}\,\cancel{oz}}{3 \cdot \cancel{4}\,\cancel{oz}} = \frac{1}{3}$

27. $\frac{24mi}{32mi} = \frac{3 \cdot \cancel{8}\,\cancel{mi}}{4 \cdot \cancel{8}\,\cancel{mi}} = \frac{3}{4}$

29. $\frac{0.3}{0.9} = \frac{0.3}{0.9} \cdot \frac{10}{10} = \frac{3}{9} = \frac{1}{3}$

31. $\frac{0.65}{0.15} = \frac{0.65}{0.15} \cdot \frac{100}{100} = \frac{65}{15} = \frac{13}{3}$

33. $\frac{3.8}{7.8} = \frac{3.8}{7.8} \cdot \frac{10}{10} = \frac{38}{78} = \frac{19}{39}$

35. $\frac{7}{24.5} = \frac{7}{24.5} \cdot \frac{10}{10} = \frac{70}{245} = \frac{2}{7}$

37. $\frac{2\frac{1}{3}}{4\frac{2}{3}} = \frac{\frac{7}{3}}{\frac{14}{3}} = \frac{7}{3} \cdot \frac{3}{14} = \frac{\cancel{7}}{\cancel{3}} \cdot \frac{\cancel{3}}{2 \cdot \cancel{7}} = \frac{1}{2}$

39. $\frac{10\frac{1}{2}}{1\frac{3}{4}} = \frac{\frac{21}{2}}{\frac{7}{4}} = \frac{21}{2} \cdot \frac{4}{7} = \frac{3 \cdot \cancel{7}}{\cancel{2}} \cdot \frac{\cancel{2} \cdot 2}{\cancel{7}} = \frac{6}{1}$

41. $\frac{12 \text{ min}}{1hr} = \frac{12 \text{ min}}{60 \text{ min}} = \frac{\cancel{12}\,\cancel{\text{min}}}{5 \cdot \cancel{12}\,\cancel{\text{min}}} = \frac{1}{5}$

43. $\frac{3d}{1wk} = \frac{3d}{7d} = \frac{3\cancel{d}}{7\cancel{d}} = \frac{3}{7}$

45. $\frac{18mo}{2yr} = \frac{18mo}{24mo} = \frac{3 \cdot \cancel{6}\,\cancel{mo}}{4 \cdot \cancel{6}\,\cancel{mo}} = \frac{3}{4}$

47. $\frac{21in}{3ft} = \frac{21in}{36in} = \frac{\cancel{3} \cdot 7\cancel{in}}{\cancel{3} \cdot 12\cancel{in}} = \frac{7}{12}$

49. $\frac{64ft}{6\sec} = \frac{\cancel{2} \cdot 32ft}{\cancel{2} \cdot 3\sec} = \frac{32ft}{3\sec}$

51. $\frac{75d}{20gal} = \frac{\cancel{5} \cdot 15d}{4 \cdot \cancel{5}gal} = \frac{15d}{4gal}$

53. $\frac{84 \text{ made}}{100 \text{ attempts}} = \frac{\cancel{4} \cdot 21 \text{ made}}{\cancel{4} \cdot 25 \text{ attempts}} = \frac{21 \text{ made}}{25 \text{ attempts}}$

55. $\dfrac{18\text{ beats}}{12\text{ measures}} = \dfrac{3\cdot \cancel{6}\text{ beats}}{2\cdot \cancel{6}\text{ measures}} = \dfrac{3\text{ beats}}{2\text{ measures}}$

57. $\dfrac{60\text{ rev}}{5\text{ min}} = \dfrac{\cancel{5}\cdot 12\text{ rev}}{\cancel{5}\text{ min}} = 12$ revolutions per minute

59. $\dfrac{\$50{,}000}{10\,yr} = \dfrac{\cancel{10}\cdot \$5{,}000}{\cancel{10}\,yr} = \$5{,}000$ per year

61. $12 \div 8 = 1.5$
1.5 errors per hour

63. $4{,}007{,}500 \div 12{,}500 = 320.6$
320.6 people per square mile

65. $48 \div 12 = 4$
\$4 per minute

67. $272 \div 4 = 68$
\$68 per item

69. $78 \div 65 = 1.2$
1.2 cents per ounce

71. $3.50 \div 50 = 0.07$
\$0.07 per foot

APPLICATIONS

73.
a. $\dfrac{12}{18} = \dfrac{2}{3}$

b. $\dfrac{18}{12} = \dfrac{3}{2}$

75. $\dfrac{5\frac{1}{4}}{1\frac{3}{4}} = \dfrac{\frac{21}{4}}{\frac{7}{4}} = \dfrac{21}{4}\cdot\dfrac{4}{7} = \dfrac{3}{1}$

77. $\dfrac{0.02}{1.1} = \dfrac{2}{110} = \dfrac{1}{55}$

79. $\dfrac{5¢}{\$1} = \dfrac{5¢}{100¢} = \dfrac{1}{20}$

81. $\dfrac{155}{25} = \dfrac{\cancel{5}\cdot 31}{\cancel{5}\cdot 5} = \dfrac{31\text{ compressions}}{5\text{ breaths}}$

83.
a. $800 + 600 + 100 + 100 + 120 + 80$
$= \$1{,}800$

b. $\dfrac{800}{1800} = \dfrac{4}{9}$

c. $\dfrac{600}{1800} = \dfrac{1}{3}$

d. $\dfrac{100}{1800} = \dfrac{1}{18}$

85. Since the height and width of a square are the same, the ratio is $\dfrac{1}{1}$.

87. $\dfrac{3.29}{1{,}000} = \dfrac{329\text{ complaints}}{100{,}000\text{ passengers}}$

89.
a. $112{,}500 - 4{,}500 = 108{,}000$

b. $\dfrac{108{,}000}{4{,}500} = 24$ browsers per buyer

91. $\dfrac{1{,}065}{35{,}500} = \dfrac{0.03\text{ ounces of gold}}{\text{cubic foot of gravel}}$

93. $\dfrac{84}{12} = 7$ cents per ounce

95. $\dfrac{25}{20} = 1.25$ cents per minute

97. $\dfrac{222.50}{50} = \$4.45$ per pound

99. $\dfrac{11{,}880}{27} = 440$ gallons per minute

101.
a. $35{,}071 - 34{,}746 = 325$ miles

b. $\dfrac{325}{5} = 65$ mph

103. The 6oz. can: $\frac{89}{6} \approx 14.8¢ / oz$

The 8oz. can: $\frac{119}{8} \approx 14.9¢ / oz$

The 6oz. can is a better buy.

105. The box of 20: $\frac{429}{20} = 21.45¢ / \text{pill}$

The box of 50: $\frac{959}{50} = 19.18¢ / \text{pill}$

The 50-tablet box is a better buy.

107. Car: $\frac{345}{6} = 57.5 \text{ mph}$

Truck: $\frac{376}{6.2} \approx 60.6 \text{ mph}$

The truck is traveling faster.

109. First car: $\frac{1,235}{51.3} \approx 24.1 mpg$

Second car: $\frac{1,456}{55.78} \approx 26.1 mpg$

The second car had better mileage.

WRITING

111. No. In the first case the first quantity is triple the second while in the second case the first quantity is a third of the second.

113. It will be easier to figure out which sales are actually bargains.

REVIEW

115. $\frac{48}{54} = \frac{\cancel{6} \cdot 8}{\cancel{6} \cdot 9} = \frac{8}{9}$

117. $18x = 63$

$\frac{18x}{18} = \frac{63}{18}$

$x = 3.5$

Section 6.2: Proportions

VOCABULARY

1. A <u>proportion</u> is a statement that two ratios (or rates) are equal.

3. The <u>cross</u> products for the proportion $\frac{4}{7} = \frac{36}{x}$ are $4 \cdot x$ and $7 \cdot 36$.

5. A letter that is used to represent an unknown number is called a <u>variable</u>.

7. We solve proportions by writing a series of steps that result in an equation of the form $x =$ a number or a number $= x$. We say that the variable x is <u>isolated</u> on one side of the equation.

CONCEPTS

9. If the cross products of a proportion are equal, the proportion is <u>true</u>. If the cross products are *not equal*, the proportion is <u>false</u>.

11. $9 \cdot 10 = 90$ $\quad$ $2 \cdot 45 = 90$

$\frac{9}{2} = \frac{45}{10}$

13. $\frac{\text{Teacher's aides} \rightarrow 12}{\text{Children} \rightarrow 100} = \frac{3 \leftarrow \text{Teacher's aides}}{25 \leftarrow \text{Children}}$

NOTATION

15. $2 \cdot 9 = 3 \cdot x$

$18 = 3 \cdot x$

$\frac{18}{3} = \frac{3 \cdot x}{3}$

$6 = x$

The solution is 6.

GUIDED PRACTICE

17. $\frac{20}{30} = \frac{2}{3}$

19. $\frac{400 \text{ sheets}}{100 \text{ beds}} = \frac{4 \text{ sheets}}{1 \text{ bed}}$

21. $\frac{70}{81} = \frac{2 \cdot 5 \cdot 7}{3 \cdot 3 \cdot 3 \cdot 3}$

The fraction does not simplify so the proportion is false.

23. $\frac{21}{14} = \frac{3\cdot\cancel{7}}{2\cdot\cancel{7}} = \frac{3}{2}$

$\frac{18}{12} = \frac{3\cdot\cancel{6}}{2\cdot\cancel{6}} = \frac{3}{2}$

The proportion is true.

25. $4\cdot 16 \stackrel{?}{=} 2\cdot 32$

$64 \stackrel{?}{=} 64$

The proportion is true.

27. $9\cdot 80 \stackrel{?}{=} 19\cdot 38$

$720 \stackrel{?}{=} 722$

The proportion is false.

29. $0.5\cdot 1.3 \stackrel{?}{=} 1.1\cdot 0.8$

$0.65 \stackrel{?}{=} 0.88$

The proportion is false.

31. $1.2\cdot 5.4 \stackrel{?}{=} 1.8\cdot 3.6$

$6.48 \stackrel{?}{=} 6.48$

The proportion is true.

33. $1\frac{4}{5}\cdot 4\frac{1}{6} \stackrel{?}{=} 2\frac{3}{16}\cdot 3\frac{3}{7}$

$\frac{9}{5}\cdot\frac{25}{6} \stackrel{?}{=} \frac{35}{16}\cdot\frac{24}{7}$

$\frac{\cancel{3}\cdot 3\cdot\cancel{5}\cdot 5}{\cancel{5}\cdot 2\cdot\cancel{3}} \stackrel{?}{=} \frac{5\cdot\cancel{7}\cdot\cancel{4}\cdot\cancel{2}\cdot 3}{2\cdot\cancel{2}\cdot\cancel{4}\cdot\cancel{7}}$

$\frac{15}{2} \stackrel{?}{=} \frac{15}{2}$

The proportion is true.

35. $\frac{1}{5}\cdot 11\frac{2}{3} \stackrel{?}{=} 1\frac{1}{7}\cdot 1\frac{1}{6}$

$\frac{1}{5}\cdot\frac{35}{3} \stackrel{?}{=} \frac{8}{7}\cdot\frac{7}{6}$

$\frac{\cancel{5}\cdot 7}{\cancel{5}\cdot 3} \stackrel{?}{=} \frac{\cancel{2}\cdot 4\cdot\cancel{7}}{\cancel{7}\cdot\cancel{2}\cdot 3}$

$\frac{7}{3} \stackrel{?}{=} \frac{4}{3}$

The proportion is false.

37. $\frac{18}{54} \stackrel{?}{=} \frac{3}{9}$

$18\cdot 9 \stackrel{?}{=} 3\cdot 54$

$162 \stackrel{?}{=} 162$

Yes the numbers are proportional.

39. $\frac{8}{6} \stackrel{?}{=} \frac{21}{16}$

$8\cdot 16 \stackrel{?}{=} 21\cdot 6$

$128 \stackrel{?}{=} 126$

No the numbers are not proportional.

41. $\frac{5}{10} = \frac{3}{x}$

$5\cdot x = 3\cdot 10$

$5x = 30$

$\frac{5x}{5} = \frac{30}{5}$

$x = 6$

Check:

$\frac{5}{10} \stackrel{?}{=} \frac{3}{6}$

$5\cdot 6 \stackrel{?}{=} 3\cdot 10$

$30 = 30$

43. $\frac{2}{3} = \frac{x}{6}$

$2\cdot 6 = x\cdot 3$

$\frac{12}{3} = \frac{x\cdot 3}{3}$

$4 = x$

Check:

$\frac{2}{3} \stackrel{?}{=} \frac{4}{6}$

$2\cdot 6 \stackrel{?}{=} 4\cdot 3$

$12 = 12$

45.

$$\frac{0.6}{9.6}=\frac{x}{4.8}$$
$$0.6\cdot 4.8 = x\cdot 9.6$$
$$\frac{2.88}{9.6}=\frac{x\cdot 9.6}{9.6}$$
$$0.3 = x$$

Check:

$$\frac{0.6}{9.6}\stackrel{?}{=}\frac{0.3}{4.8}$$
$$0.6\cdot 4.8 \stackrel{?}{=} 0.3\cdot 9.6$$
$$2.88 = 2.88$$

47.

$$\frac{2.75}{x}=\frac{1.5}{1.2}$$
$$2.75\cdot 1.2 = 1.5\cdot x$$
$$\frac{3.3}{1.5}=\frac{1.5\cdot x}{1.5}$$
$$2.2 = x$$

Check:

$$\frac{2.75}{2.2}\stackrel{?}{=}\frac{1.5}{1.2}$$
$$2.75\cdot 1.2 \stackrel{?}{=} 1.5\cdot 2.2$$
$$3.3 = 3.3$$

49.

$$\frac{x}{1\frac{1}{2}}=\frac{10\frac{1}{2}}{4\frac{1}{2}}$$
$$\frac{x}{\frac{3}{2}}=\frac{\frac{21}{2}}{\frac{9}{2}}$$
$$x\cdot\frac{9}{2}=\frac{21}{2}\cdot\frac{3}{2}$$
$$x\cdot\frac{9}{2}\cdot\frac{2}{9}=\frac{21}{2}\cdot\frac{3}{2}\cdot\frac{2}{9}$$
$$x=\frac{\not{3}\cdot 7}{\not{2}}\cdot\frac{\not{3}}{2}\cdot\frac{\not{2}}{\not{3}\cdot\not{3}}$$
$$x=\frac{7}{2}=3\frac{1}{2}$$

Check:

$$\frac{3\frac{1}{2}}{1\frac{1}{2}}\stackrel{?}{=}\frac{10\frac{1}{2}}{4\frac{1}{2}}$$
$$\frac{\frac{7}{2}}{\frac{3}{2}}\stackrel{?}{=}\frac{\frac{21}{2}}{\frac{9}{2}}$$
$$\frac{7}{2}\cdot\frac{9}{2}\stackrel{?}{=}\frac{21}{2}\cdot\frac{3}{2}$$
$$\frac{63}{4}=\frac{63}{4}$$

51.

$$\frac{x}{1\frac{1}{6}} = \frac{2\frac{5}{8}}{3\frac{1}{2}}$$

$$\frac{x}{\frac{7}{6}} = \frac{\frac{21}{8}}{\frac{7}{2}}$$

$$\frac{7}{2} \cdot x = \frac{21}{8} \cdot \frac{7}{6}$$

$$\frac{\frac{7}{2} \cdot x}{\frac{7}{2}} = \frac{\frac{21}{8} \cdot \frac{7}{6}}{\frac{7}{2}}$$

$$x = \frac{21}{8} \cdot \frac{7}{6} \cdot \frac{2}{7}$$

$$x = \frac{\not{3} \cdot \not{7} \cdot 7 \cdot \not{2}}{2 \cdot 2 \cdot 2 \cdot \not{2} \cdot \not{3} \cdot \not{7}}$$

$$x = \frac{7}{8}$$

Check:

$$\frac{\frac{7}{8}}{1\frac{1}{6}} \stackrel{?}{=} \frac{2\frac{5}{8}}{3\frac{1}{2}}$$

$$\frac{\frac{7}{8}}{\frac{7}{6}} \stackrel{?}{=} \frac{\frac{21}{8}}{\frac{7}{2}}$$

$$\frac{7}{8} \cdot \frac{7}{2} \stackrel{?}{=} \frac{21}{8} \cdot \frac{7}{6}$$

$$\frac{49}{16} = \frac{49}{16}$$

TRY IT YOURSELF

53.

$$\frac{4,000}{x} = \frac{3.2}{2.8}$$

$$4,000 \cdot 2.8 = 3.2 \cdot x$$

$$11,200 = 3.2 \cdot x$$

$$3,500 = x$$

55.

$$\frac{12}{6} = \frac{x}{\frac{1}{4}}$$

$$12 \cdot \frac{1}{4} = x \cdot 6$$

$$3 = x \cdot 6$$

$$\frac{3}{6} = x$$

$$\frac{1}{2} = x$$

57.

$$\frac{x}{800} = \frac{900}{200}$$

$$x \cdot 200 = 900 \cdot 800$$

$$x \cdot 200 = 720,000$$

$$x = 3,600$$

59.

$$\frac{x}{2.5} = \frac{3.7}{9.25}$$

$$x \cdot 9.25 = 3.7 \cdot 2.5$$

$$x \cdot 9.25 = 9.25$$

$$x = 1$$

61.

$$\frac{0.8}{2} = \frac{x}{5}$$

$$0.8 \cdot 5 = 2 \cdot x$$

$$4 = 2 \cdot x$$

$$2 = x$$

63.

$$\frac{x}{4\frac{1}{10}} = \frac{3\frac{3}{4}}{1\frac{7}{8}}$$

$$\frac{x}{\frac{41}{10}} = \frac{\frac{15}{4}}{\frac{15}{8}}$$

$$x \cdot \frac{15}{8} = \frac{15}{4} \cdot \frac{41}{10}$$

$$x = \frac{15}{4} \cdot \frac{41}{10} \cdot \frac{8}{15}$$

$$x = \frac{41}{5} = 8\frac{1}{5}$$

65. $\frac{340}{51}=\frac{x}{27}$

$340\cdot 27=x\cdot 51$

$9{,}180=51\cdot x$

$180=x$

67. $\frac{0.4}{1.2}=\frac{6}{x}$

$0.4\cdot x=6\cdot 1.2$

$0.4\cdot x=7.2$

$x=18$

69. $\frac{4.65}{7.8}=\frac{x}{5.2}$

$4.65\cdot 5.2=x\cdot 7.8$

$24.18=x\cdot 7.8$

$3.1=x$

71. $\frac{\frac{3}{4}}{\frac{1}{2}}=\frac{0.25}{x}$

$\frac{3}{4}\cdot x=\frac{1}{2}\cdot 0.25$

$\frac{3}{4}x=\frac{1}{2}\cdot\frac{1}{4}$

$x=\frac{1}{2}\cdot\frac{1}{\cancel{4}}\cdot\frac{\cancel{4}}{3}$

$x=\frac{1}{6}$

APPLICATIONS

73. $\frac{6}{1.75}=\frac{750}{x}$

$6\cdot x=1.75\cdot 750$

$6x=1{,}312.5$

$x=\$218.75$

75. $\frac{12}{57.99}=\frac{16}{x}$

$12\cdot x=57.99\cdot 16$

$12\cdot x=927.84$

$x=\$77.32$

77. $\frac{10}{4}\stackrel{?}{=}\frac{25}{10}$

$10\cdot 10\stackrel{?}{=}25\cdot 4$

$100\stackrel{?}{=}100$

Yes the ratios are equal.

79. $\frac{3}{7}=\frac{x}{56}$

$3\cdot 56=7\cdot x$

$168=7x$

$\frac{168}{7}=\frac{7x}{7}$

$x=24$ drops

81. $\frac{5}{25}=\frac{195}{x}$

$5\cdot x=25\cdot 195$

$5\cdot x=4875$

$x=975$

83. $\frac{280}{7}=\frac{x}{2}$

$280\cdot 2=7\cdot x$

$560=7\cdot x$

$x=80\,ft.$

85. $\frac{87}{1}=\frac{x}{.75}$

$1\cdot x=.75\cdot 87$

$x=65.25\,ft.$

(or 65 ft. 3 in.)

87. $\frac{160}{1}=\frac{420}{x}$

$160\cdot x=1\cdot 420$

$x=\frac{420}{160}$

$x=2.625in.$

89.
$$\frac{1\frac{1}{4}}{3\frac{1}{2}}=\frac{x}{12}$$
$$\frac{5}{4}\cdot 12=\frac{7}{2}\cdot x$$
$$15=\frac{7}{2}\cdot x$$
$$\frac{2}{7}\cdot 15=x$$
$$x=4\frac{2}{7}$$

91.
$$\frac{15}{2.85}=\frac{100}{x}$$
$$15\cdot x=2.85\cdot 100$$
$$15\cdot x=285$$
$$x=19\text{sec.}$$

93.
$$\frac{5}{6}=\frac{x}{37.5}$$
$$5\cdot 37.5=6\cdot x$$
$$187.5=6\cdot x$$
$$x=31.25in$$

95.
$$\frac{412}{40}=\frac{x}{30}$$
$$412\cdot 30=40\cdot x$$
$$12{,}360=40\cdot x$$
$$x=\$309$$

WRITING

97. A ratio is how two quantities relate to one another. A proportion is how two ratios are related.

99. If 4 ounces of cashews have 639 calories, how many calories are in 10 ounces of cashews?

REVIEW

101. $7.4+6.78+35+0.008=49.188$

103. $48.8-17.372=31.428$

105. $-3.8-(-7.9)=-3.8+7.9=4.1$

107. $-35.1-13.99=-49.09$

Section 6.3: American Units of Measurement

VOCABULARY

1. A ruler is used for measuring <u>length</u>.

3. $\frac{3\text{ft}}{1\text{yd}}, \frac{1\text{ton}}{2{,}000\text{lb}}$, and $\frac{4\text{qt}}{1\text{gal}}$ are examples of <u>unit</u> conversion factors.

5. Some examples of American units of <u>capacity</u> are cups, pints, quarts, and gallons.

CONCEPTS

7. a. 12 inches = 1 foot   b. 3 feet = 1 yard

c. 1 yard = 36 inches   d. 1 mile = 5,280 feet

9. a. 1 cup = 8 fluid ounces

b. 1 pint = 2 cups

c. 2 pints = 1 quart

d. 4 quarts = 1 gallon

11. The value of any unit conversion factor is <u>1</u>.

13. a. oz can be removed.

b. lb remains

15. a. $\frac{1\text{ton}}{2{,}000\text{lb}}$ b. $\frac{2\text{pt}}{1\text{qt}}$

17. a. iv   b. i   c. ii   d. iii

19. a. iii   b. iv   c. i   d. ii

NOTATION

21. a. pound   b. ounce   c. fluid ounce

23.
$$2yd=\frac{2yd}{1}\cdot\frac{36in}{1yd}$$
$$=2\cdot 36in$$
$$=72in$$

25.
$$1ton=\frac{1ton}{1}\cdot\frac{2{,}000lb}{1ton}\cdot\frac{16oz}{1lb}$$
$$=1\cdot 2{,}000\cdot 16oz$$
$$=32{,}000oz$$

GUIDED PRACTICE

27. a. 8

b. $\frac{5}{8}in$, $1\frac{1}{4}in$, $2\frac{7}{8}in$

29. a. 16

b. $\frac{9}{16}in$, $1\frac{3}{4}in$, $2\frac{3}{16}in$

31. $2\frac{9}{16}in$

33. $10\frac{7}{8}in$

35. $\frac{4yd}{1}\cdot\frac{3ft}{1yd}=12ft$

37. $\frac{35yd}{1}\cdot\frac{3ft}{1yd}=105ft$

39. $3\frac{1}{2}ft=\frac{7}{2}ft\cdot\frac{12in}{1ft}=\frac{84}{2}in=42in$

41. $5\frac{1}{4}ft=\frac{21}{4}ft\cdot\frac{12in}{1ft}=\frac{252}{4}in=63in$

43. $\frac{105yd}{1}\cdot\frac{3ft}{1yd}\cdot\frac{1mi}{5,280ft}=\frac{315}{5,280}mi$

$=\frac{21}{352}mi\approx 0.06mi$

45. $\frac{1,540yd}{1}\cdot\frac{3ft}{1yd}\cdot\frac{1mi}{5,280ft}$

$=\frac{4,620}{5,280}mi=\frac{7}{8}mi$

$=0.875mi$

47. $\frac{44oz}{1}\cdot\frac{1lb}{16oz}=\frac{44}{16}lb=\frac{11}{4}lb$

$=2\frac{3}{4}lb=2.75lb$

49. $\frac{72oz}{1}\cdot\frac{1lb}{16oz}=\frac{72}{16}lb=\frac{9}{2}lb$

$=4\frac{1}{2}lb=4.5lb$

51. $\frac{50lb}{1}\cdot\frac{16oz}{1lb}=50\cdot 16oz=800oz$

53. $\frac{87lb}{1}\cdot\frac{16oz}{1lb}=87\cdot 16oz=1,392oz$

55. $\frac{8pt}{1}\cdot\frac{2c}{1pt}\cdot\frac{8fl\,oz}{1c}=8\cdot 2\cdot 8fl\,oz=128fl\,oz$

57. $\frac{21pt}{1}\cdot\frac{2c}{1pt}\cdot\frac{8fl\,oz}{1c}=21\cdot 2\cdot 8fl\,oz=336fl\,oz$

59. $\frac{165\min}{1}\cdot\frac{1hr}{60\min}=\frac{165}{60}hr$

$=\frac{11}{4}hr=2\frac{3}{4}hr$

61. $\frac{330\min}{1}\cdot\frac{1hr}{60\min}=\frac{330}{60}hr$

$=\frac{11}{2}hr=5\frac{1}{2}hr$

TRY IT YOURSELF

63. $\frac{3qt}{1}\cdot\frac{2pt}{1qt}=6pt$

65. $\frac{7,200\min}{1}\cdot\frac{1hr}{60\min}\cdot\frac{1\text{day}}{24hr}$

$=\frac{7,200}{1,440}\text{ days}=5\text{days}$

67. $\frac{56in}{1}\cdot\frac{1ft}{12in}=\frac{56}{12}ft$

$=\frac{14}{3}ft=4\frac{2}{3}ft$

69. $\frac{4ft}{1}\cdot\frac{12in}{1ft}=4\cdot 12in=48in$

71. $\frac{16pt}{1}\cdot\frac{1qt}{2pt}\cdot\frac{1gal}{4qt}=\frac{16}{8}gal=2gal$

73. $\frac{80oz}{1}\cdot\frac{1lb}{16oz}=\frac{80}{16}lb=5lb$

75. $\frac{240\text{ min}}{1}\cdot\frac{1hr}{60\text{ min}}=\frac{240}{60}hr=4hr$

77. $\frac{8yd}{1}\cdot\frac{3ft}{1yd}\cdot\frac{12in}{1ft}=8\cdot3\cdot12in=288in$

79. $\frac{90in}{1}\cdot\frac{1ft}{12in}\cdot\frac{1yd}{3ft}=\frac{90}{36}yd=\frac{5}{2}yd=2\frac{1}{2}yd$

81. $\frac{5yd}{1}\cdot\frac{3ft}{1yd}=5\cdot3ft=15ft$

83. $\frac{12.4\text{ tons}}{1}\cdot\frac{2{,}000lb}{1\text{ ton}}=12.4\cdot2{,}000lb$

$=24{,}800lb$

85. $\frac{7ft}{1}\cdot\frac{1yd}{3ft}=\frac{7}{3}yd=2\frac{1}{3}yd$

87. $\frac{15{,}840ft}{1}\cdot\frac{1mi}{5{,}280ft}=\frac{15{,}840}{5{,}280}mi=3mi$

89. $\frac{1}{2}mi\cdot\frac{5{,}280ft}{1mi}=\frac{5{,}280}{2}ft=2{,}640ft$

91. $\frac{7{,}000lb}{1}\cdot\frac{1\text{ ton}}{2{,}000lb}=\frac{7{,}000}{2{,}000}\text{tons}$

$=\frac{7}{2}\text{tons}=3\frac{1}{2}\text{tons}$

93. $\frac{32\,fl\,oz}{1}\cdot\frac{1c}{8\,fl\,oz}\cdot\frac{1pt}{2c}=\frac{32}{16}pt=2pt$

APPLICATIONS

95. $\frac{450ft}{1}\cdot\frac{1yd}{3ft}=\frac{450}{3}yd=150yd$

97. $\frac{240ft}{1}\cdot\frac{12in}{1ft}=240\cdot12in=2{,}880in$

99. $\frac{1{,}454ft}{1}\cdot\frac{1mi}{5{,}280ft}=\frac{1{,}454}{5{,}280}mi\approx0.28mi$

101. $\frac{35mi}{1}\cdot\frac{5{,}280ft}{1mi}\cdot\frac{1yd}{3ft}$

$=\frac{35\cdot5{,}280}{3}yd$

$=61{,}600yd$

103. $\frac{8lb}{1}\cdot\frac{16oz}{1lb}=8\cdot16oz=128oz$

105. $\frac{9{,}900lb}{1}\cdot\frac{1\text{ton}}{2{,}000lb}=\frac{9{,}900}{2{,}000}\text{tons}=4.95\text{tons}$

107. $\frac{17gal}{1}\cdot\frac{4qt}{1gal}=17\cdot4qt=68qt$

109. $\frac{575pt}{1}\cdot\frac{1qt}{2pt}\cdot\frac{1gal}{4qt}=\frac{575}{8}gal=71.875gal$

111. $\frac{2.5gal}{1}\cdot\frac{4qt}{1gal}\cdot\frac{2pt}{1qt}\cdot\frac{2c}{1pt}\cdot\frac{8\,fl\,oz}{1c}=320\,fl\,oz$

113. $\frac{147hr}{1}\cdot\frac{1\text{day}}{24hr}=\frac{147}{24}\text{day}=6.125\text{ days}$

WRITING

115. a. Write a form of 1 that uses feet and inches.

$\frac{1ft}{12in}$ or $\frac{12in}{1ft}$

b. Write a form of 1 that involves pints and gallons. $\frac{8pt}{1gal}$ or $\frac{1gal}{8pt}$

REVIEW

117. $2.09\cdot100=209$

119. $2.83(0.001)=0.00283$

Section 6.4: Metric Units of Measurement

VOCABULARY

1. The meter, the gram, and the liter are basic units of measurement in the metric system.

3. a. *Deka* means tens.

b. *Hecto* means hundreds.

c. *Kilo* means thousands.

5. We can convert from one unit to another in the metric system using unit conversion factors or a conversion chart like the one shown below.

7. The weight of an object is determined by the Earth's gravitational pull on the object.

CONCEPTS

9. a. 1 kilometer = 1,000 meters

b. 100 centimeters = 1 meter

c. 1,000 millimeters = 1 meter

11. a. 1,000 milliliters = 1 liter

b. 1 dekaliter = 10 liters

13. a. $\frac{1km}{1,000m}$

b. $\frac{100cg}{1g}$

c. $\frac{\text{1,000 millilters}}{\text{1 liter}}$

15. a. iii b. i c. ii

17. a. ii b. iii c. i

NOTATION

19. $20cm = \frac{20cm}{1} \cdot \frac{1m}{100cm} = \frac{20}{100}m = 0.2m$

21. $0.2kg = \frac{0.2kg}{1} \cdot \frac{1,000g}{1kg} \cdot \frac{1,000mg}{1g}$

$= 0.2 \cdot 1,000 \cdot 1,000mg$

$= 200,000mg$

GUIDED PRACTICE

23. $1cm$, $3cm$, $5cm$

25. a. 10, 1 millimeter b. $27mm$, $41mm$, $55mm$

27. $156mm$

29. $280mm$

31. $\frac{380cm}{1} \cdot \frac{1m}{100cm} = \frac{380}{100}m = 3.8m$

33. $\frac{120cm}{1} \cdot \frac{1m}{100cm} = \frac{120}{100}m = 1.2m$

35. $\frac{8.7m}{1} \cdot \frac{1000mm}{1m} = 8.7 \cdot 1000mm = 8,700mm$

37. $\frac{2.89m}{1} \cdot \frac{1000mm}{1m} = 2.89 \cdot 1000mm = 2,890mm$

39. $\frac{4.5cm}{1} \cdot \frac{1m}{100cm} \cdot \frac{1km}{1000m} = \frac{4.5}{100,000}km$

$= 0.000045km$

41. $\frac{0.3cm}{1} \cdot \frac{1m}{100cm} \cdot \frac{1km}{1000m} = \frac{0.3}{100,000}km$

$= 0.000003km$

43. $\frac{1.93kg}{1} \cdot \frac{1,000g}{1kg} = 1.93 \cdot 1,000g = 1,930g$

45. $\frac{4.531kg}{1} \cdot \frac{1,000g}{1kg} = 4.531 \cdot 1,000g = 4,531g$

47. $\frac{6,000mg}{1} \cdot \frac{1g}{1,000mg} = \frac{6,000}{1,000}g = 6g$

49. $\frac{3,500mg}{1} \cdot \frac{1g}{1,000mg} = \frac{3,500}{1,000}g = 3.5g$

51. $\frac{3L}{1} \cdot \frac{1,000mL}{1L} = 3 \cdot 1,000mL = 3,000mL$

53. $\frac{26.3L}{1} \cdot \frac{1,000mL}{1L}$

$= 26.3 \cdot 1,000mL = 26,300mL$

TRY IT YOURSELF

55. Refer to a conversion chart: 1 place to the right
$0.31dm = 3.1cm$

57. Refer to a conversion chart: 3 places to the left
$500mL = 0.5L$

59. Refer to a conversion chart: 3 places to the right
$2kg = 2{,}000g$

61. Refer to a conversion chart: 1 place to the right
$0.074cm = 0.74mm$

63. Refer to a conversion chart: 3 places to the right
$1{,}000kg = 1{,}000{,}000g$

65. Refer to a conversion chart: 3 places to the left
$658.23L = 0.65823kL$

67. Refer to a conversion chart: 1 place to the left
$4.72cm = 0.472dm$

69. $10mL = 10cc$

71. Refer to a conversion chart: 3 places to the left
$1250mcg = 1.25mg$

73. Refer to a conversion chart: 3 places to the left
$5{,}689g = 5.689kg$

75. Refer to a conversion chart: 2 places to the left
$453.2cm = 4.532m$

77. Refer to a conversion chart: 1 place to the left
$0.325dL = 0.0325L$

79. Refer to a conversion chart: 3 places to the right
$675dam = 675{,}000cm$

81. Refer to a conversion chart: 3 places to the left
$0.00777cm = 0.0000077dam$

83. Refer to a conversion chart: 2 places to the left
$134m = 1.34hm$

85. Refer to a conversion chart: 2 places to the right
$65.78km = 6{,}578dam$

APPLICATIONS

87. $\frac{500}{1{,}000} = 0.5km \qquad \frac{1{,}000}{1{,}000} = 1km$

$\frac{1{,}500}{1{,}000} = 1.5km \qquad \frac{5{,}000}{1{,}000} = 5km$

$\frac{10{,}000}{1{,}000} = 10km$

89. $\frac{343m}{1} \cdot \frac{1hm}{100m} = \frac{343}{100}hm = 3.43hm$

91. $\frac{120mm}{1} \cdot \frac{1cm}{10mm} = \frac{120}{10}cm = 12cm$

$\frac{80mm}{1} \cdot \frac{1cm}{10mm} = \frac{80}{10}cm = 8cm$

93. $\frac{0.05mL}{1} \cdot \frac{1L}{1{,}000mL} = \frac{0.05}{1{,}000}L = 0.00005L$

95. $60(50) = 3{,}000mg$

$\frac{3{,}000mg}{1} \cdot \frac{1g}{1{,}000mg} = 3g$

97. $0.5 \cdot 6 = 3L$

$\frac{3L}{1} \cdot \frac{1{,}000mL}{1L} = 3{,}000mL$

99. 3 bottles would only have 0.852kg . You would need 4 bottles.

101. $1{,}500cc = 1{,}500mL$

$\frac{1{,}500mL}{1} \cdot \frac{1L}{1{,}000mL} = 1.5L$

WRITING

103. We move 3 places to the right because we are multiplying by 1,000.

105. century – 100 years

centipede – 100 legs

centennial – 100th birthday

cent – 1/100 of a dollar

centiliter – 1/100 of a liter

REVIEW

107.

$$9\overline{)8.00}\quad 0.88\ldots$$
$$\frac{-72}{80}\qquad \frac{-72}{8}$$

$8 \div 9 = 0.\overline{8}$

109.

$$90\overline{)7.000}\quad 0.077\cdots$$
$$\frac{-630}{700}\qquad \frac{-630}{700}$$

$\frac{7}{90} = .0777... = 0.0\overline{7}$

Section 6.5: Converting between American and Metric Units

VOCABULARY

1. In the American system, temperatures are measured in degrees Fahrenheit. In the metric system, temperatures are measured in degrees Celsius.

CONCEPTS

3. a. meter b. meter

c. inch d. mile

5. a. liter b. liter c. gallon

7. a. $\frac{0.30m}{1ft}$ b. $\frac{0.45kg}{1lb}$ c. $\frac{3.79L}{1gal}$

NOTATION

9.
$$4,500ft \approx \frac{4,500ft}{1}\cdot\frac{0.30m}{1ft}$$
$$\approx 1,350m$$

11.
$$3kg \approx \frac{3kg}{1}\cdot\frac{1,000g}{1kg}\cdot\frac{0.035oz}{1g}$$
$$\approx 3\cdot 1,000\cdot 0.035oz$$
$$\approx 105oz$$

GUIDED PRACTICE

13. $\frac{25cm}{1}\cdot\frac{0.39in}{1cm} \approx 9.75in \approx 10in$

15. $\frac{88cm}{1}\cdot\frac{0.39in}{1cm} \approx 34.32in \approx 34in$

17. $\frac{8,400ft}{1}\cdot\frac{0.3m}{1ft} \approx 2,520m$

19. $\frac{25,115ft}{1}\cdot\frac{0.3m}{1ft} \approx 7,534.5m$

21. $\frac{20lb}{1}\cdot\frac{16oz}{1lb}\cdot\frac{28.35g}{1oz} \approx 9,072g$

23. $\frac{75lb}{1}\cdot\frac{16oz}{1lb}\cdot\frac{28.35g}{1oz} \approx 34,020g$

25. $\frac{6.5kg}{1}\cdot\frac{2.2lb}{1kg} \approx 14.3lb$

27. $\frac{300kg}{1}\cdot\frac{2.2lb}{1kg} \approx 660lb$

29.
$$\frac{650mL}{1}\cdot\frac{1L}{1,000mL}\cdot\frac{1.06qt}{1L}$$
$$\approx \frac{689}{1000}qt \approx .689qt \approx 0.7qt$$

31.
$$\frac{1,200mL}{1}\cdot\frac{1L}{1,000mL}\cdot\frac{1.06qt}{1L}$$
$$\approx \frac{1,272}{1000}qt \approx 1.272qt \approx 1.3qt$$

33.
$$C = \frac{5}{9}(120-32)$$
$$= \frac{5}{9}(88) = \frac{440}{9}$$
$$\approx 48.8888\ldots \approx 48.9°C$$

34.
$$C = \frac{5}{9}(110-32)$$
$$= \frac{5}{9}(78) = \frac{390}{9}$$
$$\approx 43.333\ldots \approx 43.3°C$$

35.
$$C = \frac{5}{9}(35-32)$$
$$= \frac{5}{9}(3) = \frac{15}{9}$$
$$\approx 1.6666\ldots \approx 1.7°C$$

36.
$$C = \frac{5}{9}(45-32)$$
$$= \frac{5}{9}(13) = \frac{65}{9}$$
$$\approx 7.2222\ldots \approx 7.2°C$$

37.
$$F = \frac{9}{5}(75)+32$$
$$= 135+32 = 167°F$$

38.
$$F = \frac{9}{5}(85)+32$$
$$= 153+32 = 185°F$$

39.
$$F = \frac{9}{5}(10)+32$$
$$= 18+32 = 50°F$$

40.
$$F = \frac{9}{5}(20)+32$$
$$= 36+32 = 68°F$$

TRY IT YOURSELF

41.
$$\frac{25lb}{1}\cdot\frac{16oz}{1lb}\cdot\frac{28.35g}{1oz} \approx 11{,}340g$$

42.
$$\frac{7.5oz}{1}\cdot\frac{28.35g}{1oz} \approx 212.6g$$

43.
$$F = \frac{9}{5}(50)+32$$
$$= 90+32 = 122°F$$

45.
$$\frac{0.75qt}{1}\cdot\frac{1L}{1.06qt}\cdot\frac{1{,}000mL}{1L}$$
$$\approx \frac{750}{1.06}mL \approx 707.5mL$$

47.
$$\frac{0.5kg}{1}\cdot\frac{1lb}{0.45kg}\cdot\frac{16oz}{1lb}$$
$$\approx \frac{8}{0.45}oz \approx 17.8oz$$

49.
$$\frac{3.75m}{1}\cdot\frac{1yd}{0.91m}\cdot\frac{3\,ft}{1yd}\cdot\frac{12in}{1\,ft}$$
$$\approx \frac{135}{0.91}in \approx 148.4in$$

51.
$$\frac{3\,fl\,oz}{1}\cdot\frac{1L}{33.81\,fl\,oz} \approx 0.1L$$

53.
$$\frac{12km}{1}\cdot\frac{1mi}{1.61km}\cdot\frac{5{,}280\,ft}{1mi}$$
$$\approx \frac{63{,}360}{1.61}ft \approx 39{,}354\,ft$$

55.
$$\frac{37oz}{1}\cdot\frac{28.35g}{1oz}\cdot\frac{1kg}{1{,}000g}$$
$$\approx \frac{1{,}048.95}{1{,}000}kg \approx 1.0kg$$

57.
$$F = \frac{9}{5}(-10)+32$$
$$= -18+32 = 14°F$$

59.
$$\frac{17g}{1}\cdot\frac{0.035oz}{1g} \approx 0.6oz$$

61.
$$\frac{7.2L}{1}\cdot\frac{33.81\,fl\,oz}{1L} \approx 243.4\,fl\,oz$$

63.
$$\frac{3\,ft}{1}\cdot\frac{12in}{1\,ft}\cdot\frac{2.54cm}{1in} \approx 91.4cm$$

65.
$$\frac{500mL}{1}\cdot\frac{1L}{1{,}000mL}\cdot\frac{1.06qt}{1L} = \frac{530}{1{,}000}qt$$
$$= 0.53qt \approx 0.5qt$$

67.
$$C = \frac{5}{9}(50 - 32)$$
$$= \frac{5}{9}(18) = 10°C$$

69.
$$\frac{5{,}000in}{1} \cdot \frac{1ft}{12in} \cdot \frac{1m}{3.28ft}$$
$$\approx \frac{5{,}000}{39.36} m \approx 127m$$

71.
$$C = \frac{5}{9}(-5 - 32)$$
$$= \frac{5}{9}(-37) = \frac{-185}{9}$$
$$\approx -20.5555\ldots \approx -20.6°C$$

APPLICATIONS

73.
$$\frac{8km}{1} \cdot \frac{.62mi}{1km} = 4.96mi$$
About 5 miles.

75.
$$\frac{112km}{1hr} \cdot \frac{62mi}{1km} \approx 69mph$$

77.
$$\frac{6{,}288 \text{ ft}}{1} \cdot \frac{1 \text{ mi}}{5{,}280 \text{ ft}} \cdot \frac{1.61 \text{ km}}{1 \text{ mi}} \approx 1.9 \; km$$

79.
$$\frac{0.75in}{1} \cdot \frac{2.54cm}{1in} \approx 1.9cm$$

81.
$$\frac{182.5kg}{1} \cdot \frac{2.2lb}{1kg} \approx 402lb$$
$$\frac{363kg}{1} \cdot \frac{2.2lb}{1kg} \approx 799lb$$

83.
$$\frac{8oz}{1} \cdot \frac{28.35g}{1oz} \approx 226.8g$$
$$\frac{8fl\,oz}{1} \cdot \frac{29.57mL}{1fl\,oz} \cdot \frac{1L}{1{,}000mL} \approx 0.24L$$

85.
$$\frac{32kg}{1} \cdot \frac{2.2lb}{1kg} \approx 70.4lb. \text{ No.}$$

87.
$$C = \frac{5}{9}(143 - 32) = \frac{5}{9}(111) = 61.\overline{6}$$
About 62°C

89. 15°C = 59°F

50°C=122°F

Choose the 38°C (approximately 100°F.)

91. -5°C and 0°C only: 10°C is too warm for snow.

93. The three quart purchase: $\frac{\$4.50}{3qt} = \$1.50/qt$

The two liter purchase: $\frac{\$3.6}{2L} \cdot \frac{1L}{1.06} \approx \$1.70/qt$

The 3 quart purchase is the better buy.

WRITING

95. Convert to miles by multiplying the number of kilometers by 0.62 or by dividing by 1.61.

97. It is possibly costing money because people can't easily compare the prices to see if the United States prices are a better deal.

REVIEW

99.
$$\frac{3}{5} + \frac{4}{3} = \frac{9}{15} + \frac{20}{15} = \frac{29}{15}$$

101.
$$\frac{3}{5} \cdot \frac{4}{3} = \frac{4}{5}$$

103.
$$\begin{array}{r} \overset{1}{3}.25 \\ +4.80 \\ \hline 8.05 \end{array}$$

105.
$$\begin{array}{r} 325 \\ \times 48 \\ \hline 2{,}600 \\ +13{,}000 \\ \hline 15{,}600 \end{array}$$

Since the original factors had a total of 3 places right of the decimal, this becomes 15.6.

Chapter 6 Review

1. $\frac{7}{25}$

2. $\frac{15}{16}$

3. $\frac{24}{36}=\frac{2\cdot\cancel{12}}{3\cdot\cancel{12}}=\frac{2}{3}$

4. $\frac{21}{14}=\frac{3\cdot\cancel{7}}{2\cdot\cancel{7}}=\frac{3}{2}$

5. $\frac{4in}{12in}=\frac{\cancel{4}\,\cancel{in}}{3\cdot\cancel{4}\,\cancel{in}}=\frac{1}{3}$

6. $\frac{63m}{72m}=\frac{7\cdot\cancel{9}\,\cancel{m}}{8\cdot\cancel{9}\,\cancel{m}}=\frac{7}{8}$

7. $\frac{0.28}{0.35}=\frac{28}{35}=\frac{4\cdot\cancel{7}}{5\cdot\cancel{7}}=\frac{4}{5}$

8. $\frac{5.1}{1.7}=\frac{51}{17}=\frac{3\cdot\cancel{17}}{\cancel{17}}=\frac{3}{1}$

9. $\frac{2\frac{1}{3}}{2\frac{2}{3}}=\frac{\frac{7}{3}}{\frac{8}{3}}=\frac{7}{\cancel{3}}\cdot\frac{\cancel{3}}{8}=\frac{7}{8}$

10. $\frac{4\frac{1}{6}}{3\frac{1}{3}}=\frac{\frac{25}{6}}{\frac{10}{3}}=\frac{25}{6}\cdot\frac{3}{10}$

$=\frac{\cancel{5}\cdot 5\cdot\cancel{3}}{2\cdot\cancel{3}\cdot 2\cdot\cancel{5}}=\frac{5}{4}$

11. $\frac{15\,\text{min}}{3hr}=\frac{15\,\text{min}}{180\,\text{min}}=\frac{\cancel{15}\,\cancel{\text{min}}}{12\cdot\cancel{15}\,\cancel{\text{min}}}=\frac{1}{12}$

12. $\frac{8oz}{2lb}=\frac{8oz}{32oz}=\frac{\cancel{8}\,\cancel{oz}}{4\cdot\cancel{8}\,\cancel{oz}}=\frac{1}{4}$

13. $\frac{64cm}{12yr}=\frac{\cancel{4}\cdot 16cm}{3\cdot\cancel{4}\,yr}=\frac{16cm}{3yr}$

14. $\frac{\$15}{25\,\text{min}}=\frac{\$3\cdot\cancel{5}}{5\cdot\cancel{5}\,\text{min}}=\frac{\$3}{5\,\text{min}}$

15. $\frac{600t}{20\,\text{min}}=\frac{\cancel{20}\cdot 30t}{\cancel{20}\,\text{min}}=30$ tickets per minute

16. $\frac{45in}{3t}=\frac{\cancel{3}\cdot 15in}{\cancel{3}t}=15$ inches per turn

17. $\frac{195\,ft}{6r}=32.5$ feet per roll

18. $\frac{48cal}{15p}=3.2$ calories per piece

19. $\frac{\$11.45}{5p}=\$2.29$ per pair

20. $\frac{\$3b}{12mo}=\$0.25$ billion per month

21. $\frac{185}{160}=\frac{\cancel{5}\cdot 37}{\cancel{5}\cdot 32}=\frac{37}{32}$

22. $\frac{\$333.25}{43hr}=\$7.75/hr$

The student makes \$7.75 per hour.

23. $\frac{54{,}000\,people}{48\,\text{min}}=1{,}125\,people/\text{min}$

1,125 people per minute

24. 12 oz. can: $\frac{\$4.95}{12oz}=\$0.4125/oz$

8 oz. can: $\frac{\$3.25}{8oz}=\$0.40625/oz$

The 8 ounce can is the better buy.

25. a. $\frac{20}{30}=\frac{2}{3}$

b. $\frac{6\,\text{buses}}{100\,\text{cars}}=\frac{36\,\text{buses}}{600\,\text{cars}}$

26. $2\cdot 27=54 \qquad 9\cdot 6=54$

27. $\frac{8}{12}=\frac{2}{3}\neq\frac{3}{7}$

The proportion is false.

28. $\frac{4}{18}=\frac{2}{9}$

$\frac{10}{45}=\frac{2}{9}$

The proportion is true.

29. $9\cdot 6\stackrel{?}{=}2\cdot 27$

$54\stackrel{?}{=}54$

The proportion is true.

30. $17\cdot 21\stackrel{?}{=}7\cdot 51$

$357\stackrel{?}{=}357$

The proportion is true.

31. $3.5\cdot 3\stackrel{?}{=}1.2\cdot 9.3$

$10.5\stackrel{?}{=}11.16$

The proportion is false.

32. $\left(1\frac{1}{2}\right)\left(1\frac{1}{7}\right)\stackrel{?}{=}\left(3\frac{1}{3}\right)\left(\frac{1}{4}\right)$

$\frac{3}{2}\cdot\frac{8}{7}\stackrel{?}{=}\frac{10}{3}\cdot\frac{1}{4}$

$\frac{24}{14}\stackrel{?}{=}\frac{10}{12}$

$\frac{12}{7}\stackrel{?}{=}\frac{5}{6}$

The proportion is false.

33. $\frac{20}{36}=\frac{\cancel{4}\cdot 5}{\cancel{4}\cdot 9}=\frac{5}{9}$

The numbers are proportional.

34. $\frac{29}{54}$ does not reduce to $\frac{7}{13}$ so the numbers are not proportional.

35. $\frac{12}{18}=\frac{3}{x}$

$12x=3\cdot 18$

$x=\frac{54}{12}$

$x=4.5$

36. $\frac{4}{x}=\frac{2}{8}$

$4\cdot 8=2x$

$\frac{32}{2}=x$

$16=x$

37. $\frac{4.8}{6.6}=\frac{x}{9.9}$

$4.8\cdot 9.9=x\cdot 6.6$

$\frac{47.52}{6.6}=x$

$7.2=x$

38. $\frac{0.08}{x}=\frac{0.04}{0.06}$

$0.08\cdot 0.06=0.04\cdot x$

$\frac{0.0048}{0.04}=x$

$0.12=x$

39. $\frac{1\frac{9}{11}}{x}=\frac{3\frac{1}{3}}{2\frac{3}{4}}$

$\left(\frac{20}{11}\right)\left(\frac{11}{4}\right)=\left(\frac{10}{3}\right)x$

$5=\frac{10}{3}x$

$\frac{3}{10}\cdot 5=\frac{3}{10}\cdot\frac{10}{3}x$

$\frac{3}{2}=x=1\frac{1}{2}$

40.
$$\frac{\frac{4}{5}}{1\frac{1}{20}}=\frac{2\frac{2}{3}}{x}$$
$$\frac{4}{5}x=\frac{8}{3}\cdot\frac{21}{20}$$
$$\frac{5}{4}\cdot\frac{4}{5}x=\frac{5}{4}\cdot\frac{8}{3}\cdot\frac{21}{20}$$
$$x=\frac{7}{2}=3\frac{1}{2}$$

41.
$$\frac{\frac{2}{3}}{\frac{1}{2}}=\frac{x}{0.25}$$
$$\frac{2}{3}\cdot\frac{1}{4}=x\cdot\frac{1}{2}$$
$$\frac{1}{6}=x\cdot\frac{1}{2}$$
$$2\cdot\frac{1}{6}=x\cdot\frac{1}{2}\cdot 2$$
$$\frac{1}{3}=x$$

42.
$$\frac{x}{300}=\frac{5{,}000}{1{,}500}$$
$$1{,}500x=5{,}000\cdot 300$$
$$x=\frac{1{,}500{,}000}{1{,}500}$$
$$x=1{,}000$$

43.
$$\frac{35}{2}=\frac{x}{11}$$
$$35\cdot 11=x\cdot 2$$
$$\frac{385}{2}=x$$
$$192.5mi=x$$

44.
$$\frac{66}{12}=\frac{1{,}650}{x}$$
$$66x=1{,}650\cdot 12$$
$$x=\frac{19{,}800}{66}$$
$$x=300\text{ parts}$$

45.
$$\frac{\frac{1}{8}}{1}=\frac{1\frac{1}{2}}{x}$$
$$\frac{1}{8}x=1\cdot 1\frac{1}{2}$$
$$8\cdot\frac{1}{8}x=8\cdot\frac{3}{2}$$
$$x=12\,ft$$

46.
$$\frac{10}{8\frac{1}{2}}=\frac{x}{25\frac{1}{2}}$$
$$10\cdot\frac{51}{2}=\frac{17}{2}\cdot x$$
$$10\cdot\frac{51}{2}\cdot\frac{2}{17}=\frac{2}{17}\cdot\frac{17}{2}\cdot x$$
$$30in=x$$

47. a. 16 b. $\frac{7}{16}in$, $1\frac{1}{2}in$, $1\frac{3}{4}in$, $2\frac{5}{8}in$

48. $1\frac{1}{2}in$

49. $\frac{1mi}{5{,}280\,ft}=1$; $\frac{5{,}280\,ft}{1mi}=1$

50. a. min b. sec

51. $$\frac{5yd}{1}\cdot\frac{3\,ft}{1yd}=15\,ft$$

52. $$\frac{6yd}{1}\cdot\frac{3\,ft}{1yd}\cdot\frac{12in}{1\,ft}=216in$$

53. $$\frac{66in}{1}\cdot\frac{1\,ft}{12in}=\frac{11}{2}\,ft=5\frac{1}{2}\,ft$$

54. $$\frac{9{,}240\,ft}{1}\cdot\frac{1mi}{5{,}280\,ft}=\frac{7}{4}mi=1\frac{3}{4}mi$$

55. $$\frac{4\frac{1}{2}\,ft}{1}\cdot\frac{12in}{1\,ft}=\frac{9}{2}\cdot 12in=54in$$

56. $$\frac{1mi}{1}\cdot\frac{5{,}280\,ft}{1mi}\cdot\frac{1yd}{3\,ft}=1{,}760\,yd$$

57. $\frac{32oz}{1} \cdot \frac{1lb}{16oz} = 2lb$

58. $\frac{17.2lb}{1} \cdot \frac{16oz}{1lb} = 275.2oz$

59. $\frac{3\text{tons}}{1} \cdot \frac{2{,}000lb}{1\text{ton}} \cdot \frac{16oz}{1lb} = 96{,}000oz$

60. $\frac{4{,}500lb}{1} \cdot \frac{1\text{ton}}{2{,}000lb} = \frac{9}{4}\text{ tons} = 2\frac{1}{4}\text{ tons}$

61. $\frac{5pt}{1} \cdot \frac{2c}{1pt} \cdot \frac{8fl\,oz}{1c} = 80\,fl\,oz$

62. $\frac{8c}{1} \cdot \frac{1pt}{2c} \cdot \frac{1qt}{2pt} \cdot \frac{1gal}{4qt} = \frac{8}{16}gal = \frac{1}{2}gal$

63. $\frac{17qt}{1} \cdot \frac{2pt}{1qt} \cdot \frac{2c}{1pt} = 68c$

64. $\frac{176fl\,oz}{1} \cdot \frac{1c}{8fl\,oz} \cdot \frac{1pt}{2c} \cdot \frac{1qt}{2pt}$

$= \frac{176}{32}qt = \frac{11}{2}qt = 5.5qt$

65. $\frac{5gal}{1} \cdot \frac{4qt}{1gal} \cdot \frac{2pt}{1qt} = 40pt$

66. $\frac{3.5gal}{1} \cdot \frac{4qt}{1gal} \cdot \frac{2pt}{1qt} \cdot \frac{2c}{1pt}$

$= 3.5 \cdot 16c = 56c$

67. $\frac{20\text{min}}{1} \cdot \frac{60\text{sec}}{1\text{min}} = 1{,}200\text{sec}$

68. $\frac{900\text{sec}}{1} \cdot \frac{1\text{min}}{60\text{sec}} = 15\text{min}$

69. $\frac{200hr}{1} \cdot \frac{1d}{24hr} = \frac{25}{3}d = 8\frac{1}{3}d$

70. $\frac{6hr}{1} \cdot \frac{60\text{min}}{1hr} = 360\text{min}$

71. $\frac{4.5d}{1} \cdot \frac{24hr}{1d} = 108hr$

72. $\frac{1d}{1} \cdot \frac{24hr}{1d} \cdot \frac{60\text{min}}{1hr} \cdot \frac{60\text{sec}}{1\text{min}} = 86{,}400\text{sec}$

73. $\frac{210yd}{1} \cdot \frac{3ft}{1yd} \cdot \frac{1mi}{5{,}280ft} = \frac{21}{176}mi$

$\approx 0.12mi$

74. $\frac{40{,}500lb}{1} \cdot \frac{1\text{ton}}{2{,}000lb} = \frac{81}{4}\text{tons} = 20\frac{1}{4}\text{tons}$

75. $\frac{1{,}454ft}{1} \cdot \frac{1yd}{3ft} = \frac{1{,}454}{3}yd = 484\frac{2}{3}yd$

76. $\frac{50gal}{1} \cdot \frac{4qt}{1gal} \cdot \frac{1\ bottle}{2qt} = 100\text{ bottles}$

77. a. 10, 1mm b. 19mm, 3cm, 45mm, 62mm

78. 4cm

79. a. $\frac{1km}{1{,}000m} = 1$; $\frac{1{,}000m}{1km} = 1$

b. $\frac{1g}{100cg} = 1$; $\frac{100cg}{1g} = 1$

80. 5 places to the left

81. Refer to a conversion chart: 2 places left
$475cm = 4.75m$

82. Refer to a conversion chart: 3 places right
$8m = 8{,}000mm$

83. Refer to a conversion chart: 3 places right
$165.7km = 165{,}700m$

84. Refer to a conversion chart: 1 place left
$6{,}789cm = 678.9dm$

85. Refer to a conversion chart: 3 places left
$25mcg = 0.025mg$

86. Refer to a conversion chart: 2 places left
$800cg = 8g$

87. Refer to a conversion chart: 3 places left
$5{,}425g = 5.425kg$

88. Refer to a conversion chart: 3 places right
$5,425g = 5,425,000mg$

89. Refer to a conversion chart: 2 places left
$150cL = 1.5L$

90. Refer to a conversion chart: 3 places left
$3,250L = 3.25kL$

91. Refer to a conversion chart: 1 place left
$400mL = 40cL$

92. Refer to a conversion chart: 3 places right
$1hL = 1,000dL$

93. $\frac{1,350g}{1} \cdot \frac{1kg}{1,000g} = \frac{1,350}{1,000} kg = 1.35kg$

94. $12 \cdot 20mL = 240mL$
$\frac{240mL}{1} \cdot \frac{1L}{1,000mL} = \frac{240}{1000} L = 0.24L$

95. $100 \cdot 500mg = 50,000mg$
$\frac{50,000mg}{1} \cdot \frac{1g}{1,000mg} = \frac{50,000}{1,000} g = 50g$

96. $1L = 1,000mL$

97. $\frac{50m}{1} \cdot \frac{3.28ft}{1m} \approx 164ft$

98. $\frac{443m}{1} \cdot \frac{3.28ft}{1m} \approx 1,453.04ft$
The Willis tower is taller.

99. $\frac{1,930mi}{1} \cdot \frac{1.61km}{mi} \approx 3,107km$

100. $6ft\ 4in = 76in$
$\frac{76in}{1} \cdot \frac{2.54cm}{1in} \approx 193cm$

101. $\frac{30oz}{1} \cdot \frac{28.35g}{1oz} \approx 850.5g$

102. $\frac{15kg}{1} \cdot \frac{2.2lb}{1kg} \approx 33lb$

103. $\frac{50lb}{1} \cdot \frac{0.45kg}{1lb} \cdot \frac{1,000g}{1kg} \approx 22,500g$

104. $\frac{2,000lb}{1} \cdot \frac{0.45kg}{1lb} \approx 900kg$

105. $\frac{910g}{1} \cdot \frac{1kg}{1,000g} \cdot \frac{2.2lb}{1kg} \approx \frac{2002}{1,000} lb \approx 2lb$

106. $\frac{0.5L}{1} \cdot \frac{33.81fl\,oz}{1L} \approx 16.905fl\,oz$
The LaCroix bottles are bigger.

107. $\frac{42gal}{1barrel} \cdot \frac{4qt}{1gal} \cdot \frac{0.95L}{1qt} \approx 159.6L/barrel$

108.
$$F = \frac{9}{5}(105) + 32$$
$$= 9(21) + 32$$
$$= 189 + 32$$
$$= 221°F$$

109.
$$C = \frac{5}{9}(77 - 32) = \frac{5}{9}(45)$$
$$= \frac{225}{9} = 25°C$$

110. (Use the chart on page 618.)

$10°C = 50°F$ - too cold

$30°C = 86°F$ - just great

$50°C = 122°F$ - too hot

$70°C = 158°F$ - too hot

Chapter 6 Test

1. a. A ratio is the quotient of two numbers or the quotient of two quantities that have the same units.

b. A rate is the quotient of two quantities that have different units.

c. A proportion is a statement that two ratios (or rates) are equal.

d. The cross products for the proportion $\frac{3}{8}=\frac{6}{16}$ are $3\cdot16$ and $8\cdot6$.

e. *Deci* means tenths, *centi* means hundredths, and *milli* means thousandths.

f. The meter, the gram, and the liter are basic units of measurement in the metric system.

g. In the American system, temperatures are measured in degrees Fahrenheit. In the metric system, temperatures are measured in degrees Celsius.

2. $\frac{36}{52}=\frac{9}{13}$; $9:13$; $9\text{ to }13$

3. $\frac{6\,ft}{8\,ft}=\frac{\cancel{2}\cdot3\,\cancel{ft}}{\cancel{2}\cdot4\,\cancel{ft}}=\frac{3}{4}$

4. $\frac{8oz}{3lb}=\frac{8oz}{48oz}=\frac{\cancel{8}\,\cancel{oz}}{6\cdot\cancel{8}\,\cancel{oz}}=\frac{1}{6}$

5. $\frac{0.26}{0.65}=\frac{26}{65}=\frac{2\cdot\cancel{13}}{5\cdot\cancel{13}}=\frac{2}{5}$

6. $\frac{3\frac{1}{3}}{3\frac{8}{9}}=\frac{\frac{10}{3}}{\frac{35}{9}}=\frac{10}{3}\cdot\frac{9}{35}$

$=\frac{2\cdot\cancel{5}\cdot\cancel{3}\cdot3}{\cancel{3}\cdot\cancel{5}\cdot7}=\frac{6}{7}$

7. $\frac{54\,ft}{36\sec}=\frac{3\cdot18\,ft}{2\cdot18\sec}=\frac{3\,ft}{2\sec}$

8. 2 pound can: $\frac{\$3.38}{2lb}=\$1.69/lb$

5 pound can: $\frac{\$8.50}{5lb}=\$1.70/lb$

The 2 pound can is a better buy.

9. $\frac{675kwh}{30d}=22.5$ kilowatt hours per day

10. $\frac{15\text{ billboards}}{50\text{ miles}}=\frac{3\text{ billboards}}{10\text{ miles}}$

11. a. $25\cdot3\stackrel{?}{=}33\cdot2$

$75\neq66$

Not a proportion

b. $2.2\cdot2.8\stackrel{?}{=}3.5\cdot1.76$

$6.16=6.16$

Is a proportion

12. $\frac{35}{75}=\frac{\cancel{5}\cdot7}{\cancel{5}\cdot15}=\frac{7}{15}$

The numbers are proportional.

13. $\frac{x}{3}=\frac{35}{7}$

$7x=35\cdot3$

$x=\frac{35\cdot3}{7}=\frac{105}{7}$

$x=15$

14. $\frac{15.3}{x}=\frac{3}{12.4}$

$15.3\cdot12.4=3x$

$\frac{189.72}{3}=x$

$63.24=x$

15. $$\frac{2\frac{2}{9}}{\frac{4}{3}} = \frac{x}{1\frac{1}{2}}$$
$$\left(\frac{20}{9}\right)\left(\frac{3}{2}\right) = x\left(\frac{4}{3}\right)$$
$$\left(\frac{3}{4}\right)\left(\frac{20}{9}\right)\left(\frac{3}{2}\right) = x$$
$$x = \frac{5}{2} = 2\frac{1}{2}$$

16. $$\frac{25}{\frac{1}{10}} = \frac{50}{x}$$
$$25x = \frac{1}{10} \cdot 50$$
$$x = \frac{5}{25}$$
$$x = \frac{1}{5}$$

17. $$\frac{13oz}{\$2.79} = \frac{16oz}{x}$$
$$13x = 2.79 \cdot 16$$
$$x = \frac{44.64}{13}$$
$$x \approx \$3.43$$

18. $$\frac{1\frac{2}{3}}{5} = \frac{x}{6}$$
$$\frac{5}{3} \cdot 6 = 5x$$
$$\frac{10}{5} = x$$
$$2c = x$$

19. a. 16

b. $\frac{5}{16}in$, $1\frac{3}{8}in$, $2\frac{3}{4}in$

20. introduce

eliminate

21. $$\frac{180in}{1} \cdot \frac{1ft}{12in} = 15ft$$

22. $$\frac{25ft}{1} \cdot \frac{1yd}{3ft} = \frac{25}{3}yd = 8\frac{1}{3}yd$$

23. $$\frac{10\frac{3}{4}lb}{1} \cdot \frac{16oz}{1lb} = \frac{43}{4} \cdot 16oz = 172oz$$

24. $$\frac{1.6\,\text{tons}}{1} \cdot \frac{2000lbs}{1\,\text{ton}} = 3,200lbs$$

25. $$\frac{1gal}{1} \cdot \frac{4qt}{1gal} \cdot \frac{2pt}{1qt} \cdot \frac{2c}{1pt} \cdot \frac{8fl\,oz}{1c} = 128fl\,oz$$

26. $$\frac{80d}{1} \cdot \frac{24hr}{1d} \cdot \frac{60\,\text{min}}{1hr} = 115,200\,\text{min}$$

27. a. the larger one b. the longer one

c. the right side

28. 12mm, 5cm, 65mm

29. $$\frac{500m}{1} \cdot \frac{1km}{1,000m} = \frac{1}{2}km = 0.5km$$

30. $$\frac{5m}{1} \cdot \frac{100cm}{1m} = 500cm$$

31. $$\frac{5,000mcg}{1} \cdot \frac{1g}{1,000,000mcg} \cdot \frac{1,000mg}{1g} = 5mg$$

32. $$\frac{70L}{1} \cdot \frac{1,000mL}{1L} = 70,000mL$$

33. $$\frac{50t}{1} \cdot \frac{150mg}{1t} \cdot \frac{1g}{1,000mg} = 7.5g$$

34. $$\frac{100yd}{1} \cdot \frac{0.91m}{1yd} = 91m$$

The 100 yard race is longer.

35. $$\frac{160lb}{1} \cdot \frac{0.45kg}{1lb} \approx 72kg$$

Jim weighs more than Ricardo does.

36. $\frac{810mL}{1}\cdot\frac{1L}{1{,}000mL}\cdot\frac{1.06qt}{1L}=\frac{858.6}{1{,}000}qt$
$=0.8586qt\approx 0.9qt$

37. $\frac{16.5in}{1}\cdot\frac{2.54cm}{1in}\approx 41.91cm\approx 42cm$

38.
$F=\frac{9}{5}(83)+32$
$F=149.4+32$
$F=181.4$
$F\approx 182°F$

39. A scale is a ratio comparing the size of a drawing to the size of the actual object. For example, a map could be on the scale 1:6, where 1 inch represents 6 miles.

40. It is easier to convert between units in the metric system since multiplying and dividing by powers of 10 is simply a matter of moving the decimal point.

Chapters 1 – 6 Cumulative Review

1. a. five million, seven hundred sixty-four thousand, five hundred two

b. 5,000,000 + 700,000 + 60,000 + 4,000 + 500 + 2

2. a. Detroit:
$38+36+34+37+14+12+15=186$

Denver:
$34+40+39+32+14+12+13=184$

b. Detroit won.

c. $186+184=370$ total points

3.
$$\begin{array}{r} \overset{6\ 10\ 9\ 10\ 9\ 10\ 16}{\cancel{7}\cancel{0},\cancel{0}\cancel{0}\cancel{6}} \\ -\quad 348 \\ \hline 69{,}658 \end{array}$$

4.
$$\begin{array}{r} 504 \\ \times 729 \\ \hline 4{,}536 \\ 10{,}080 \\ +352{,}800 \\ \hline 367{,}416 \end{array}$$

5.
$$\begin{array}{r} 20 \\ 37\overline{)743} \\ -74 \\ \hline 03 \\ -0 \\ \hline 3 \end{array}$$
$20R3$

6. 1, 2, 3, 5, 6, 10, 15, 30

7. $360=36\cdot 10=6\cdot 6\cdot 10$
$=2\cdot 3\cdot 2\cdot 3\cdot 2\cdot 5=2^3\cdot 3^2\cdot 5$

8. $20=2\cdot 2\cdot 5$
$28=2\cdot 2\cdot 7$
$LCM(20,28)=2\cdot 2\cdot 5\cdot 7=140$
$GCF(20,28)=2\cdot 2=4$

9. $81+9\left[7^2-7(11-4)\right]=81+9\left[49-7(7)\right]$
$=81+9[49-49]=81-9(0)=81$

10. a. $m+158=203$

$m+158-158=203-158$

$m=45$

Check: $m+158=203$

$45+158\stackrel{?}{=}203$

$203=203$

The solution checks.

b. $\frac{n}{57}=300$

$57\cdot\frac{n}{57}=57\cdot 300$

$n=17{,}100$

Check: $\frac{n}{57}=300$

$\frac{17{,}100}{57}\stackrel{?}{=}300$

$300=300$

11. $-(-10)\square|-11|$

$10\square 11$

$-(-10)\boxed{<}|-11|$

12. $(-12+6)+(-6+8)$

$=-6+2=-4$

13. $3-(-12)=3+12=15$ shots

14. $-3^2=-3\cdot 3=-9$

$(-3)^2=(-3)(-3)=9$

15. a. -8 b. undefined c. -8

d. 0 e. 8 f. 0

16. $\frac{3+3[5(-6)-(1-10)]}{-1+(-1)}$

$=\frac{3+3[-30-(-9)]}{-2}=\frac{3+3[-30+9]}{-2}$

$=\frac{3+3[-21]}{-2}=\frac{3-63}{-2}=\frac{-60}{-2}=30$

17. $-3{,}900+(-5{,}800)+4{,}700$

$=-9{,}700+4{,}700=-5{,}000$

18. x = the gain in the polls

$-18+x=3$

$-18+x+18=3+18$

$x=21$

The candidate gained 21 points in support.

19. $t-4=-8-(-2)$

$t-4=-8+2$

$t-4=-6$

$t-4+4=-6+4$

$t=-2$

Check: $t-4=-8-(-2)$

$-2-4\stackrel{?}{=}-8-(-2)$

$-2-4\stackrel{?}{=}-8+2$

$-6=-6$

20. $-3a+(-2)=16$

$-3a+(-2)+2=16+2$

$-3a=18$

$\frac{-3a}{-3}=\frac{18}{-3}$

$a=-6$

Check: $-3a+(-2)=16$

$-3(-6)+(-2)\stackrel{?}{=}16$

$18+(-2)\stackrel{?}{=}16$

$16=16$

21. a. $w-29$ b. $\frac{10}{m^3}$

22. $\frac{-x+3(1+x)}{x^2-3}=\frac{-(-2)+3[1+(-2)]}{(-2)^2-3}$

$=\frac{2+3(-1)}{4-3}=\frac{2-3}{1}=\frac{-1}{1}=-1$

23. a. $-6(15a)=(-6\cdot 15)a=-90a$

b. $5(4x-2y+7)=20x-10y+35$

24. a. $5x-11x=(5-11)x=-6x$

b. $-4(x-3y)+5x-2y$
$=-4x+12y+5x-2y$
$=x+10y$

25.
$z+12+z=8z-4-2z$
$2z+12=6z-4$
$2z+12-2z=6z-4-2z$
$12=-4+4z$
$12+4=-4+4z+4$
$16=4z$
$\frac{16}{4}=\frac{4z}{4}$
$4=z$

Check: $z+12+z=8z-4-2z$
$(4)+12+(4)\stackrel{?}{=}8(4)-4-2(4)$
$20\stackrel{?}{=}32+4-8$
$20=20$

26.
$2y+7=2-(4y+7)$
$2y+7=2-4y-7$
$2y+7=-5-4y$
$2y+7+4y=-5-4y+4y$
$6y+7=-5$
$6y+7-7=-5-7$
$6y=-12$
$\frac{6y}{6}=\frac{-12}{6}$
$y=-2$

Check: $2(-2)+7=2-[4(-2)+7]$
$-4+7=2-(-8+7)$
$3=2-(-1)$
$3=3$

27. w = the width of the lawn
$2w+2(3w)=144$
$2w+6w=144$
$8w=144$
$\frac{8w}{8}=\frac{144}{8}$
$w=18$

The lawn is 18 feet wide and $3\cdot 18=54$ feet long.

28. $-\frac{16}{20}=-\frac{\not{4}\cdot 4}{\not{4}\cdot 5}=-\frac{4}{5}$

29. $\frac{9}{10}=\frac{9}{10}\cdot\frac{6a}{6a}=\frac{54a}{60a}$

30. $\frac{3}{10}\cdot 197{,}000{,}000=3\cdot 19{,}700{,}000$
$=59{,}100{,}000mi^2$

31. $A=\frac{1}{2}bh$

32. $\frac{5}{10b^3}\div\frac{1}{2b^2}=\frac{5}{10b^3}\cdot\frac{2b^2}{1}$
$=\frac{\not{5}\cdot\not{2}\cdot\not{b}\cdot\not{b}}{\not{2}\cdot\not{5}\cdot\not{b}\cdot\not{b}\cdot b}=\frac{1}{b}$

33. $\frac{11}{12}-\frac{7}{15}=\frac{11}{12}\cdot\frac{5}{5}-\frac{7}{15}\cdot\frac{4}{4}$
$=\frac{55}{60}-\frac{28}{60}=\frac{27}{60}=\frac{9}{20}$

34. $\frac{19}{5}\square\frac{5}{4}$
$\frac{19}{5}\cdot\frac{4}{4}\square\frac{5}{4}\cdot\frac{5}{5}$
$\frac{76}{20}>\frac{25}{20}$
$\frac{19}{15}$ is the larger fraction.

35. $\frac{5}{m}+\frac{7}{8}=\frac{5}{m}\cdot\frac{8}{8}+\frac{7}{8}\cdot\frac{m}{m}$
$=\frac{40}{8m}+\frac{7m}{8m}=\frac{40+7m}{8m}$

36. $$\frac{5}{32}+\frac{5}{16}+\frac{1}{2}=\frac{5}{32}+\frac{5}{16}\cdot\frac{2}{2}+\frac{1}{2}\cdot\frac{16}{16}$$
$$=\frac{5}{32}+\frac{10}{32}+\frac{16}{32}=\frac{31}{32}in$$

37. $$-15\frac{1}{3}\left(-\frac{9}{20}\right)=\frac{46}{3}\cdot\frac{9}{20}$$
$$=\frac{\cancel{2}\cdot 23\cdot\cancel{3}\cdot 3}{\cancel{3}\cdot\cancel{2}\cdot 10}=\frac{69}{10}=6\frac{9}{10}$$

38. $8\frac{1}{2}\div\frac{1}{4}=\frac{17}{2}\cdot\frac{4}{1}=\frac{68}{2}=34$ strips.

39. $1\frac{1}{2}-\frac{3}{4}=\frac{3}{2}-\frac{3}{4}=\frac{6}{4}-\frac{3}{4}=\frac{3}{4}$ hp

40. $$\frac{4-\frac{3}{4}}{-1\frac{7}{8}}=\frac{\frac{16}{4}-\frac{3}{4}}{-1\frac{7}{8}}=\frac{\frac{13}{4}}{-\frac{15}{8}}=-\frac{13}{4}\cdot\frac{8}{15}$$
$$=-\frac{13\cdot 2\cdot\cancel{4}}{\cancel{4}\cdot 15}=-\frac{26}{15}=-1\frac{11}{15}$$

41. $$\frac{7}{8}t=-28$$
$$\frac{8}{7}\cdot\frac{7}{8}t=\frac{8}{7}(-28)$$
$$t=8(-4)$$
$$t=-32$$
Check: $\frac{7}{8}t=-28$
$$\frac{7}{8}(-32)\stackrel{?}{=}-28$$
$$7(-4)\stackrel{?}{=}-28$$
$$-28=-28$$

42. $$\frac{4}{5}x=\frac{3}{4}x+\frac{1}{2}$$
$$20\cdot\frac{4}{5}x=20\cdot\frac{3}{4}x+20\cdot\frac{1}{2}$$
$$4\cdot 4x=5\cdot 3x+10$$
$$16x=15x+10$$
$$16x-15x=15x+10-15x$$
$$x=10$$
Check: $\frac{4}{5}x=\frac{3}{4}x+\frac{1}{2}$
$$\frac{4}{5}(10)\stackrel{?}{=}\frac{3}{4}(10)+\frac{1}{2}$$
$$8\stackrel{?}{=}7.5+0.5$$
$$8=8$$

43. t = the number of tickets sold
$$\frac{7}{8}t+200=t$$
$$\frac{7}{8}t+200-\frac{7}{8}t=t-\frac{7}{8}t$$
$$200=\frac{1}{8}t$$
$$8\cdot 200=8\cdot\frac{1}{8}t$$
$$1{,}600=t$$
There were 1,600 tickets sold for the concert.

44. $-64.22>-64.238$

45.

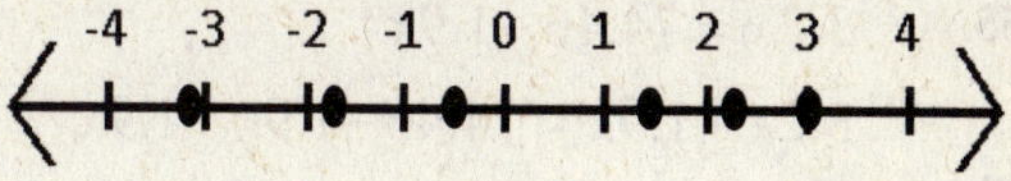

46. $-20.04+2.4=-17.64$

47. $-8.08-15.3=-8.08+(-15.3)=-23.38$

48. Since there are 2 zeros in 100, move the decimal place 2 units right.
$$2.5\cdot 100=250$$

49.

$$\begin{array}{r} 8.33 \\ \times\ \underline{55} \\ 4165 \\ +\underline{41650} \\ 458.15 \end{array}$$

50. Since there are 2 zeros in 100, move the decimal place 2 units left.

$$2.5 \div 100 = 0.025$$

51. $t = \dfrac{107.95}{8.5} = 12.7$

52.

$$\begin{array}{r} 0.0833 \\ 12\overline{)\,1.000} \\ -\underline{96} \\ 40 \\ -\underline{36} \\ 40 \\ -\underline{36} \\ 4 \end{array}$$

$$\frac{1}{12} = 0.08\overline{3}$$

53. $\frac{3}{4}(8.60) + \frac{2}{3}(5.25) = 6.45 + 3.50 = \$9.95$

54. $3\sqrt{25} + 4\sqrt{4} = 3 \cdot 5 + 4 \cdot 2 = 15 + 8 = 23$

55.

$$3.2x = 74.46 - 1.9x$$
$$3.2x + 1.9x = 74.46 - 1.9x + 1.9x$$
$$5.1x = 74.46$$
$$\frac{5.1x}{5.1} = \frac{74.46}{5.1}$$
$$x = 14.6$$

Check: $3.2x = 74.46 - 1.9x$

$$3.2(14.6) \stackrel{?}{=} 74.46 - 1.9(14.6)$$
$$46.72 \stackrel{?}{=} 74.46 - 27.74$$
$$46.72 = 46.72$$

56.

$$-5.2x = 108 - 6.1x$$
$$-5.2x + 6.1x = 108 - 6.1x + 6.1x$$
$$0.9x = 108$$
$$\frac{0.9x}{0.9} = \frac{108}{0.9}$$
$$x = 120$$

Check: $-5.2x = 108 - 6.1x$

$$-5.2(120) \stackrel{?}{=} 108 - 6.1(120)$$
$$-624 \stackrel{?}{=} 108 - 732$$
$$-624 = -624$$

57.

$$-2(x - 2.1) = -2.4$$
$$-2x + 4.2 = -2.4$$
$$-2x + 4.2 - 4.2 = -2.4 - 4.2$$
$$-2x = -6.6$$
$$\frac{-2x}{-2} = \frac{-6.6}{-2}$$
$$x = 3.3$$

Check: $-2(x - 2.1) = -2.4$

$$-2(3.3 - 2.1) \stackrel{?}{=} -2.4$$
$$-2(1.2) \stackrel{?}{=} -2.4$$
$$-2.4 = -2.4$$

58.

$$\frac{1}{5}x - 2.5 = -17.2$$
$$\frac{1}{5}x - 2.5 + 2.5 = -17.2 + 2.5$$
$$\frac{1}{5}x = -14.7$$
$$5 \cdot \frac{1}{5}x = 5(-14.7)$$
$$x = -73.5$$

Check: $\frac{1}{5}x - 2.5 = -17.2$

$$\frac{1}{5}(-73.5) - 2.5 \stackrel{?}{=} -17.2$$
$$-14.7 - 2.5 \stackrel{?}{=} -17.2$$
$$-17.2 = -17.2$$

59. s = the number of needed signatures

$20+0.05s=60$

$20+0.05s-20=60-20$

$0.05s=40$

$\frac{0.05s}{0.05}=\frac{40}{0.05}$

$s=800$

She needs to collect 800 signatures.

60. $\frac{3in}{15in}=\frac{\cancel{3}\,\cancel{in}}{\cancel{3}\cdot 5\cancel{in}}=\frac{1}{5}$

61. 94 pound bag: $\frac{\$4.48}{94lb}=\$0.0477/lb$

100 pound bag: $\frac{\$4.80}{100lb}=\$0.048/lb$

The 94 pound bag is a better deal.

62. $25\cdot 17\overset{?}{=}12\cdot 33$

$425\overset{?}{=}396$

The proportion is false.

63. $\frac{55}{12}=\frac{x}{44}$

$12\cdot x=55\cdot 44$

$12x=2420$

$x\approx 202mg$

64. $\frac{x}{3}=\frac{35}{7}$

$x\cdot 7=35\cdot 3$

$x=\frac{105}{7}$

$x=15$

65.

a. $\frac{40d}{1}\cdot\frac{24hr}{1d}=960hr$

b. $\frac{3d}{1}\cdot\frac{24h}{1d}\cdot\frac{60\min}{1h}=4{,}320\min$

c. $\frac{8\min}{1}\cdot\frac{60\sec}{1\min}=480\sec$

66. $\frac{40oz}{1}\cdot\frac{1lb}{16oz}=\frac{40}{16}lb=2.5lb$

67. $\frac{2.4m}{1}\cdot\frac{1{,}000mm}{1m}=2{,}400mm$

68. $\frac{320g}{1}\cdot\frac{1kg}{1{,}000g}=\frac{320}{1{,}000}kg=0.32kg$

69.

a. $\frac{2L}{1}\cdot\frac{1.06qt}{L}\approx 2.12qt$

$1gal=4qt$

A 1-gallon bottle is bigger.

b. a meterstick

70. $\frac{92cm}{1}\cdot\frac{0.39in}{cm}\approx 35.88in\approx 36in$

CHAPTER 7 Percents, Decimals, and Fractions

Section 7.1: Percents, Decimals, and Fractions

VOCABULARY

1. Percent means parts per one hundred.

CONCEPTS

3. To write a percent as a fraction, drop the % symbol and write the given number over 100. Then simplify the fraction, if possible.

5. To write a decimal as a percent, multiply the decimal by 100 by moving the decimal point two places to the right and then insert a % symbol.

NOTATION

7. percent

GUIDED PRACTICE

9. 84% shaded, 16% not shaded

11. 107%

13. $\frac{99}{100}=99\%$

15.

a. $\frac{15}{100}=15\%$

b. $100-15=85\%$

17. $\frac{17}{100}$

19. $\frac{91}{100}$

21. $\frac{4}{100}=\frac{\cancel{4}}{\cancel{4}\cdot 25}=\frac{1}{25}$

23. $\frac{60}{100}=\frac{3\cdot\cancel{20}}{5\cdot\cancel{20}}=\frac{3}{5}$

25. $1.9\%=\frac{1.9}{100}=\frac{1.9}{100}\cdot\frac{10}{10}=\frac{19}{1{,}000}$

27. $54.7\%=\frac{54.7}{100}=\frac{54.7}{100}\cdot\frac{10}{10}=\frac{547}{1{,}000}$

29. $12.5\%=\frac{12.5}{100}=\frac{12.5}{100}\cdot\frac{10}{10}=\frac{125}{1{,}000}$

$=\frac{\cancel{125}}{8\cdot\cancel{125}}=\frac{1}{8}$

31. $6.8\%=\frac{6.8}{100}=\frac{6.8}{100}\cdot\frac{10}{10}=\frac{68}{1{,}000}$

$=\frac{\cancel{4}\cdot 17}{\cancel{4}\cdot 250}=\frac{17}{250}$

33. $1\frac{1}{3}\%=\frac{\frac{4}{3}}{100}=\frac{4}{3}\cdot\frac{1}{100}$

$=\frac{\cancel{4}}{3}\cdot\frac{1}{\cancel{4}\cdot 25}=\frac{1}{75}$

35. $14\frac{1}{6}\%=\frac{\frac{85}{6}}{100}=\frac{85}{6}\cdot\frac{1}{100}$

$=\frac{\cancel{5}\cdot 17}{6}\cdot\frac{1}{\cancel{5}\cdot 20}=\frac{17}{120}$

37. $130\%=\frac{130}{100}=\frac{13\cdot\cancel{10}}{10\cdot\cancel{10}}=\frac{13}{10}$

39. $220\%=\frac{220}{100}=\frac{11\cdot\cancel{20}}{5\cdot\cancel{20}}=\frac{11}{5}$

41. $0.35\%=\frac{0.35}{100}=\frac{0.35}{100}\cdot\frac{100}{100}$

$=\frac{35}{10{,}000}=\frac{\cancel{5}\cdot 7}{\cancel{5}\cdot 2{,}000}=\frac{7}{2{,}000}$

43. $0.25\%=\frac{0.25}{100}=\frac{0.25}{100}\cdot\frac{100}{100}$

$=\frac{25}{10{,}000}=\frac{\cancel{25}}{\cancel{25}\cdot 400}=\frac{1}{400}$

45. $16\%=0.16$

47. $81\%=0.81$

49. $34.12\% = 0.3412$

51. $50.033\% = 0.50033$

53. $6.99\% = 0.0699$

55. $1.3\% = 0.013$

57. $7\frac{1}{4}\% = 7.25\% = 0.0725$

59. $18\frac{1}{2}\% = 18.5\% = 0.185$

61. $460\% = 4.6$

63. $316\% = 3.16$

65. $0.5\% = 0.005$

67. $0.03\% = 0.0003$

69. $0.362 = 36.2\%$

71. $0.98 = 98\%$

73. $1.71 = 171\%$

75. $4 = 400\%$

77. $\frac{2}{5} = 0.4 = 40\%$

79. $\frac{4}{25} = 0.16 = 16\%$

81. $\frac{5}{8} = 0.625 = 62.5\%$

83. $\frac{7}{16} = 0.4375 = 43.75\%$

85. $\frac{9}{4} = 2.25 = 225\%$

87. $\frac{21}{20} = 1.05 = 105\%$

89. $\frac{1}{6} = 0.1666... = 16\frac{2}{3}\% \approx 16.7\%$

91. $\frac{5}{3} = 1.6666... = 166\frac{2}{3}\% \approx 166.7\%$

	Fraction	Decimal	Percent
93.	$\frac{157}{5,000}$	0.0314	3.14%
95.	$\frac{51}{125}$	0.408	40.8%
97.	$\frac{21}{400}$	0.0525	5.25% $= 5\frac{1}{4}\%$
99.	$\frac{7}{3}$	$2.33\overline{3}$	$233\frac{1}{3}\%$ $\approx 233.3\%$

APPLICATIONS

101. $\frac{91}{100} = 91\%$

103.

a. $\frac{6}{50} = \frac{12}{100} = 12\%$

b. $\frac{4}{50} = \frac{8}{100} = 8\%$

c. $\frac{2}{50} = \frac{4}{100} = 4\%$ (Alaska, Hawaii)

105.

a. $3.75\% = 0.0375$

b. $2\% = 0.02$

c. $14.25\% = 0.1425$

107. Torso: 100% -72.5% = 27.5%

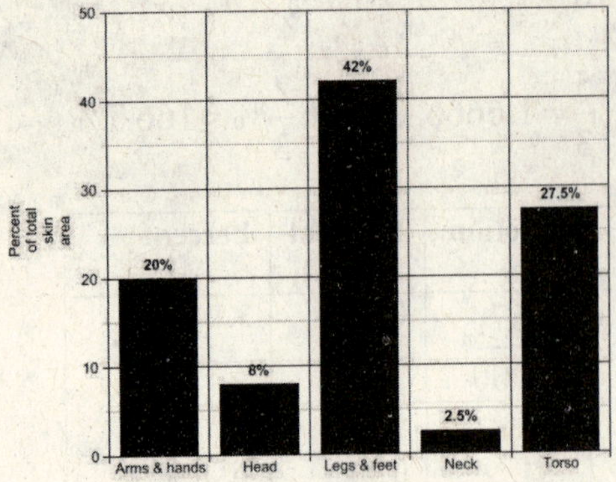

109.

a. $\frac{15}{193}$

b. $\frac{15}{193} = 0.07777... \approx 0.0777$

c. $0.0777 = 7.77\%$

111. $33\frac{1}{3}\% = \frac{1}{3} = 0.\bar{3}$

113.

a. $\frac{39}{45} = \frac{3 \cdot 13}{3 \cdot 15} = \frac{13}{15}$

b. $13 \div 15 = 86.\bar{6} = 86\frac{2}{3}\% \approx 86.7\%$

115.

a. $\frac{1}{4}\%$

b. $\frac{0.25}{100} = \frac{25}{10{,}000} = \frac{1}{400}$

c. $\frac{0.25}{100} = 0.0025$

117. $1 \div 365 = .002739... \approx 0.27\%$

WRITING

119. Even though the numbers represent the same quantity, 25% *seems* like more since 25 is larger than $\frac{1}{4}$.

121. It could happen if there are more people there than are safely allowed to be there.

REVIEW

123. a. $P = 2L + 2W$

$P = 2(6.5) + 2(10.5) = 13 + 21 = 34cm$

b. $A = LW$

$A = (6.5)(10.5) = 68.25cm^2$

Section 7.2: Solving Percent Problems Using Percent Equations and Proportions

VOCABULARY

1. We call "What number is 15% of 25?" a percent sentence. It translates to the percent equation $x = 15\% \cdot 25$.

3. When we find the value of the variable that makes a percent equation true, we say that we have solved the equation.

5. The amount is part of the base. The base is the standard of comparison – it represents the whole of some quantity.

7. The cross products for the proportion $\frac{24}{x} = \frac{36}{100}$ are $24 \cdot 100$ and $x \cdot 36$

CONCEPTS

9. Amount = percent · base

Part = percent · whole

11. $100\% - 51\% - 18\% - 8\% - 3\% - 6\% = 14\%$

NOTATION

13. a. 12% = 0.12

b. 5.6% = 0.056

c. 125% = 1.25

d. $\frac{1}{4}\% = 0.25\% = 0.0025$

GUIDED PRACTICE

15. a. $x = 7\% \cdot 16$

$\frac{x}{16} = \frac{7}{100}$

b. $125 = x \cdot 800$

$\frac{125}{800} = \frac{x}{100}$

c. $1 = 94\% \cdot x$

$\frac{1}{x} = \frac{94}{100}$

17. a. $5.4\% \cdot 99 = x$

$\frac{x}{99} = \frac{5.4}{100}$

b. $75.1\% \cdot x = 15$

$\frac{15}{x} = \frac{75.1}{100}$

c. $x \cdot 33.8 = 3.8$

$\frac{3.8}{33.8} = \frac{x}{100}$

19. $\frac{34}{100} = \frac{x}{200}$

$34 \cdot 200 = x \cdot 100$

$\frac{6{,}800}{100} = \frac{x \cdot 100}{100}$

$68 = x$

21. $\frac{88}{100} = \frac{x}{150}$

$88 \cdot 150 = x \cdot 100$

$\frac{13{,}200}{100} = \frac{x \cdot 100}{100}$

$132 = x$

23. $\frac{224}{100} = \frac{x}{7.9}$

$224 \cdot 7.9 = x \cdot 100$

$\frac{1{,}769.6}{100} = \frac{x \cdot 100}{100}$

$17.696 = x$

25. $\frac{105}{100} = \frac{x}{23.2}$

$105 \cdot 23.2 = x \cdot 100$

$\frac{2{,}436}{100} = \frac{x \cdot 100}{100}$

$24.36 = x$

27. $\frac{8}{32} = \frac{x}{100}$

$8 \cdot 100 = x \cdot 32$

$\frac{800}{32} = \frac{x \cdot 32}{32}$

$25\% = x$

29. $\frac{51}{60} = \frac{x}{100}$

$51 \cdot 100 = x \cdot 60$

$\frac{5{,}100}{60} = \frac{x \cdot 60}{60}$

$85\% = x$

31. $\frac{5}{8} = \frac{x}{100}$

$5 \cdot 100 = x \cdot 8$

$\frac{500}{8} = \frac{x \cdot 8}{8}$

$62.5\% = x$

33. $\frac{7}{16} = \frac{x}{100}$

$7 \cdot 100 = x \cdot 16$

$\frac{700}{16} = \frac{x \cdot 16}{16}$

$43.75\% = x$

35. $\frac{66}{60} = \frac{x}{100}$

$66 \cdot 100 = x \cdot 60$

$\frac{6{,}600}{60} = \frac{x \cdot 60}{60}$

$110\% = x$

37.

$$\frac{84}{24}=\frac{x}{100}$$
$$84\cdot 100=x\cdot 24$$
$$\frac{8,400}{24}=\frac{x\cdot 24}{24}$$
$$350\%=x$$

39.

$$\frac{9}{x}=\frac{30}{100}$$
$$\frac{9}{x}=\frac{3}{10}$$
$$9\cdot 10=3\cdot x$$
$$\frac{90}{3}=\frac{3\cdot x}{3}$$
$$30=x$$

41.

$$\frac{36}{x}=\frac{24}{100}$$
$$\frac{36}{x}=\frac{6}{25}$$
$$36\cdot 25=6\cdot x$$
$$\frac{900}{6}=\frac{6\cdot x}{6}$$
$$150=x$$

43.

$$\frac{19.2}{x}=\frac{33\frac{1}{3}}{100}$$
$$19.2\cdot 100=\frac{100}{3}\cdot x$$
$$19.2\cdot \cancel{100}\cdot\frac{3}{\cancel{100}}=\frac{\cancel{3}}{\cancel{100}}\cdot\frac{\cancel{100}}{\cancel{3}}\cdot x$$
$$19.2\cdot 3=x$$
$$57.6=x$$

45.

$$\frac{48.4}{x}=\frac{66\frac{2}{3}}{100}$$
$$48.4\cdot 100=\frac{200}{3}\cdot x$$
$$48.4\cdot \cancel{100}\cdot\frac{3}{\cancel{200}_{2}}=\frac{\cancel{3}}{\cancel{200}}\cdot\frac{\cancel{200}}{\cancel{3}}\cdot x$$
$$48.4\cdot\frac{3}{2}=x$$
$$72.6=x$$

TRY IT YOURSELF

47.

$$\frac{0.5}{40}=\frac{x}{100}$$
$$0.5\cdot 100=x\cdot 40$$
$$\frac{50}{40}=\frac{x\cdot 40}{40}$$
$$1.25\%=x$$

49.

$$\frac{7.8}{x}=\frac{12}{100}$$
$$7.8\cdot 100=x\cdot 12$$
$$\frac{780}{12}=\frac{x\cdot 12}{12}$$
$$65=x$$

51.

$$\frac{33}{x}=\frac{33\frac{1}{3}}{100}$$
$$33\cdot 100=\frac{100}{3}\cdot x$$
$$33\cdot \cancel{100}\cdot\frac{3}{\cancel{100}}=\frac{\cancel{3}}{\cancel{100}}\cdot\frac{\cancel{100}}{\cancel{3}}\cdot x$$
$$33\cdot 3=x$$
$$99=x$$

53.

$$\frac{36}{100}=\frac{x}{250}$$
$$36\cdot 250=x\cdot 100$$
$$\frac{9,000}{100}=\frac{x\cdot 100}{100}$$
$$90=x$$

55. $\dfrac{16}{20}=\dfrac{x}{100}$

$16\cdot 100 = x\cdot 20$

$\dfrac{1,600}{20}=\dfrac{x\cdot 20}{20}$

$80\% = x$

57. $\dfrac{0.8}{100}=\dfrac{x}{12}$

$0.8\cdot 12 = x\cdot 100$

$\dfrac{9.6}{100}=\dfrac{x\cdot 100}{100}$

$0.096 = x$

59. $\dfrac{3.3}{x}=\dfrac{7.5}{100}$

$3.3\cdot 100 = 7.5\cdot x$

$\dfrac{330}{7.5}=\dfrac{7.5\cdot x}{7.5}$

$44 = x$

61. $\dfrac{1.25}{0.05}=\dfrac{x}{100}$

$1.25\cdot 100 = x\cdot 0.05$

$\dfrac{125}{0.05}=\dfrac{x\cdot 0.05}{0.05}$

$2,500\% = x$

63. $\dfrac{x}{105}=\dfrac{102}{100}$

$100\cdot x = 102\cdot 105$

$\dfrac{100\cdot x}{100}=\dfrac{10,710}{100}$

$x = 107.1$

65. $\dfrac{5.7}{x}=\dfrac{9\frac{1}{2}}{100}$

$5.7\cdot 100 = \dfrac{19}{2}\cdot x$

$\overset{30}{\cancel{570}}\cdot\dfrac{2}{\cancel{19}}=\dfrac{\cancel{2}}{\cancel{19}}\cdot\dfrac{\cancel{19}}{\cancel{2}}\cdot x$

$30\cdot 2 = x$

$60 = x$

67. $\dfrac{2,500}{8,000}=\dfrac{x}{100}$

$\dfrac{25}{80}=\dfrac{x}{100}$

$25\cdot 100 = x\cdot 80$

$\dfrac{2,500}{80}=\dfrac{x\cdot 80}{80}$

$31.25\% = x$

69. $\dfrac{7\frac{1}{4}}{100}=\dfrac{x}{600}$

$7.25\cdot 600 = x\cdot 100$

$\dfrac{4350}{100}=\dfrac{x\cdot 100}{100}$

$43.5 = x$

APPLICATIONS

71. $\dfrac{24}{100}\cdot 50 = 12$

12K bytes = 12,000 bytes

73. a. $3.80+7.50+9.45=\$20.75$

b. $\dfrac{20}{100}\cdot 20.75 = 4.15$

$4.15 rebate

75. $\dfrac{180}{100}\cdot 1.5 = \dfrac{\overset{90}{\cancel{180}}}{100}\cdot\dfrac{3}{\underset{1}{\cancel{2}}}=\dfrac{270}{100}=2.7in$

77. $\dfrac{28}{40}=\dfrac{x}{100}$

$28\cdot 100 = 40\cdot x$

$\dfrac{2,800}{40}=\dfrac{40\cdot x}{40}$

$70 = x$

Yes, he passed.

79. $\dfrac{200}{4,000}=\dfrac{1}{20}=0.05=5\%$

The driver paid 5% of the cost.

81. $\frac{70}{100}=\frac{84}{x}$

$70\cdot x=84\cdot 100$

$\frac{70x}{70}=\frac{8,400}{70}$

$x=120$

The center can enroll a maximum of 120 children.

83. $\frac{33\frac{1}{3}}{100}=\frac{4,500}{x}$

$33\frac{1}{3}\cdot x=4500\cdot 100$

$\frac{100}{3}x=450,000$

$\frac{3}{100}\left(\frac{100}{3}x\right)=\frac{3}{100}(450,000)$

$x=13,500km$

The river and its tributaries are 13,500 km in length.

85. $23\%+37\%=60\%$

$\frac{60}{100}=\frac{x}{3.603\text{ trillion}}$

$100x=60(3.603)$

$100x=216.18$

$x=2.1618$

\$2.1618 trillion was spent on those programs.

87. $0.25x=6$

$\frac{0.25x}{0.25}=\frac{6}{0.25}$

$x=24$

There is 24 oz. in the larger bottle.

89. Tank 1: $0.5\cdot 60=30$ gal

Tank 2: $0.3\cdot 40=12$ gal

91. $10,000=x\cdot 25$

$\frac{10,000}{25}=x$

$400=x$

$40,000\%=x$

93.

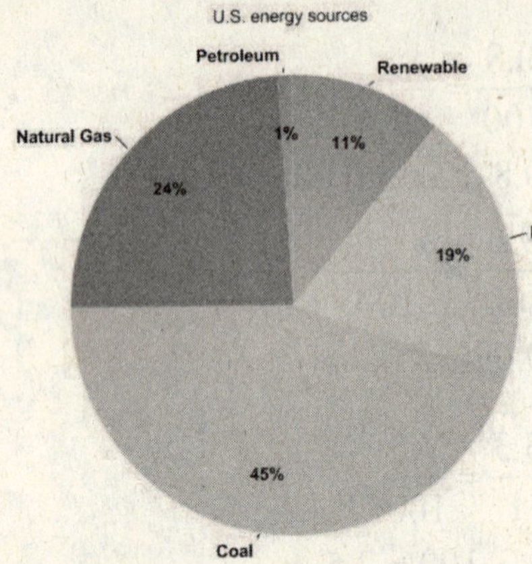

95. $\frac{529}{2,300}=0.23=23\%;\frac{690}{2,300}=0.3=30\%$

$\frac{115}{2,300}=0.05=5\%;\frac{138}{2,300}=0.06=6\%$

$\frac{828}{2,300}=0.36=36\%$

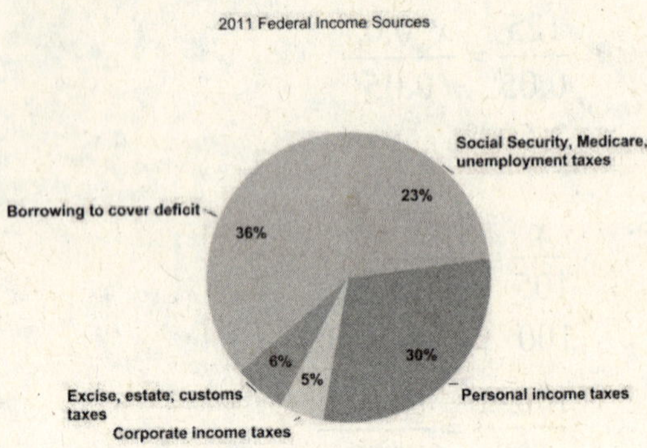

WRITING

97. Out of 20 students in my class, 9 of them are 19 years old. What percent of the students are 19 years old?

99. 150% of a number is larger because it is more than 100%, which is the number itself.

101. It is best to write the number as a fraction when it can be easily converted to a denominator of 100.

REVIEW

103. $2.68+6+9.09+0.3=18.17$

105. The distance from 4.9 is 0.1, the distance from 5.001 is 0.001, so 5.001 is closer.

107. $(0.2)^3=(0.2)(0.2)(0.2)=0.008$

Section 7.3: Applications of Percent

VOCABULARY

1. Instead of working for a salary or getting paid at an hourly rate, some salespeople are paid on <u>commission</u>. They earn a certain percent of the total dollar amount of the goods or services they sell.

3. a. When we use percent to describe how a quantity has increased compared to its original value we are finding its percent of <u>increase</u>.

b. When we use percent to describe how a quantity has decreased compared to its <u>original</u> value, we are finding the percent of decrease.

CONCEPTS

5. Sales tax = sales tax rate · <u>purchase price</u>

7. Commission = commission rate · <u>sales</u>

9. a. $59.32+4.75=\$64.07$

b. $150.99-15.99=\$135.00$

11. To find the percent decrease, <u>subtract</u> the smaller number from the larger number to find the amount of decrease. The find what percent that difference is of the <u>original</u> amount.

GUIDED PRACTICE

13. $0.04\cdot\$92.70=\$3.708\approx\$3.71$

15. $0.05\cdot\$83.90=\$4.195\approx\$4.20$

17. $0.038\cdot\$68.24=\$2.59312\approx\$2.59$

$\$68.24+\$2.59=\$70.83$

19. $0.064\cdot\$60.18=\$3.85852\approx\$3.85$

$\$60.18+\$3.85=\$64.03$

21. $7.28=x\cdot 140$

$$\frac{7.28}{140}=\frac{x\cdot 140}{140}$$

$0.052=x$

$x=5.2\%$

23. $4{,}590=x\cdot 30{,}000$

$$\frac{4{,}590}{30{,}000}=\frac{x\cdot 30{,}000}{30{,}000}$$

$0.153=x$

$x=15.3\%$

25. $0.12\cdot\$95=\$11.40$

27. $0.35\cdot\$480=\$168$

29. $15=x\cdot 750$

$$\frac{15}{750}=\frac{x\cdot 750}{750}$$

$0.02=x$

$x=2\%$

31. $20=x\cdot 500$

$$\frac{20}{500}=\frac{x\cdot 500}{500}$$

$0.04=x$

$x=4\%$

33. $88-80=8$

$8=x\cdot 80$

$$\frac{8}{80}=\frac{x\cdot 80}{80}$$

$0.1=x$

$x=10\%$

35. $345-300=45$

$45=x\cdot 300$

$$\frac{45}{300}=\frac{x\cdot 300}{300}$$

$0.15=x$

$x=15\%$

37. $30 - 24 = 6$

$6 = x \cdot 30$

$\frac{6}{30} = \frac{x \cdot 30}{30}$

$0.2 = x$

$x = 20\%$

39. $40 - 36 = 4$

$4 = x \cdot 40$

$\frac{4}{40} = \frac{x \cdot 40}{40}$

$0.1 = x$

$x = 10\%$

41. $d = 0.33 \cdot \$90 = \$29.70$

$\$90 - \$29.70 = \$60.30$

43. $d = 0.15 \cdot \$58 = \$8.70$

$\$58 - \$8.70 = \$49.30$

45. $d = \$79.95 - \$64.95 = \$15$

$15 = x \cdot 79.95$

$\frac{15}{79.95} = \frac{x \cdot 79.95}{79.95}$

$0.1876 \approx x$

$x \approx 19\%$

47. $\$209 - \$179 = \$30$

$30 = x \cdot 209$

$\frac{30}{209} = \frac{x \cdot 209}{209}$

$x \approx 0.1435$

$x \approx 14\%$

APPLICATIONS

49. $0.0625 \cdot \$900 = \$56.25$

51. Subtotal: $\$8.97 + \$9.87 + \$28.50 = \$47.34$

Sales Tax: $0.06 \cdot \$47.34 \approx \$2.84$

Total: $\$47.34 + \$2.84 = \$50.18$

53. $10.32 = x \cdot 129$

$\frac{10.32}{129} = x$

$0.08 = x$

$x = 8\%$

55. $5 = x \cdot 2{,}000$

$\frac{5}{2{,}000} = \frac{x \cdot 2{,}000}{2{,}000}$

$0.0025 = x$

$x = 0.25\%$

57. $0.01 \cdot \$24{,}950 = \$249.50$

59. Fed. Tax: $28.8 = x \cdot 360$

$\frac{28.8}{360} = \frac{x \cdot 360}{360}$

$0.08 = x$

$x = 8\%$

Work. Comp: $13.50 = x \cdot 360$

$\frac{13.50}{360} = x$

$0.0375 = x$

$x = 3.75\%$

Medicare: $5.22 = x \cdot 360$

$\frac{5.22}{360} = x$

$0.0145 = x$

$x = 1.45\%$

Social Security: $22.32 = x \cdot 360$

$\frac{22.32}{360} = x$

$0.062 = x$

$x = 6.2\%$

61. $84 - 80 = 4$

$4 = x \cdot 80$

$\frac{4}{80} = \frac{x \cdot 80}{80}$

$0.05 = x$

$x = 5\%$

63. $7.6-5.8=1.8$

$1.8=x\cdot 5.8$

$\frac{1.8}{5.8}=\frac{x\cdot 5.8}{5.8}$

$x\approx 0.3103$

$x\approx 31\%$

65. $106-42=64$

$64=x\cdot 42$

$\frac{64}{42}=\frac{x\cdot 42}{42}$

$1.5238\approx x$

$x\approx 152\%$

67. $150-96=54$

$54=x\cdot 150$

$\frac{54}{150}=\frac{x\cdot 150}{150}$

$0.36=x$

$x=36\%$

69. $\$112-\$98=\$14$

$14=x\cdot 112$

$\frac{14}{112}=\frac{x\cdot 112}{112}$

$0.125=x$

$x=12.5\%$

71. a. $1.25-1=0.25$

$0.25=1\cdot x$

$x=25\%$

b. $1.25-0.8=0.45$

$0.45=x\cdot 1.25$

$\frac{0.45}{1.25}=\frac{x\cdot 1.25}{1.25}$

$0.36=x$

$x=36\%$

73. $0.06\cdot \$198{,}500=\$11{,}910$

$\frac{1}{2}\cdot \$11{,}910=\$5{,}955$

75. $37{,}500=x\cdot 2{,}500{,}000$

$\frac{37{,}500}{2{,}500{,}000}=\frac{x\cdot 2{,}500{,}000}{2{,}500{,}000}$

$0.015=x$

$x=1.5\%$

77. $144=x\cdot 160$

$\frac{144}{160}=\frac{x\cdot 160}{160}$

$0.9=x$

$x=90\%$

79. $6\cdot 6{,}000=\$36{,}000$

$\frac{33\frac{1}{3}}{100}\cdot \$36{,}000=\frac{100}{3}\cdot\frac{360}{1}=\$12{,}000$

81. a. $0.2\cdot \$39.95=\$7.99$

b. $\$39.95-\$7.99=\$31.96$

83. $5{,}700-5{,}350=350$

$350=x\cdot 5{,}700$

$\frac{350}{5{,}700}=\frac{x\cdot 5{,}700}{5{,}700}$

$x\approx 0.0614$

$x\approx 6\%$

85. $\$399.97-\$50=\$349.97$

$50=x\cdot 399.97$

$\frac{50}{399.97}=\frac{x\cdot 399.97}{399.97}$

$x\approx 0.125$

$x\approx 13\%$

87. $\$15.48-\$3.60=\$11.88$

$3.6=x\cdot 15.48$

$\frac{3.6}{15.48}=\frac{x\cdot 15.48}{15.48}$

$x\approx 0.2326$

$x\approx 23\%$

89. $0.55\cdot \$170=\$93.50$

$\$170-\$93.50=\$76.50$

91. $149.99 = 0.8 \cdot x$

$\frac{149.99}{0.8} = \frac{0.8 \cdot x}{0.8}$

$x \approx \$187.49$

WRITING

93. The sales tax is the amount of money you pay for an item in addition to the price of the item. The sales tax rate is what percent of the price of your item is used to calculate the sales tax.

95. The base price is incorrect in the equation.

REVIEW

97. $-5(-5)(-2) = 25(-2) = -50$

99. $-4-(-7) = -4+7 = 3$

101. $|-5-8| = |-13| = 13$

Section 7.4: Estimation with Percent

VOCABULARY

1. Estimation can be used to find approximations when exact answers aren't necessary.

CONCEPTS

3. To find 1% of a number, move the decimal point in the number two places to the left.

5. To find 20% of a number, find 10% of the number by moving the decimal point one place to the left, and then double (multiply by 2) the result.

7. To find 25% of a number, divide the number by 4.

9. To find 15% of a number, find the sum of 10% of the number and 5% of the number.

GUIDED PRACTICE

11. a. 2.751

b. 1% of 300 = 3

13. a. 0.1267

b. 1% of 10 = 0.1

15. a. 405.9 lb

b. 10% of 4,000 = 400 lb

17. a. 69.14 min

b. 10% of 700 = 70 min

19. $0.1 \cdot 350 = 35$

$2 \cdot 35 = 70$

21. $0.1 \cdot 70 = 7$

$2 \cdot 7 = 14$

23. $\frac{4,200,000}{2} = 2,100,000$

25. $\frac{400,000}{2} = 200,000$

27. $\frac{16}{4} = 4$

29. $\frac{48}{4} = 12$

31. $0.1 \cdot 16,400 = 1,640$

$\frac{1,640}{2} = 820$

33. $0.1 \cdot 400 = 40$

$\frac{40}{2} = 20$

35. $0.1 \cdot 60 = 6$

$\frac{6}{2} = 3$

$6+3 = \$9$

37. $0.1 \cdot 30 = 3$

$\frac{3}{2} = 1.5$

$3+1.5 = \$4.50$

39. $0.1 \cdot 120 = 12$

$\frac{12}{2} = 6$

$12+6 = \$18$

41. $0.1 \cdot 10 = 1$

$\frac{1}{2} = 0.5$

$1 + 0.5 = \$1.50$

43. $2 \cdot 4 = 8$

45. $2 \cdot 36 = 72$

TRY IT YOURSELF

47. $0.01 \cdot 600 = 6$

$2 \cdot 6 = 12$

49. $0.1 \cdot 18 = 1.8$

$3 \cdot 1.8 = 5.4$

51. $3 \cdot 60 = 180$

53. $0.1 \cdot 4{,}600 = 460$

$\frac{460}{2} = 230$

55. $0.01 \cdot 600 = 6$

57. $0.1 \cdot 120 = 12$

$\frac{12}{2} = 6$

$12 + 6 = 18$

59. $0.1 \cdot 70 = 7$

61. $\frac{280}{4} = 70$

63. $\frac{24{,}000}{2} = 12{,}000$

65. $2 \cdot 0.9 = 1.8$

67. $0.5 \cdot 98 = 49$

$0.01 \cdot 49 = 0.49$

69. $0.2 \cdot 400 = 80$

$0.15 \cdot 80 = 12$

APPLICATIONS

71. $0.1 \cdot 820 = 82$

$2 \cdot 82 = 164$

164 students

73. $0.1 \cdot 200 = 20$

$3 \cdot 20 = \$60$

75. $0.1 \cdot 40 = 4$

$\frac{4}{2} = 2$

$4 + 2 = \$6$

77. $\$28.55 + \$19.75 = \$48.30$

$0.1 \cdot 50 = 5$

$\frac{5}{2} = 2.5$

$5 + 2.5 = \$7.50$

79. $\frac{120{,}000}{4} = \$30{,}000$

81. $2 \cdot 160 = 320\ lb$

83. $0.1 \cdot 700 = 70$

$3 \cdot 70 = 210$ motorists

85. $24\% \approx 25\%$

$\frac{880}{4} = 220$ people

87. $58\% \approx 60\%$

$0.1 \cdot 30{,}000 = 3{,}000$

$6 \cdot 3{,}000 = 18{,}000$ people

89. $48\% \approx 50\%$

$\frac{6{,}200}{2} = 3{,}100$ volunteers

WRITING

91. 200% of a number is twice the number since 100% is the number, 2(100%) is the number 2 times, or twice the number.

93. Since 5% is half of 10%, find 10% of the number and divide by 2.

REVIEW

95.

a. $\frac{5}{6}+\frac{1}{2}=\frac{5}{6}+\frac{1}{2}\cdot\frac{3}{3}=\frac{5}{6}+\frac{3}{6}=\frac{8}{6}=\frac{4}{3}=1\frac{1}{3}$

b. $\frac{5}{6}-\frac{1}{2}=\frac{5}{6}-\frac{1}{2}\cdot\frac{3}{3}=\frac{5}{6}-\frac{3}{6}=\frac{2}{6}=\frac{1}{3}$

c. $\frac{5}{6}\cdot\frac{1}{2}=\frac{5\cdot 1}{6\cdot 2}=\frac{5}{12}$

d. $\frac{5}{6}\div\frac{1}{2}=\frac{5}{6}\cdot\frac{2}{1}=\frac{5\cdot \cancel{2}}{\cancel{2}\cdot 3\cdot 1}=\frac{5}{3}=1\frac{2}{3}$

Section 7.5: Interest

VOCABULARY

1. In general, <u>interest</u> is money that is paid for the use of money.

3. The percent that is used to calculate the amount of interest to be paid is called the interest <u>rate</u>.

5. The <u>total</u> amount in an investment account is the sum of the principal and the interest.

CONCEPTS

7. a. $125,000

b. 5%

c. 30 years

9. a. $7\%=0.07$

b. $9.8\%=0.098$

c. $6\frac{1}{4}\%=6.25\%=0.0625$

11. $10,000\cdot 0.06\cdot 3=600\cdot 3=\$1,800$

13. a. compound interest b. $1,000

c. 4 d. $50 e. 4 quarters is 1 year.

NOTATION

15. $I=Prt$

GUIDED PRACTICE

17. $I=Prt$

$I=\$2,000\cdot 0.05\cdot 1$

$=\$100\cdot 1$

$=\$100$

19. $I=Prt$

$I=\$700\cdot 0.09\cdot 4$

$=\$63\cdot 4$

$=\$252$

21. $I=Prt$

$I=\$500\cdot 0.025\cdot 2$

$=\$12.5\cdot 2$

$=\$25$

$\$500+\$25=\$525$

23. $I=Prt$

$I=\$1,500\cdot 0.012\cdot 5$

$=\$18\cdot 5$

$=\$90$

$\$1,500+\$90=\$1,590$

25. $\frac{9mo}{1}\cdot\frac{1yr}{12\ mo}=\frac{9}{12}yr=0.75yr$

$I=Prt$

$I=\$550\cdot 0.04\cdot 0.75$

$=\$22\cdot 0.75$

$=\$16.50$

27. $\frac{4mo}{1}\cdot\frac{1yr}{12\ mo}=\frac{4}{12}yr=\frac{1}{3}yr$

$I=Prt$

$I=\$1,320\cdot 0.07\cdot\frac{1}{3}$

$=\$92.4\cdot\frac{1}{3}$

$=\$30.80$

29. $\frac{90d}{1}\cdot\frac{1yr}{365d}=\frac{90}{365}yr=\frac{18}{73}yr$

$I=Prt$

$I=\$12,600\cdot 0.18\cdot\frac{18}{73}$

$=\$2,268\cdot\frac{18}{73}$

$\approx \$559.23$

$\$12,600+\$559.23=\$13,159.23$

31. $\frac{45d}{1}\cdot\frac{1yr}{365d}=\frac{45}{365}yr=\frac{9}{73}yr$

$I=Prt$

$I=\$40,000\cdot 0.1\cdot\frac{9}{73}$

$=\$4,000\cdot\frac{9}{73}$

$\approx \$493.15$

$\$40,000+\$493.15=\$40,493.15$

33.

$A=P\left(1+\frac{r}{n}\right)^{nt}$

$A=2,000\left(1+\frac{0.03}{4}\right)^{4\cdot 1}$

$A=2,000(1.0075)^4$

$A\approx \$2,060.68$

35.

$A=P\left(1+\frac{r}{n}\right)^{nt}$

$A=5,400\left(1+\frac{.04}{4}\right)^{4\cdot 1}$

$A=5,400(1.01)^4$

$A\approx \$5,619.26$

37.

$A=P\left(1+\frac{r}{n}\right)^{nt}$

$A=30,000\left(1+\frac{0.048}{365}\right)^{365\cdot 6}$

$A=\$40,011.96$

$\$40,011.96-\$30,000=\$10,011.96$ can be withdrawn, leaving the balance intact.

39.

$A=P\left(1+\frac{r}{n}\right)^{nt}$

$A=55,250\left(1+\frac{0.0855}{365}\right)^{365\cdot 4}$

$A\approx \$77,775.64$

APPLICATIONS

41. $I=\$5,000\cdot 0.06\cdot 1=\$300$

$\$5,000+\$300=\$5,300$

43. $I=\$1,200\cdot 0.055\cdot 3$

$=\$66\cdot 3=\$198$

45. $I=\$4,500\cdot 0.12\cdot 2$

$=\$540\cdot 2=\$1,080$

$\$4,500+\$1,080=\$5,580$

47. $\frac{3mo}{1}\cdot\frac{1yr}{12mo}=\frac{1}{4}yr=0.25yr$

$I=\$1,500\cdot 0.125\cdot 0.25$

$=\$187.5\cdot 0.25\approx \$46.88$

49. $\frac{30d}{1}\cdot\frac{1yr}{365d}=\frac{6}{73}yr$

$I=\$4,200\cdot 0.18\cdot\frac{6}{73}$

$=\$756\cdot\frac{6}{73}\approx \$62.14$

$\$4,200+\$62.14=\$4,262.14$

51.

P	*r*	*t*	*I*
$10,000	0.0725	2 years	$1,450

53. 4. $\$1,200 \cdot 0.08 \cdot 2 = \$192$

5. $\$1,200 + \$192 = \$1,392$

6. $\dfrac{\$1,392}{24} = \$58$

Loan Application Worksheet

1. Amount of loan (principal) $1,200.00
2. Length of loan (time) 2 YEARS
3. Annual percentage rate (simple interest) 8%
4. Interest charged $192
5. Total amount to be repaid $1,392
6. Check method of repayment: ☐ 1 lump sum ☑ monthly payments

Borrower agrees to pay 24 equal payments of $58 to repay loan.

55. $I = \$18m \cdot 0.023 \cdot 3.5 = \$1.449m$

$\$18m + \$1.449m = \$19.449$ million

57.
$$A = P\left(1 + \frac{r}{n}\right)^{nt}$$
$$A = 600\left(1 + \frac{.08}{1}\right)^{3} \approx \$755.83$$

59.
$$A = P\left(1 + \frac{r}{n}\right)^{nt}$$
$$A = 1,000\left(1 + \frac{0.06}{365}\right)^{365 \cdot 4} \approx \$1,271.22$$

61.
$$A = 545\left(1 + \frac{0.046}{365}\right)^{365 \cdot 1} \approx \$570.65$$

63.
$$A = 500,000\left(1 + \frac{0.06}{365}\right)^{365 \cdot 1} \approx \$30,915.66$$

65.
$$A = 90,000\left(1 + \frac{0.051}{365}\right)^{365 \cdot 20} \approx \$249,569.75$$

She can withdraw

$\$249,569.75 - \$90,000 = \$159,569.75$ in interest.

WRITING

67. Simple interest is only calculated once. Compound interest is calculated based on previous interest earned / paid, in addition to the principal amount.

69. They would do that because then they do not make interest off of the loan amount.

REVIEW

71. $\sqrt{\dfrac{1}{4}} = \dfrac{1}{2}$

73. $\dfrac{3}{7} + \dfrac{2}{5} = \dfrac{3}{7} \cdot \dfrac{5}{5} + \dfrac{2}{5} \cdot \dfrac{7}{7} = \dfrac{15}{35} + \dfrac{14}{35} = \dfrac{29}{35}$

75. $2\dfrac{1}{2} \cdot 3\dfrac{1}{3} = \dfrac{5}{\cancel{2}_1} \cdot \dfrac{\cancel{10}^5}{3} = \dfrac{25}{3} = 8\dfrac{1}{3}$

77. $-6^2 = -6 \cdot 6 = -36$

Chapter 7 Review

1. 39% , 0.39 , $\dfrac{39}{100}$

2. 111% , 1.11 , $\dfrac{111}{100}$

3. 100% - 39% = 61%

4. a. 54%

b. 100% – 54% = 46%

5. $\dfrac{15}{100} = \dfrac{3 \cdot \cancel{5}}{\cancel{5} \cdot 20} = \dfrac{3}{20}$

6. $\dfrac{120}{100} = \dfrac{6 \cdot \cancel{20}}{5 \cdot \cancel{20}} = \dfrac{6}{5}$

7. $\dfrac{9\frac{1}{4}}{100} = \dfrac{9.25}{100} = \dfrac{925}{10{,}000} = \dfrac{\cancel{25} \cdot 37}{\cancel{25} \cdot 400} = \dfrac{37}{400}$

8. $\dfrac{0.2}{100} = \dfrac{2}{1{,}000} = \dfrac{\cancel{2}}{\cancel{2} \cdot 500} = \dfrac{1}{500}$

9. $27\% = 0.27$

10. $8\% = 0.08$

11. $655\% = 6.55$

12. $1.8\% = 0.018$

13. $0.75\% = 0.0075$

14. $0.23\% = 0.0023$

15. $0.83 = 83\%$

16. $1.625 = 162.5\%$

17. $0.051 = 5.1\%$

18. $6 = 600\%$

19. $0.5 = 50\%$

20. $0.8 = 80\%$

21. $0.875 = 87.5\%$

22. $0.0625 = 6.25\%$

23. $\dfrac{1}{3} = 0.3333... = 33.3333...\% = 33\dfrac{1}{3}\% \approx 33.3\%$

24. $\dfrac{5}{6} = 0.83333... = 83.3333...\% = 83\dfrac{1}{3}\% \approx 83.3\%$

25. $\dfrac{11}{12} = 0.91666... = 91.666...\% = 91\dfrac{2}{3}\% \approx 91.7\%$

26. $\dfrac{15}{9} = 1.6666.... = 166.6666...\% = 166\dfrac{2}{3}\% \approx 166.7\%$

27. a. $97.2\% = 0.972$

b. $\dfrac{97.2}{100} = \dfrac{972}{1{,}000} = \dfrac{\cancel{4} \cdot 243}{\cancel{4} \cdot 250} = \dfrac{243}{250}$

28. $\dfrac{17}{27} = 0.629... \approx 63\%$

29. a. $\dfrac{1}{4}\% = 0.25\% = 0.0025$

b. $\dfrac{0.25}{100} = \dfrac{25}{10{,}000} = \dfrac{\cancel{25}}{\cancel{25} \cdot 400} = \dfrac{1}{400}$

30. $\dfrac{1}{15} = 0.0666... = 6.666...\% = 6\dfrac{2}{3}\% \approx 6.7\%$

31. a. amount: 15, base: 45, percent: $33\dfrac{1}{3}\%$

b. Amount = percent·base
Part = percent·whole

32. a. 0.13

b. 0.071

c. 1.95

d. $0.25\% = 0.0025$

e. $33\dfrac{1}{3}\% = \dfrac{100}{3}\% = \dfrac{\frac{100}{3}}{100} = \dfrac{100}{3} \cdot \dfrac{1}{100} = \dfrac{1}{3}$

f. $66\dfrac{2}{3}\% = \dfrac{200}{3}\% = \dfrac{\frac{200}{3}}{100} = \dfrac{\cancel{200}^{2}}{3} \cdot \dfrac{1}{\cancel{100}} = \dfrac{2}{3}$

g. $16\dfrac{2}{3}\% = \dfrac{50}{3}\% = \dfrac{\frac{50}{3}}{100} = \dfrac{\cancel{50}}{3} \cdot \dfrac{1}{\cancel{100}^{2}} = \dfrac{1}{6}$

33. a. $x = 32\% \cdot 96$

b. $64 = x \cdot 135$

c. $9 = 47.2\% \cdot x$

34.

a. $\frac{x}{96}=\frac{32}{100}$

b. $\frac{64}{135}=\frac{x}{100}$

c. $\frac{9}{x}=\frac{47.2}{100}$

35. $\frac{x}{500}=\frac{40}{100}$

$100\cdot x=40\cdot 500$

$\frac{\cancel{100}x}{\cancel{100}}=\frac{20,000}{100}$

$x=200$

36. $\frac{16}{100}=\frac{20}{x}$

$16\cdot x=20\cdot 100$

$\frac{\cancel{16}x}{\cancel{16}}=\frac{2,000}{16}$

$x=125$

37. $\frac{1.4}{80}=\frac{x}{100}$

$1.4\cdot 100=80\cdot x$

$\frac{140}{80}=\frac{\cancel{80}x}{\cancel{80}}$

$1.75=x$

$x=1.75\%$

38. $\frac{x}{3,150}=\frac{66\frac{2}{3}}{100}$

$\frac{3,150}{1}\cdot\frac{x}{3,150}=\frac{66\frac{2}{3}}{100}\cdot\frac{3,150}{1}$

$x=2,100$

39. $A=2.20\cdot 55=121$

40. $A=0.0005\cdot 60,000=30$

41. $\frac{2.5}{100}=\frac{3.88}{x}$

$2.5\cdot x=3.88\cdot 100$

$\frac{2.5x}{2.5}=\frac{388}{2.5}$

$x=155.2$

42. $\frac{x}{100}=\frac{4.24}{0.08}$

$\cancel{100}\cdot\frac{x}{\cancel{100}}=100\cdot 53$

$x=5,300\%$

43. $0.04\cdot 15=0.6gal$

44. $51=0.75\cdot x$

$\frac{51}{0.75}=\frac{\cancel{0.75}x}{\cancel{0.75}}$

$68=x$

45. $96=110\cdot x$

$\frac{96}{110}=\frac{\cancel{110}x}{\cancel{110}}$

$x\approx 0.87$

$x=87\%$

46. $T=0.15\cdot 36.20=\$5.43$

47.

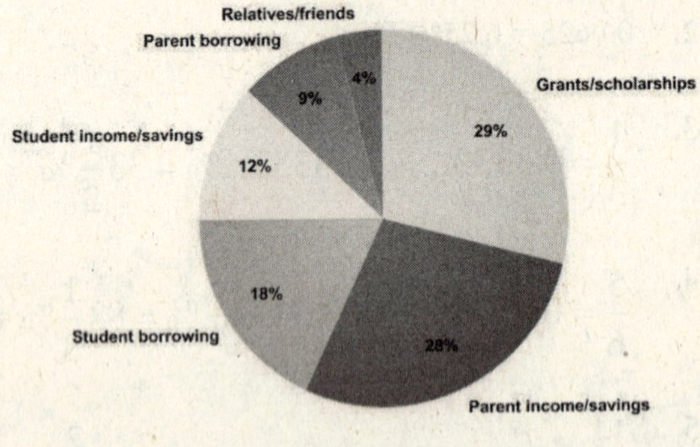

48. $0.709\cdot 196,800,000=139,531,200mi^2$

49. $0.055\cdot 59.99\approx\$3.30$

$59.99+3.30=\$63.29$

50. $492 = x \cdot 12{,}300$

$\frac{492}{12{,}300} = \frac{\cancel{12{,}300}x}{\cancel{12{,}300}}$

$0.04 = x$

$x = 4\%$

51. Total sales: $\$369.97 + \$299.97 = \$669.94$

$0.06 \cdot \$669.94 \approx \$40.20$

52. $646 = x \cdot 15{,}200$

$\frac{646}{15{,}200} = \frac{\cancel{15{,}200}x}{\cancel{15{,}200}}$

$0.0425 = x$

$x = 4.25\%$

53. $33\frac{1}{3}\% = \frac{1}{3}$

$\frac{1}{3}(12{,}000)(25) = \$100{,}000$

54. The percent of increase (or decrease) is a percent of the <u>original</u> number, that is, the number before the change occurred.

55. $\frac{19{,}500 - 17{,}000}{17{,}000} = 0.1470... \approx 15\%$

56. $\frac{18.8 - 17.0}{18.8} = 0.09574... \approx 9.6\%$

57. a. Sales tax = sales tax rate · <u>purchase price</u>

b. Total cost = purchase price + <u>sales tax</u>

c. Commission <u>= commission rate</u> · sales

58. a. Amount of discount = original price – <u>sale price</u>

b. Amount of discount = discount rate · <u>original price</u>

c. Sale price = original price – <u>discount</u>

59. discount: $\$180$

original price: $\$2{,}320 + \$180 = \$2{,}500$

discount rate: $\frac{180}{2{,}500} = 0.072 = 7.2\%$

60. There is a discount of \$49.

$\frac{49}{980} = 0.05 = 5\%$

61. 3.4023 ; 3

62. 86.87 ; 90

63. 4.34sec ; 4sec

64. 1,090L ; 1,000L

65. 10% is roughly 6, so 20% is roughly 12.

66. 10% is roughly 60, so 20% is roughly 120.

67. 50% of approximately 280,000 is 140,000.

68. 50% of approximately 300 is 150.

69. 25% of approximately 12 is 3.

70. 25% of approximately 40 is 10.

71. 10% is roughly 700, so 5% is roughly 350.

72. 10% is roughly 2,000, so 5% is roughly 1,000.

73. 200% of approximately 30 is 60.

74. 200% of approximately 1 is 2.

75. 10% is about \$24. 5% is about \$12.

\$24 + \$12 = \$36.

76. 10% is about \$5. 5% is about \$2.50.

\$5 + \$2.50 = \$7.50.

77. $50\,fl\,oz \approx 48\,fl\,oz$

25% of 48 is $12\,fl\,oz$

78. 785 people is close to 800 people.

10 % of 800 is 80, and 5% of 800 is 40.

80 + 40 = 120 people.

79. Estimate the number of cars with 4,000 and the percent with 5%.

10% of 4,000 is 400 so 5% of 4,000 is 200 cars.

80. Estimate the cost at \$300,000 and percent at 10%.

10% of \$300,000 is \$30,000.

81.
$$I = P \cdot r \cdot t$$
$$I = 6{,}000 \cdot 0.08 \cdot 2$$
$$I = 480 \cdot 2 = \$960$$

82.
$$I = P \cdot r \cdot t$$
$$I = 24{,}000 \cdot 0.045 \cdot 3$$
$$I = 1{,}080 \cdot 3 = \$3{,}240$$
$$\$24{,}000 + \$3{,}240 = \$27{,}240$$

83.
$$3mo = \frac{1}{4}yr$$
$$I = P \cdot r \cdot t$$
$$I = 2{,}750 \cdot 0.11 \cdot \frac{1}{4} = 302.5 \cdot \frac{1}{4}$$
$$I = \$75.63$$

84.
$$I = 10{,}000 \cdot 0.125 \cdot \frac{90}{365}$$
$$I = 1{,}250 \cdot \frac{90}{365} = \$308.22$$
$$\$10{,}000 + \$308.22 = \$10{,}308.22$$

85. a. $I = 1{,}500 \cdot 0.0775 \cdot 1 = \$116.25$

b. $\$1{,}500 + \$116.25 = \$1{,}616.25$

c. $\frac{1{,}616.25}{12} = \$134.69$

86.
$$A = P\left(1 + \frac{r}{n}\right)^{nt}$$
$$A = 2{,}000\left(1 + \frac{0.07}{2}\right)^{2 \cdot 1}$$
$$A = \$2{,}142.45$$

87.
$$A = P\left(1 + \frac{r}{n}\right)^{nt}$$
$$A = 5{,}000\left(1 + \frac{0.065}{365}\right)^{365 \cdot 3}$$
$$A \approx \$6{,}076.45$$

88.
$$A = P\left(1 + \frac{r}{n}\right)^{nt}$$
$$A = 500{,}000\left(1 + \frac{0.083}{365}\right)^{365 \cdot 1}$$
$$A \approx \$543{,}265.78$$
$$\$543{,}265.78 - \$500{,}000 = \$43{,}265.78$$

Chapter 7 Test

1. a. Percent means parts per one hundred.

b. The key words in a percent sentence translate as follows:
* is translates to an equal symbol =
* of translates to multiplication that is shown with a raised dot ·
* what *number* or what *percent* translates to an unknown number that is represented by a variable.

c. In the percent sentence "5 is 25% of 20", 5 is the amount, 25% is the percent and 20 is the base.

d. When we use percent to describe how a quantity has increased compared to its original value, we are finding the percent of increase.

e. Simple interest is interest earned only on the original principal. Compound interest is interest paid on the principal and previously earned interest.

2. a. 61% , $\frac{61}{100}$, 0.61

b. $100\% - 61\% = 39\%$

3. 199% , $\frac{199}{100}$, 1.99

4. a. 0.67 b. 0.123 c. 9.75% = 0.0975

5. a. 0.0006 b. 2.1 c. 0.55375

6. a. 0.25 = 25%

b. 0.625 = 62.5%

c. 1.12 = 112%

7. a. 19% b. 347% c. 0.5%

8. a. 66.7% b. 200% c. 90%

9.

a. $55\% = \frac{55}{100} = \frac{\cancel{5} \cdot 11}{\cancel{5} \cdot 20} = \frac{11}{20}$

b. $0.01\% = \frac{0.01}{100} = \frac{1}{10,000}$

c. $125\% = \frac{125}{100} = \frac{5 \cdot \cancel{25}}{4 \cdot \cancel{25}} = \frac{5}{4}$

10.

a. $6\frac{2}{3}\% = \frac{\frac{20}{3}}{100} = \frac{\cancel{20}}{3} \cdot \frac{1}{\cancel{100}_5} = \frac{1}{15}$

b. $37.5\% = \frac{37.5}{100} = \frac{375}{1,000} = \frac{3 \cdot \cancel{125}}{8 \cdot \cancel{125}} = \frac{3}{8}$

c. $8\% = \frac{8}{100} = \frac{2 \cdot \cancel{4}}{\cancel{4} \cdot 25} = \frac{2}{25}$

11.

a. $\frac{1}{30} = \frac{1}{30} \cdot \frac{100}{1}\% = \frac{10}{3}\% = 3\frac{1}{3}\% \approx 3.3\%$

b.

$\frac{16}{9} = \frac{16}{9} \cdot \frac{100}{1}\% = \frac{1,600}{9}\% = 177\frac{7}{9}\% \approx 177.8\%$

12.

$$\frac{65}{1,000} = \frac{x}{100}$$
$$65 \cdot 100 = 1,000 \cdot x$$
$$\frac{6,500}{1,000} = \frac{\cancel{1000}x}{\cancel{1000}}$$
$$6.5 = x$$

65 is 6.5% of 1000.

13.

$$\frac{35}{14} = \frac{x}{100}$$
$$35 \cdot 100 = 14 \cdot x$$
$$\frac{3,500}{14} = \frac{\cancel{14}x}{\cancel{14}}$$
$$250 = x$$

35 is 250% of 14.

14. $\frac{467}{497} = 0.93963... \approx 93.96\%$

15.

$$\frac{18}{x} = \frac{20}{100}$$
$$18 \cdot 100 = x \cdot 20$$
$$\frac{1,800}{20} = \frac{\cancel{20}x}{\cancel{20}}$$
$$90 = x$$

16. $A = 0.03 \cdot 700 = 21$

17. $A = 2.24 \cdot 60 = 134.4$

18.

$$33\frac{1}{3}\% = \frac{1}{3}$$
$$2.6 = \frac{1}{3}x$$
$$3(2.6) = \cancel{3}\left(\frac{1}{\cancel{3}}x\right)$$
$$7.8 = x$$

19. a. $S = 0.03 \cdot 34 = 1.02in$

b. $34 - 1.02 = 32.98in$

20. $T = 0.029 \cdot 25.50 = \$0.74$

$\$25.50 + \$0.74 = \$26.24$

21.

$$2.7 = x \cdot 90$$
$$\frac{2.7}{90} = \frac{90x}{90}$$
$$0.03 = x$$
$$x = 3\%$$

22. There was an increase of 644 people.

$$\frac{644}{2,800}=0.23=23\%$$

23. $A=0.04\cdot 898=\$35.92$

24. $528=x\cdot 4,800$

$$\frac{528}{4,800}=\frac{\cancel{4,800}x}{\cancel{4,800}}$$

$0.11=x$

$x=11\%$

25. $I=0.036\cdot 40,000=\$1,440$

$\$40,000+\$1,440=\$41,440$

26. discount: \$9

sale price: \$75 – \$9 = \$66

discount rate: $\frac{9}{75}=0.12=12\%$

27. discount: $0.33\cdot 20=\$6.60$

sales price: $\$20-\$6.6=\$13.40$

28. a. To find 1% of a number, move the decimal point in the number <u>two</u> places to the <u>left</u>.

b. To find 10% of a number, move the decimal point in the number <u>one</u> place to the <u>left</u>.

29. a. 20% of 400 is 80.

b. 50% of 6,000,000 is 3,000,000.

c. 200% of 21 is 42.

30. 100% – 4% – 8% – 9% – 10% – 13% – 14% – 16%

= 26%

31.

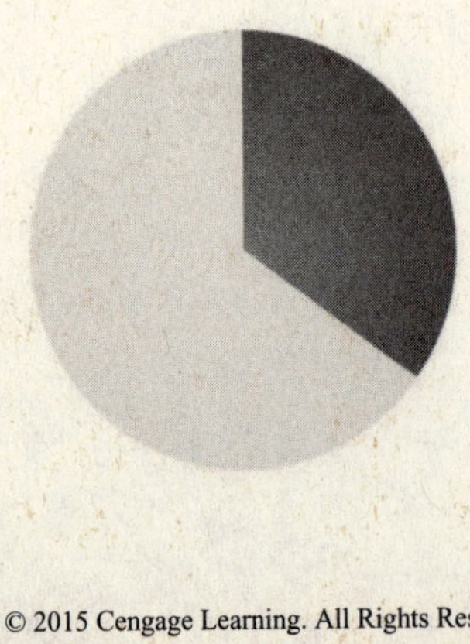

32. 10% of 2,000 is 200, so 5% is 100.

33. Use \$30 to estimate the bill.

10% of 30 is 3, so 5% is 1.5.

\$3 + \$1.5 = \$4.50

34. Use 25% of 64,000 as an approximation: 16,000 females.

35. $I=P\cdot r\cdot t$

$I=3,000\cdot 0.05\cdot 1=\$150$

36. $I=P\cdot r\cdot t$

$I=23,000\cdot 0.045\cdot 5$

$I=1,035\cdot 5=\$5,175$

$\$23,000+\$5,175=\$28,175$

37. $I=P\cdot r\cdot t$

$$I=2,000\cdot 0.08\cdot\frac{90}{365}$$

$$I=160\cdot\frac{90}{365}=\$39.45$$

38.

$$A=P\left(1+\frac{r}{n}\right)^{nt}$$

$$A=24,000\left(1+\frac{0.064}{365}\right)^{365\cdot 3}\approx\$29,079.60$$

$\$29,079.60-\$24,000=\$5,079.60$

Chapters 1 – 7 Cumulative Review

1. a. six million, fifty-four thousand, three hundred forty-six

b. 6,000,000 + 50,000 + 4,000 + 300 + 40 + 6

2. $23+21+23+\ldots+24+25=279$ days

3.

$$\begin{array}{r} 50,055 \\ -\ 7,899 \\ \hline 42,156 \end{array}$$

4.

$$\begin{array}{r} 308 \\ \times \quad 75 \\ \hline 1\,540 \\ +21\,560 \\ \hline 23{,}100 \end{array}$$

5. $A = s^2 = (8\,ft)^2 = 64\,ft^2$

6.

$$\begin{array}{r} 15 \\ 37\overline{)561} \\ -37 \\ 191 \\ -185 \\ 6 \end{array}$$

$15R6$

7. a. 1 , 2 , 4 , 5 , 8 , 10 , 20 , 40

b. $294 = 2\cdot 147 = 2\cdot 3\cdot 49 = 2\cdot 3\cdot 7\cdot 7 = 2\cdot 3\cdot 7^2$

8.

$24 = 2\cdot 2\cdot 2\cdot 3$

$30 = 2\cdot 3\cdot 5$

$LCM(24,30) = 2\cdot 2\cdot 2\cdot 3\cdot 5 = 120$

$GCF(24,30) = 2\cdot 3 = 6$

9.

$$\frac{39+3\left[4^3-2\left(2^2-3\right)\right]}{4\cdot 2^2-1}$$

$$=\frac{39+3\left[64-2(4-3)\right]}{4\cdot 4-1}=\frac{39+3\left[64-2(1)\right]}{16-1}$$

$$=\frac{39+3[62]}{15}=\frac{39+186}{15}=\frac{225}{15}=15$$

10.

$$\frac{2{,}672+1{,}680+2{,}485+1{,}370+2{,}737+1{,}692}{6}$$

$$=\frac{12{,}636}{6}=\$2{,}106$$

11. a. $23+a=207$

$23+a-23=207-23$

$a=184$

Check: $23+a=207$

$23+184 \stackrel{?}{=} 207$

$207=207$

b. $23a=207$

$\frac{23a}{23}=\frac{207}{23}$

$a=9$

Check: $23a=207$

$23(9) \stackrel{?}{=} 207$

$207=207$

12.

$|-8| \square -(-5)$

$8>5$

$|-8|>-(-5)$

13. $(-20+9)+(-13+24)=-11+11=0$

14.

$75+60+25+3(10)=\$190$

$\$55-\$190=-\$135$

15.

$-6^2=-6\cdot 6=-36$

$(-6)^2=(-6)(-6)=36$

16. a. undefined b. 0

c. 0 . $0-(-14)=0+14=14$

17.

$24\div 2(3)-3^3=12(3)-27$

$=36-27=9$

18. $-5{,}700+(-2{,}300)+3{,}400+2{,}700=-1{,}900$

19.
$$6 = 2 - 2x$$
$$6 - 2 = 2 - 2x - 2$$
$$4 = -2x$$
$$\frac{4}{-2} = \frac{-2x}{-2}$$
$$-2 = x$$
Check: $6 = 2 - 2x$
$$6 \stackrel{?}{=} 2 - 2(-2)$$
$$6 \stackrel{?}{=} 2 + 4$$
$$6 = 6$$

20. x = the gain in market share
$$-51 + x = -15$$
$$-51 + x + 51 = -15 + 51$$
$$x = 36$$
The company's market share increased 36 points.

21. $2t - 16$

22.
$$C = \frac{5(F - 32)}{9} = \frac{5(59 - 32)}{9}$$
$$= \frac{5(27)}{9} = 5(3) = 15°C$$

23. a. $4(8m) = (4 \cdot 8)m = 32m$

b. $(4d + 8)2 = 4d \cdot 2 + 8 \cdot 2 = 8d + 16$

24. a. $5x - 4x = (5 - 4)x = 1x = x$

b. $l + w + l + w = 2l + 2w$

25.
$$-(3x - 3) = 6(2x - 7)$$
$$-3x + 3 = 12x - 42$$
$$-3x + 3 + 3x = 12x - 42 + 3x$$
$$3 = 15x - 42$$
$$3 + 42 = 15x - 42 + 42$$
$$45 = 15x$$
$$\frac{45}{15} = \frac{15x}{15}$$
$$3 = x$$
Check: $-(3x - 3) = 6(2x - 7)$
$$-(3(3) - 3) \stackrel{?}{=} 6(2(3) - 7)$$
$$-(9 - 3) \stackrel{?}{=} 6(6 - 7)$$
$$-6 \stackrel{?}{=} 6(-1)$$
$$-6 = -6$$

26. m = the number of months needed
$$34 + 6m = 100$$
$$34 + 6m - 34 = 100 - 34$$
$$6m = 66$$
$$\frac{6m}{6} = \frac{66}{6}$$
$$m = 11$$
She needs 11 months to reach her goal.

27. $\frac{4}{11}$

28.
$$\frac{10y}{15y} = \frac{2 \cdot 5 \cdot y}{3 \cdot 5 \cdot y} = \frac{2 \cdot \cancel{5} \cdot \cancel{y}}{3 \cdot \cancel{5} \cdot \cancel{y}} = \frac{2}{3}$$

29.
$$\frac{4}{5} = \frac{4}{5} \cdot \frac{9}{9} = \frac{36}{45}$$

30.
$$\frac{1}{4}(-240) = \frac{\cancel{4}(-60)}{\cancel{4}} = -60$$

31.
$$A = \frac{1}{2}bh = \frac{1}{2} \cdot 50 \cdot 26$$
$$= 25 \cdot 26 = 650 in^2$$

32. $-\frac{16a}{35}\cdot\frac{25}{48a^2}=-\frac{16\cdot a\cdot 5\cdot 5}{5\cdot 7\cdot 3\cdot 16\cdot a\cdot a}$

$=-\frac{\cancel{16}\cdot\cancel{a}\cdot\cancel{5}\cdot 5}{\cancel{5}\cdot 7\cdot 3\cdot\cancel{16}\cdot\cancel{a}\cdot a}=-\frac{5}{7\cdot 3\cdot a}=-\frac{5}{21a}$

33. $\frac{4}{9}\div\left(-\frac{16}{27}\right)=-\frac{4}{9}\cdot\frac{27}{16}=-\frac{\cancel{4}\cdot 3\cdot\cancel{9}}{\cancel{9}\cdot\cancel{4}\cdot 4}=-\frac{3}{4}$

34. $\frac{9}{10}-\frac{3}{14}=\frac{9}{10}\cdot\frac{7}{7}-\frac{3}{14}\cdot\frac{5}{5}$

$=\frac{63}{70}-\frac{15}{70}=\frac{48}{70}=\frac{\cancel{2}\cdot 24}{\cancel{2}\cdot 35}=\frac{24}{35}$

35. $\frac{4}{m}+\frac{2}{7}=\frac{4}{m}\cdot\frac{7}{7}+\frac{2}{7}\cdot\frac{m}{m}=\frac{28}{7m}+\frac{2m}{7m}=\frac{28+2m}{7m}$

36. $\frac{23}{20}=\frac{23}{20}\cdot\frac{3}{3}=\frac{69}{60}$

$\frac{7}{6}=\frac{7}{6}\cdot\frac{10}{10}=\frac{70}{60}$

$\frac{7}{6}>\frac{23}{20}$

37. $\frac{1}{3}-\frac{1}{4}=\frac{1}{3}\cdot\frac{4}{4}-\frac{1}{4}\cdot\frac{3}{3}=\frac{4}{12}-\frac{3}{12}=\frac{1}{12}lb$

38. $-3\frac{3}{4}(8)=-\frac{15}{4}\cdot\frac{8}{1}=\frac{-15\cdot 2\cdot\cancel{4}}{\cancel{4}}=-30$

39. $32+\frac{3}{4}\cdot 5=32+\frac{15}{4}=32+3\frac{3}{4}=35\frac{3}{4}in$

40.

$$\begin{array}{r} 34\frac{1}{9} \\ -13\frac{5}{6} \\ \hline \end{array}\qquad \begin{array}{r} 34\frac{2}{18} \\ -13\frac{15}{18} \\ \hline \end{array}\qquad \begin{array}{r} 33\frac{20}{18} \\ -13\frac{15}{18} \\ \hline 20\frac{5}{18} \end{array}$$

41. $\dfrac{\frac{1}{3}-\frac{3}{4}}{\frac{1}{6}+\frac{1}{3}}=\dfrac{(12)\frac{1}{3}-(12)\frac{3}{4}}{(12)\frac{1}{6}+(12)\frac{1}{3}}=\frac{4-9}{2+4}=\frac{-5}{6}=-\frac{5}{6}$

42. p = the number of pages

$\frac{2}{3}p=350$

$\frac{3}{2}\cdot\frac{2}{3}p=\frac{3}{2}\cdot 350$

$p=3(175)$

$p=525$

The phone book has 525 pages.

43. $\frac{5}{6}y=-25$

$\frac{6}{5}\cdot\frac{5}{6}y=\frac{6}{5}(-25)$

$y=6(-5)$

$y=-30$

Check: $\frac{5}{6}y=-25$

$\frac{5}{6}(-30)\stackrel{?}{=}-25$

$5(-5)\stackrel{?}{=}-25$

$-25=-25$

44. $\frac{y}{6}=\frac{y}{12}+\frac{2}{3}$

$12\cdot\frac{y}{6}=12\cdot\frac{y}{12}+12\cdot\frac{2}{3}$

$2y=y+8$

$2y-y=y+8-y$

$y=8$

Check: $\frac{y}{6}=\frac{y}{12}+\frac{2}{3}$

$\frac{8}{6}\stackrel{?}{=}\frac{8}{12}+\frac{2}{3}$

$\frac{4}{3}\stackrel{?}{=}\frac{2}{3}+\frac{2}{3}$

$\frac{4}{3}=\frac{4}{3}$

45. a. 452.03

b. 452.030

46. $3.4-(6.6+7.3)+5=3.4-13.9+5$

$=-10.5+5=-5.5$

47. $40(15.90)+4(1.5)(15.90)$

$=\$636+\$95.4=\$731.40$

48. Start by moving the decimals two units right.

$$\begin{array}{r} 0.27 \\ 58\overline{)15.66} \\ -\underline{116} \\ 406 \\ -\underline{406} \\ 0 \end{array}$$

49.

$$\begin{array}{r} 0.733 \\ 15\overline{)11.000} \\ -\underline{105} \\ 50 \\ -\underline{45} \\ 50 \\ -\underline{45} \\ 5 \end{array}$$

$\frac{11}{15}=0.7\overline{3}$

50. $3\sqrt{81}-8\sqrt{49}=3\cdot 9-8\cdot 7$

$=27-56=-29$

51. $37.1n+12.2=12.4n+93.4+4.4n$

$37.1n+12.2=16.8n+93.4$

$37.1n+12.2-16.8n=16.8n+93.4-16.8n$

$20.3n+12.2=93.4$

$20.3n+12.2-12.2=93.4-12.2$

$20.3n=81.2$

$\frac{20.3n}{20.3n}=\frac{81.2}{20.3}$

$n=4$

Check: $37.1n+12.2=12.4n+93.4+4.4n$

$37.1(4)+12.2\stackrel{?}{=}12.4(4)+93.4+4.4(4)$

$148.4+12.2\stackrel{?}{=}49.6+93.4+17.6$

$160.6=160.6$

52. $2(a-4.3)+1.2=-6.2$

$2a-8.6+1.2=-6.2$

$2a-7.4=-6.2$

$2a-7.4+7.4=-6.2+7.4$

$2a=1.2$

$a=0.6$

Check: $2(a-4.3)+1.2=-6.2$

$2(0.6-4.3)+1.2\stackrel{?}{=}-6.2$

$2(-3.7)+1.2\stackrel{?}{=}-6.2$

$-7.4+1.2\stackrel{?}{=}-6.2$

$-6.2=-6.2$

53. x = the number of hours for repair

$175+35x=297.50$

$175+35x-175=297.5-175$

$35x=122.5$

$\frac{35x}{35}=\frac{122.5}{35}$

$x=3.5$

It took 3.5 hours to fix the car.

54. $\frac{1\frac{1}{4}}{1\frac{1}{2}}=\frac{\frac{5}{4}}{\frac{3}{2}}=\frac{5}{4}\cdot\frac{2}{3}=\frac{5\cdot \cancel{2}}{2\cdot \cancel{2}\cdot 3}=\frac{5}{6}$

55. $\frac{7}{14}=\frac{2}{x}$

$7\cdot x=2\cdot 14$

$\frac{\cancel{7}x}{\cancel{7}}=\frac{28}{7}$

$x=4$

56. m = the number of minutes needed

$\frac{385}{7}=\frac{495}{m}$

$385\cdot m=495\cdot 7$

$\frac{385m}{385}=\frac{3,465}{385}$

$m=9$

It would take her 9 minutes.

57. $\frac{960hr}{1}\cdot\frac{1\text{day}}{24hr}=\frac{960}{24}=40\text{days}$

58. $\frac{2{,}400mm}{1}\cdot\frac{1m}{1{,}000mm}=\frac{2{,}400}{1{,}000}m=2.4m$

59. $\frac{6.5kg}{1}\cdot\frac{2.2lb}{1kg}=6.5\cdot 2.2lb=14.3lb$

60.

Percent	Decimal	Fraction
29%	0.29	$\frac{29}{100}$
47.3%	0.473	$\frac{47.3}{100}=\frac{473}{1{,}000}$
87.5%	0.875	$\frac{875}{1{,}000}=\frac{7\cdot\cancel{125}}{8\cdot\cancel{125}}=\frac{7}{8}$

61. $20=0.16\cdot x$

$\frac{20}{0.16}=\frac{\cancel{0.16}x}{\cancel{0.16}}$

$125=x$

62. $\frac{180}{10{,}000{,}000}=0.000018=0.0018\%$

63. There was an increase of 234. $\frac{234}{300}=0.78=78\%$

64. sale price: $321

discount: $107

regular price: $321 + $107 = $428

discount rate: $\frac{107}{428}=0.25=25\%$

65. Estimate $80 for the bill.

10% of 80 is 8 so 5% of 80 is 4.

$8 + $4 = $12

Total check: $78.18 + $12 = $90.18

66. $I=P\cdot r\cdot t$

$I=10{,}000\cdot 0.0725\cdot 2$

$I=725\cdot 2=\$1{,}450$

CHAPTER 8 Graphs and Statistics

Section 8.1: Reading Graphs and Tables

VOCABULARY

1. Graph (a) is a bar graph.

3. Graph (c) is a pictograph.

5. Graph (d) is a histogram.

7. A horizontal or vertical line used for reference in a graph is called an axis.

CONCEPTS

9. To read a table, we must find the intersection of the row and column that contains the desired information.

11. A pictograph is like a bar graph, but the bars are made from pictures or symbols.

13. A histogram is a bar graph with three important features:
    *The bars of a histogram touch.
    *Data values never fall at the edge of a bar.
    *The widths of the bars of a histogram are equal and represent a range of values.

NOTATION

15. About 500 buses

GUIDED PRACTICE

17. From Row 8: $13.30

19. Only the 2 pound 9 ounce package: $12.00
    Only the 3 pound 8 ounce package: $15.40
    Shipped separately: $12.00 + $15.40 = $27.40

    Combined, the packages weigh 6 pounds 1 ounce.
    Shipped together: $21.80

    She saves $27.40 –$ 21.80 = $5.60.

21. fish, cat, dog

23. dogs and cats

25. Yes

27. 3,000,000 + 3,000,000 + 5,000,000 = 11,000,000 metric tons

29. 1990, 2000, 2011

31. 13,000,000 – 7,000,000 = 6,000,000 metric tons

33. Seniors

35. $7.5\cdot 100 - 7\cdot 100 = 750 - 700 = \$50$

37. Chinese

39. Combined, those languages account for 14%, which is not more than the 18% who speak Chinese.

41. 5% + 2% + 18% + 3% + 3% + 5% + 1% + 1% = 38%
    100% – 38% = 62%

43. $$\frac{18}{100} = \frac{x}{7{,}064{,}000{,}000}$$
$$100x = 18\cdot 7{,}064{,}000{,}000$$
$$100x = 127{,}152{,}000{,}000$$
$$x = 1{,}271{,}520{,}000$$
$$x \approx 1{,}272{,}000{,}000$$

45. 494

47. 2002 to 2003; 2004 to 2005; 2005 to 2006; 2007 to 2008; 2008 to 2009; 2009 to 2010

49. 2001 and 2003

51. 2005 and 2006: decrease of 492 – 478 =14 resorts

53. Runner number 1

55. At time B.

57. Runner number 1

59. Runner number 1was running and runner number 2 was stopped.

61. a. 27 b. 13 + 9 = 22

TRY IT YOURSELF

63. $4,991.25 + 0.25(79,250 - 36,250)$
$= 4,991.25 + 0.25(43,000)$
$= 4,991.25 + 10,750 = \$15,741.25$

65. $125,846.00 + 0.396(157,000,000 - 450,000)$
$= 125,846 + 0.396(156,550,000)$
$= 125,846 + 61,993,800 = \$62,119,646$

67. a. $4,991.25 + 0.25(53,000 - 36,250)$
$= 4,991.25 + 0.25(16,750)$
$= 4,991.25 + 4,187.50 = \$9,178.75$

b. $1,785 + 0.15[(53,000 - 3,900) - 17,850)]$
$= 1,785 + 0.15(31,250)$
$= 1,785 + 4,687.50 = \$6,472.50$

c. $\$9,178.75 - \$6,472.50 = \$2,706.25$

69. 2007; about 2.1%

71. Increased; about 0.5%

73. It increased.

75. At time D.

77. Reckless driving and failure to yield

79. Reckless driving

81. About \$440

83. No

85. The miner's

87. The miners

89. About \$42

91. Boston - San Francisco = 5(33) - 5(28)
=165 - 140 = \$25

93. $\frac{8.3}{97.2} = \frac{x}{100}$
$97.2x = 8.3(100)$
$97.2x = 830$
$x = \frac{830}{97.2} = 8.53...$
About 9% comes from nuclear energy.

95. $\frac{9.1 + 8.3}{97.2} = \frac{x}{100}$
$\frac{17.4}{97.2} = \frac{x}{100}$
$97.2x = 17.4(100)$
$97.2x = 1740$
$x = \frac{1740}{97.2} = 17.90...$
About 18% comes from both.

97.

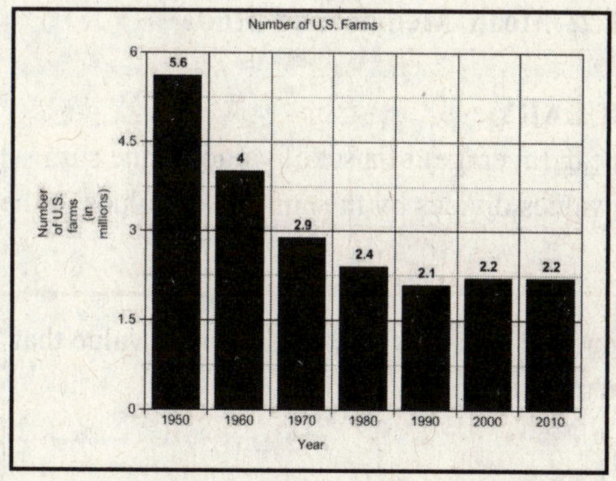

99.

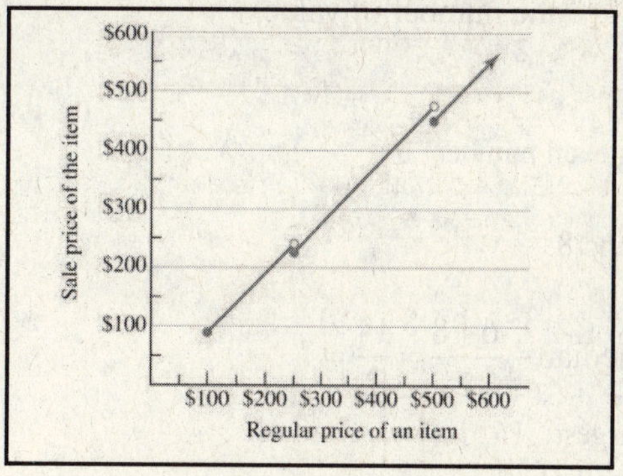

101. * A circle graph would make it easy to make multiple comparisons about student's choice of major.

* A line graph would show the trend over the time period.

* A histogram would show the time intervals spent studying for final exams.

* A table would allow comparisons of the different cities.

* A bar graph would allow the comparisons between the average salaries easy to see.

103. 11, 13, 17, 19, 23, 29

105. 0 and 4

Section 8.2: Mean, Median, and Mode

VOCABULARY

1. The mean (average) of a set of values is the sum of the values divides by the number of values in the set.

3. The mode of a set of values is the single value that occurs most often.

CONCEPTS

5. $\text{Mean} = \dfrac{\text{the sum of the values}}{\text{the number of values}}$

7. a. An even number

b. 6 and 8

c. $\text{Median} = \dfrac{6+8}{2} = \dfrac{14}{2} = 7$

d. Largest: 15
Smallest: 4
Range: $15 - 4 = 11$

9. $\text{Mean} = \dfrac{3+4+7+7+8+11+16}{7} = \dfrac{56}{7} = 8$

11. $\text{Mean} = \dfrac{5+9+12+35+37+45+60+77}{8}$
$= \dfrac{280}{8} = 35$

13. $\text{Mean} = \dfrac{15+7+12+19+27+17+19+35+20}{9}$
$= \dfrac{171}{9} = 19$

15. $\text{Mean} = \dfrac{4.2+3.6+7.1+5.9+8.2}{5}$
$= \dfrac{29}{5} = 5.8$

17. 1 2 5 **9** 11 17 29 Median = 9

19. 1 3 4 4 **5** 6 7 7 7 Median = 5

21. 13.6 15.1 **17.2** 19.7 44.9 Median = 17.2

23. $\dfrac{1}{100}\ \dfrac{1}{3}\ \dfrac{5}{8}\ \dfrac{999}{1{,}000}\ \dfrac{16}{15}$ $\text{Median} = \dfrac{5}{8}$

25. 6 7 **8** **10** 16 63

$\text{Median} = \dfrac{8+10}{2} = \dfrac{18}{2} = 9$

27. 1 39 **41** **47** 50 51

$\text{Median} = \dfrac{41+47}{2} = \dfrac{88}{2} = 44$

29. 1.7 1.8 2.0 **2.0** **2.1** 2.1 2.3 9.0

$\text{Median} = \dfrac{2.0+2.1}{2} = \dfrac{4.1}{2} = 2.05$

31. $\dfrac{1}{5}\ \dfrac{2}{5}\ \dfrac{3}{5}\ \dfrac{7}{5}\ \dfrac{11}{5}\ \dfrac{13}{5}$

$\text{Median} = \dfrac{\frac{3}{5}+\frac{7}{5}}{2} = \dfrac{\frac{10}{5}}{2} = \dfrac{2}{2} = 1$

33. 3

35. -6

37. 22.7

39. Both $\frac{1}{3}$ and $\frac{1}{2}$ (bimodal)

APPLICATIONS

41.

a. $\frac{75+80+90+85}{4}=\frac{330}{4}=82.5$

b. $\frac{75+80+90+2\cdot 85}{5}=\frac{415}{5}=83$

43.

a. $\frac{98,790}{37}=2,670mi$

b. $\frac{2,670mi}{30d}=89mi/d$

45.

a. $2,500+4\cdot 500+35\cdot 150+85\cdot 25$
$=2,500+2,000+5,250+2,125$
$=\$11,875$

b. $1+4+35+85=125$

c. $\frac{11,875}{125}=\$95$

47.

a. $\frac{\text{sum}}{20}=\frac{1300}{20}=65¢$

b. 45,50,50,50,50,50,50,50,60,60,
60,70,75,75,75,75,75,80,100,100

Median = 60¢

c. Mode = 50¢

d. Range = $100-45=55$¢

49. $\frac{53+54+\cdots 68+71}{12}=\frac{732}{12}=61°$

51. $\frac{5\cdot 2+3\cdot 4+1\cdot 3+4\cdot 1}{13}$
$=\frac{10+12+3+4}{13}=\frac{29}{13}\approx 2.23$

53. $\frac{3\cdot 1+4\cdot 3+1\cdot 4+3\cdot 2+5\cdot 3}{16}$
$=\frac{3+12+4+6+15}{16}=\frac{40}{16}=2.5$

55. 85 85 85 85 85
Mean, Median, and Mode = 85

57. $B:\frac{37+53+78}{3}=\frac{168}{3}=56$
$S:\frac{53+57+58}{3}=\frac{168}{3}=56$

They had the same average, but the sister was more consistent.

59. 11.2 17.5 18 **24** **26** 27 28 28.5

Median = $\frac{24+26}{2}=\frac{50}{2}=25oz$

Mean = $\frac{11.2+17.5+\cdots 28.5}{8}$
$=\frac{180.2}{8}=22.525oz$

Range = $28.5-11.2=17.3oz$

61. Mean = $\frac{7.0+7.0+\cdots+7.1}{17}=\frac{125}{17}\approx 7.4$

7.0 7.0 7.0 7.0 7.1 7.1 7.1 7.2 **7.2**
7.2 7.2 7.3 7.4 7.6 7.7 7.9 9.0

Median = 7.2

Range = 9.0 – 7.0 = 2.0

63. 3 3 4 4 4 4 4 5 **5** **5** 6 6 6 7 8 9 9 9 13

Median = 5lb

Mode = 4lb

Range = 13 – 3 = 10lb

WRITING

65. The mean is the sum of the values divided by the number of the values. The mode is the value which appears the most often. The median is the middle value (or the average of the two middle values) when the values are written in numerical order. The range is the difference between the largest value and the smallest value.

67. Median would be the best way to evaluate it so that the high CEO salaries do not skew the results.

REVIEW

69.
$$\frac{x}{100}=\frac{52}{80}$$
$$x\cdot 80=52\cdot 100$$
$$\frac{x\cdot 80}{80}=\frac{5,200}{80}$$
$$x=65$$

52 is 65% of 80.

71.
$$\frac{28}{x}=\frac{66\frac{2}{3}}{100}$$
$$28\cdot 100=\frac{200}{3}\cdot x$$
$$28\cdot \cancel{100}\cdot\frac{3}{\cancel{200}_{2}}=\frac{\cancel{3}}{\cancel{200}}\cdot\frac{\cancel{200}}{\cancel{3}}\cdot x$$
$$28\cdot\frac{3}{2}=x$$
$$42=x$$

73.
$$\frac{5}{8}=\frac{x}{100}$$
$$5\cdot 100=x\cdot 8$$
$$\frac{500}{8}=\frac{x\cdot 8}{8}$$
$$62.5=x$$

5 is 62.5% of 8.

75.
$$\frac{7\frac{1}{4}}{100}=\frac{x}{600}$$
$$7.25\cdot 600=x\cdot 100$$
$$\frac{4,350}{100}=\frac{x\cdot 100}{100}$$
$$43.5=x$$

Section 8.3: Equations in Two Variables; The Rectangular Coordinate System

VOCABULARY

1. $x+y=4$ is an equation in two variables.

3. When we substitute 1 for x and 3 for y in the equation $x+y=4$, the result is the true statement $4=4$. We say that $(1,3)$ satisfies the equation.

5. Solutions of equations in two variables can be listed in a table of solutions like the one shown below.

7. The point with coordinates $(0,0)$ is called the origin.

9. In the ordered pair $(-2,4)$, -2 is the x - coordinate and 4 is the y - coordinate.

CONCEPTS

11.

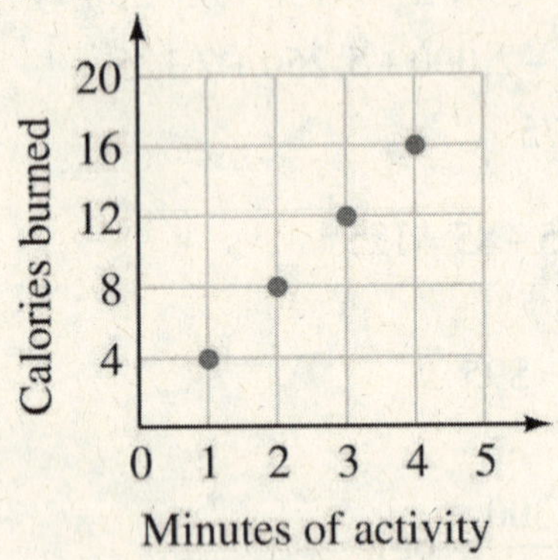

13. a. $2x+5=10$ is an equation in one variable.

b. $2x+5y=10$ is an equation in two variables.

15. a. On the $x-$axis, the positive direction is to the right.

b. On the $y-$axis, the positive direction is upward.

17. To plot the point $(3,-4)$, we start at the origin and move 3 units to the right on the x – axis and then move 4 units down.

19. A: x – axis; B: QI; C: QIII; D: origin; E: QII; F: QIV

NOTATION

21. $(3,6)$ — $\nearrow$ x, $\nwarrow$ y

23. $1\frac{1}{3}=\frac{4}{3}$, $-\frac{5}{2}=-2.5$, so yes

25.
$$4x+3y=14$$
$$4(2)+3y=14$$
$$8+3y=14$$
$$8+3y-8=14-8$$
$$3y=6$$
$$\frac{3y}{3}=\frac{6}{3}$$
$$y=2$$

GUIDED PRACTICE

27.
$$2x+3y=13$$
$$2(2)+3(3)\stackrel{?}{=}13$$
$$4+9\stackrel{?}{=}13$$
$$13=13$$
$(2,3)$ is a solution.

29.
$$x+5y=-7$$
$$9+5(-3)\stackrel{?}{=}-7$$
$$9-15\stackrel{?}{=}-7$$
$$-6=-7$$
$(9,-3)$ is not a solution.

31.
$$-3x+4y=-11$$
$$-3(-9)+4(-9)\stackrel{?}{=}-11$$
$$27-36\stackrel{?}{=}-11$$
$$-9=-11$$
$(-9,-9)$ is not a solution.

33.
$$10x-y=0$$
$$10(0)-(0)\stackrel{?}{=}0$$
$$0-0\stackrel{?}{=}0$$
$$0=0$$
$(0,0)$ is a solution.

35.
$$y=5x-4$$
$$1\stackrel{?}{=}5(2)-4$$
$$1\stackrel{?}{=}10-4$$
$$1=6$$
$(2,1)$ is not a solution.

37.
$$y=-x+4$$
$$6\stackrel{?}{=}-(-2)+4$$
$$6\stackrel{?}{=}2+4$$
$$6=6$$
$(-2,6)$ is a solution.

39.
$$y=-3x+13$$
$$25\stackrel{?}{=}-3(-4)+13$$
$$25\stackrel{?}{=}12+13$$
$$25=25$$
$(-4,25)$ is a solution.

41.
$$y=7x-9$$
$$-9\stackrel{?}{=}7(0)-9$$
$$-9\stackrel{?}{=}0-9$$
$$-9=-9$$
$(0,-9)$ is a solution.

43.

a.
$$2x+y=8$$
$$2(0)+y=8$$
$$0+y=8$$
$$y=8$$
$(0,8)$

b.
$$2x+y=8$$
$$2x+2=8$$
$$2x=6$$
$$x=3$$
$(3,2)$

45. a. $x-3y=5$
$2-3y=5$
$-3y=3$
$y=-1$
$(2,-1)$

b. $x-3y=5$
$x-3(3)=5$
$x-9=5$
$x=14$
$(14,3)$

47. a. $5x+3y=15$
$5(-6)+3y=15$
$-30+3y=15$
$3y=45$
$y=15$
$(-6,15)$

b. $5x+3y=15$
$5x+3(4)=15$
$5x+12=15$
$5x=3$
$x=\frac{3}{5}$
$\left(\frac{3}{5},4\right)$

49. a. $8x-15y=-4$
$8(1)-15y=-4$
$8-15y=-4$
$-15y=-12$
$y=\frac{-12}{-15}=\frac{4}{5}$
$\left(1,\frac{4}{5}\right)$

b. $8x-15y=-4$
$8x-15(-4)=-4$
$8x+60=-4$
$8x=-64$
$x=-8$
$(-8,-4)$

51. $4(0)+3y=24$
$3y=24$
$y=8$

$4x+3(0)=24$
$4x=24$
$x=6$

$4(-3)+3y=24$
$-12+3y=24$
$3y=36$
$y=12$

x	*y*	*(x, y)*
0	8	(0, 8)
6	0	(6, 0)
-3	12	(-3, 12)

53. $5(0)-4y=20$
$-4y=20$
$y=-5$

$5x-4(0)=20$
$5x=20$
$x=4$

$5(-4)-4y=20$
$-20-4y=20$
$-4y=40$
$y=-10$

x	y	(x,y)
0	-5	$(0,-5)$
4	0	$(4,0)$
-4	-10	$(-4,-10)$

55.

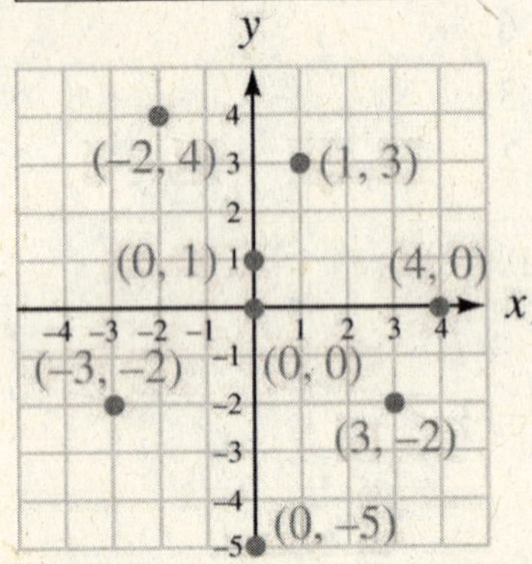

57.

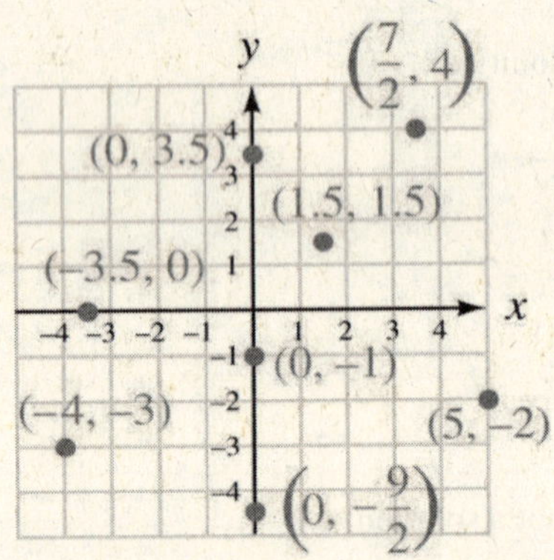

59. A: $(2,4)$ B: $(-3,3)$ C: $(-2,-3)$

D: $(4,-3)$ E: $(3,0)$ F: $(0,-1)$

61. A: $(-3,-4)$ B: $(2.5,3.5)$ C: $(-2.5,0)$

D: $(2.5,0)$ E: $(3,-3)$ F: $(0,3)$

TRY IT YOURSELF

63. $3x-6y=12$

$3(-3.6)-6(-3.8)\stackrel{?}{=}12$

$-10.8+22.8\stackrel{?}{=}12$

$12=12$

$(-3.6,-3.8)$ is a solution.

65. $y-6x=12$

$7-6\left(\frac{5}{6}\right)\stackrel{?}{=}12$

$7-5\stackrel{?}{=}12$

$2=12$

$\left(\frac{5}{6},7\right)$ is not a solution.

APPLICATIONS

67. Rockford: $(5,B)$ Forreston: $(2,C)$

Harvard : $(7,A)$ intersection: $(5,E)$

69. a. $(2,-1)$ b. no c. yes d. no

71. New Delhi, Kampala, Coats Land, Reykjavik, Buenos Aires, Havana

73.

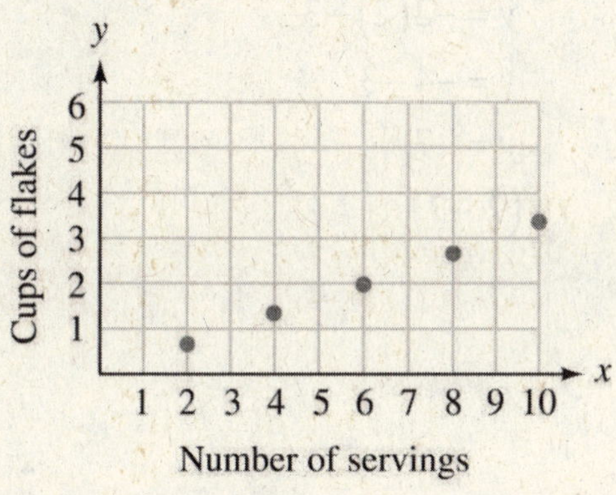

WRITING

75. To get the first point you need to go left, then up. To get the second point you go to the right, then down – you don't end at the same place.

77. To plot $(1,-7)$, first go 1 unit to the right, then go 7 units down.

79. The origin is $(0,0)$ because it is 0 units from the x-axis and 0 units from the y-axis.

REVIEW

81. $(-8-5)-3=-13-3=-16$

83. $-4^2-3^2=-16-9=-25$

85.

$\frac{x}{3}+3=10$

$\frac{x}{3}+3-3=10-3$

$\frac{x}{3}=7$

$3\cdot\frac{x}{3}=3\cdot 7$

$x=21$

Check: $\frac{x}{3}+3=10$

$\frac{21}{3}+3\stackrel{?}{=}10$

$7+3\stackrel{?}{=}10$

$10=10$

87.

$5-(7-x)=-5$

$5-7+x=-5$

$-2+x=-5$

$-2+x+2=-5+2$

$x=-3$

Check: $5-(7-x)=-5$

$5-(7-(-3))\stackrel{?}{=}-5$

$5-(7+3)\stackrel{?}{=}-5$

$5-10\stackrel{?}{=}-5$

$-5=-5$

Section 8.4: Graphing Linear Equations

VOCABULARY

1. $y = 2x + 3$ is an equation in two variables, x and y.

3. The graph of a linear equation is a mathematical "picture" of all its solutions.

5. The point where the graph of a linear equation crosses the x-axis is called the x-intercept.

7. To graph $y = 2x + 3$, we can use the following steps:

Step 1. Find three ordered pairs that are solutions of $y = 2x + 3$ by selecting three values for x and calculating the corresponding values of y.

Step 2. Plot/graph the solutions on a rectangular coordinate system.

Step 3. Draw a straight line passing through the points. If the points do not lie on a line, check your calculations.

9. a. $(0,1)$ b. $(-2,0)$ c. yes

11. To graph $3x + 4y = 12$, we can use the following two-step process:

Step 1. To find the x-intercept, substitute 0 in for y in $3x + 4y = 12$ and solve for x.

Step 2. To find the y-intercept, substitute 0 in for x in $3x + 4y = 12$ and solve for y.

13. a. The graph of the equation $y = 3$ is a horizontal line.

b. The graph of the equation $x = -2$ is a vertical line.

15. a. $(-3,4)\ (0,4)\ (2,4)$

b. $(-1,-5)\ (-1,0)\ (-1,3)$

NOTATION

17.

$$2x - 4y = 8$$
$$2x - 4(0) = 8$$
$$2x - 0 = 8$$
$$2x = 8$$
$$\frac{2x}{2} = \frac{8}{2}$$
$$x = 4$$
$$(4,0)$$

GUIDED PRACTICE

19.

$x = -2$	$x = 0$	$x = 2$
$y = 3(-2) + 3$	$y = 3(0) + 3$	$y = 3(2) + 3$
$y = -6 + 3$	$y = 0 + 3$	$y = 6 + 3$
$y = -3$	$y = 3$	$y = 9$
$(-2,-3)$	$(0,3)$	$(2,9)$

21.

$x = -2$	$x = 0$
$y = -2(-2) - 3$	$y = -2(0) - 3$
$y = 4 - 3$	$y = 0 - 3$
$y = 1$	$y = -3$
$(-2,1)$	$(0,-3)$

$$x = 2$$
$$y = -2(2) - 3$$
$$y = -4 - 3$$
$$y = -7$$
$$(2,-7)$$

23.

$x=0$	$x=3$
$y=2(0)-5$	$y=2(3)-5$
$y=0-5$	$y=6-5$
$y=-5$	$y=1$
$(0,-5)$	$(3,1)$

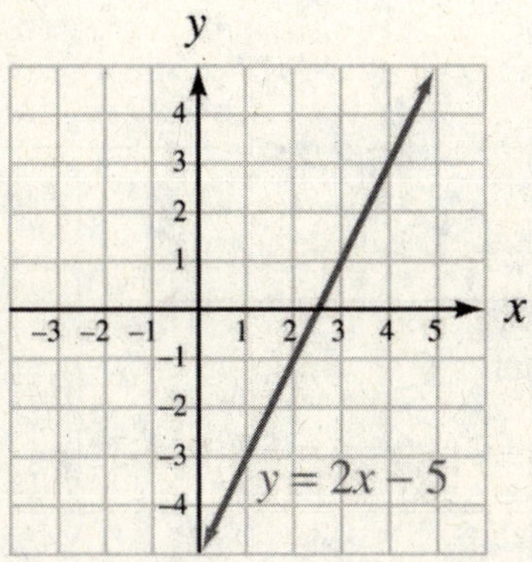

25.

$x=0$	$x=2$
$y=-3(0)+2$	$y=-3(2)+2$
$y=0+2$	$y=-6+2$
$y=2$	$y=-4$
$(0,2)$	$(2,-4)$

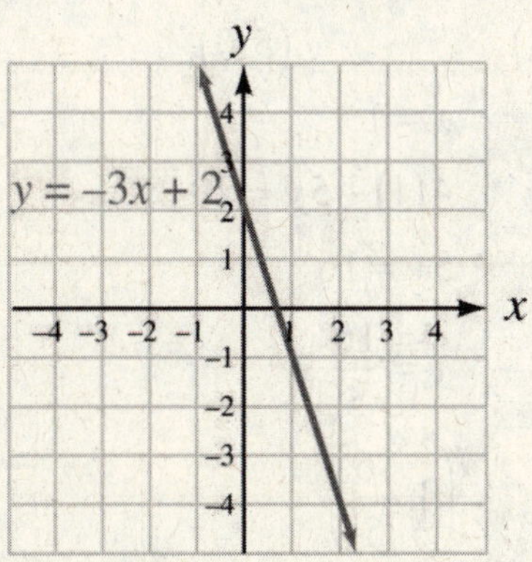

27.

$x=0$	$x=3$
$y=-\frac{1}{3}(0)+1$	$y=-\frac{1}{3}(3)+1$
$y=0+1$	$y=-1+1$
$y=1$	$y=0$
$(0,1)$	$(3,0)$

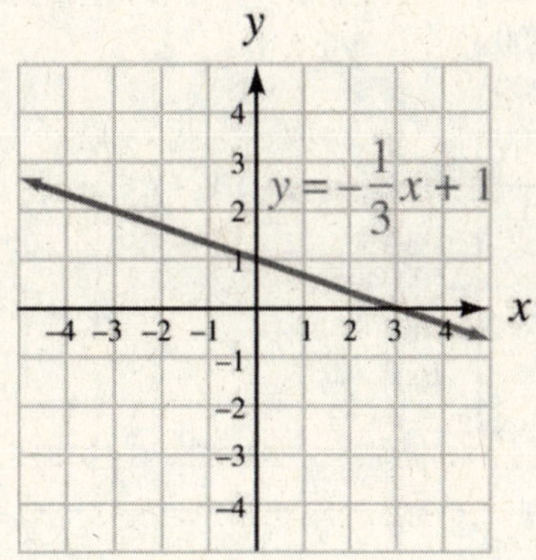

29.

$x=-4$	$x=2$
$y=-\frac{1}{2}(-4)+1$	$y=-\frac{1}{2}(2)+1$
$y=2+1$	$y=-1+1$
$y=3$	$y=0$
$(-4,3)$	$(2,0)$

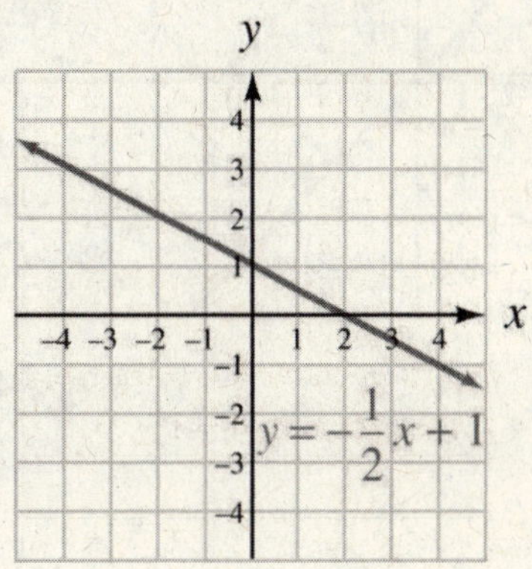

31.

$x=0$ | $x=3$

$y=100(0)$ | $y=100(3)$

$y=0$ | $y=300$

$(0,0)$ | $(3,300)$

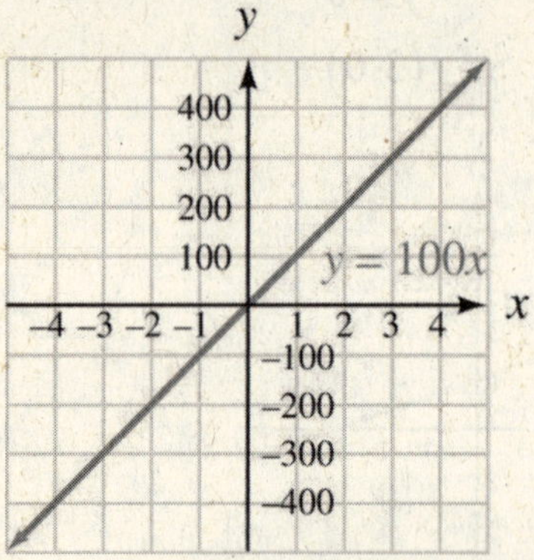

33.

$x=-3$ | $x=2$

$y=-30(-3)$ | $y=-30(2)$

$y=90$ | $y=-60$

$(-3,90)$ | $(2,-60)$

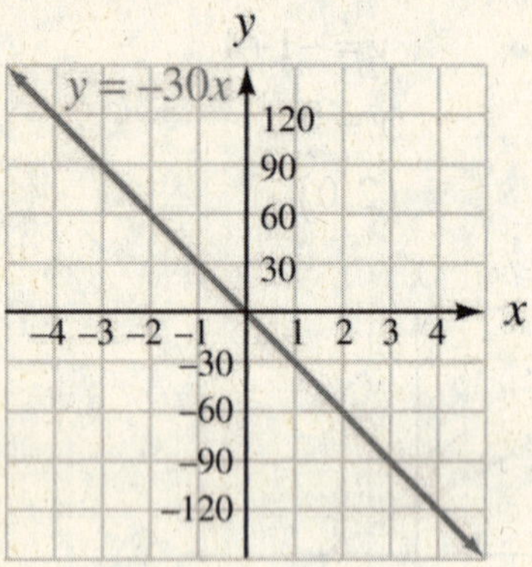

35.

$x=0$

$0-2y=-4$

$-2y=-4$

$y=2$

$(0,2)$

$y=0$

$x-2(0)=-4$

$x=-4$

$(-4,0)$

$x=4$

$4-2y=-4$

$-2y=-8$

$y=4$

$(4,4)$

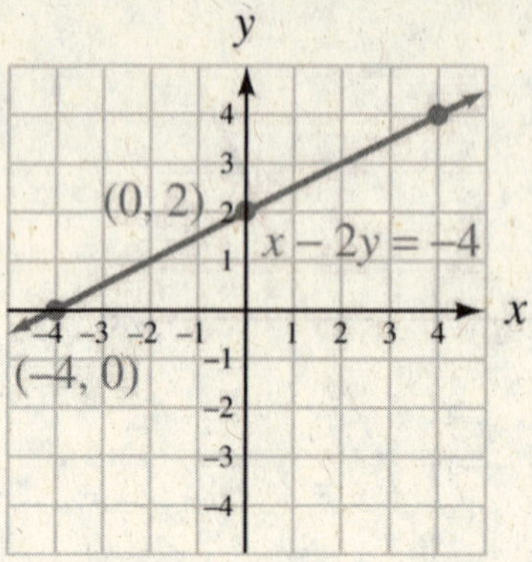

37.

$x=0:$ | $y=0$

$4(0)+5y=20$ | $4x+5(0)=20$

$5y=20$ | $4x=20$

$y=4$ | $x=5$

$(0,4)$ | $(5,0)$

$x=1$

$4(1)+5y=20$

$5y=16$

$y=\dfrac{16}{5}$

$\left(1,\dfrac{16}{5}\right)$

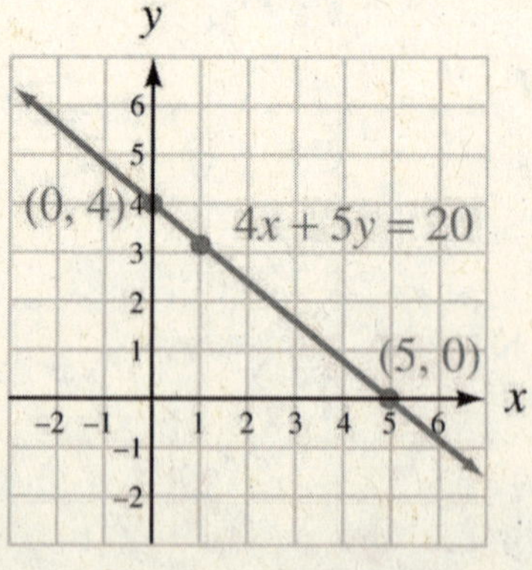

39.

$x = 0$ $y = 0$

$30(0) + y = -30$ $30x + 0 = -30$

$0 + y = -30$ $30x = -30$

$y = -30$ $x = -1$

$(0, -30)$ $(-1, 0)$

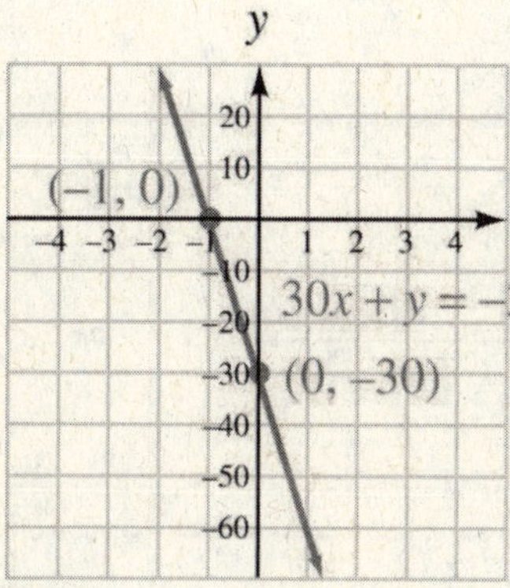

41.

$x = 0$ $y = 0$

$4(0) - 20y = 60$ $4x - 20(0) = 60$

$-20y = 60$ $4x = 60$

$y = -3$ $x = 15$

$(0, -3)$ $(15, 0)$

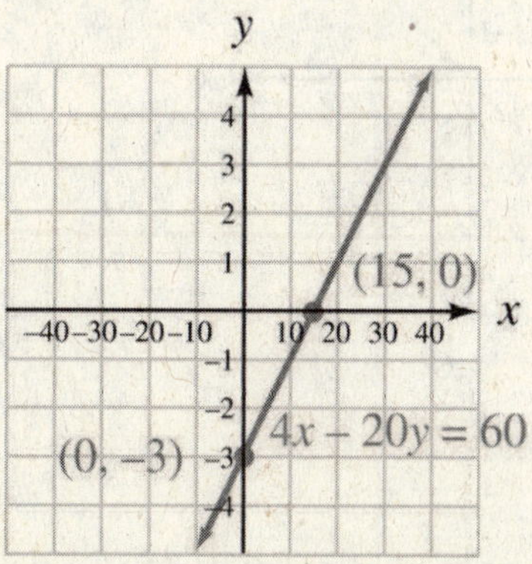

43.

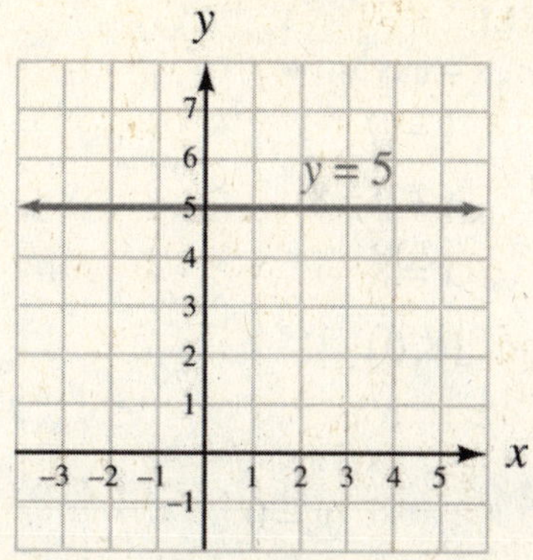

45.

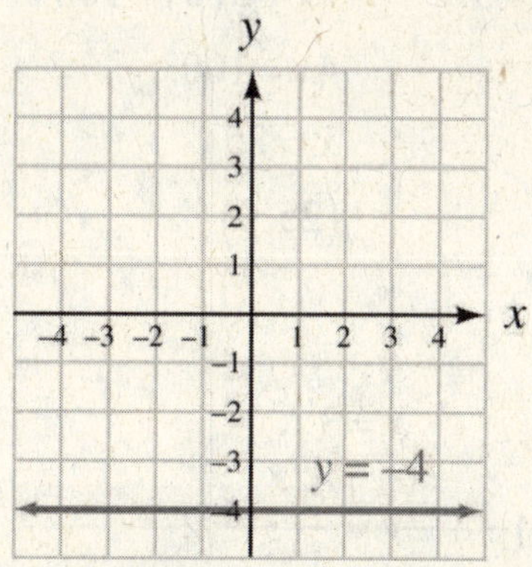

47.

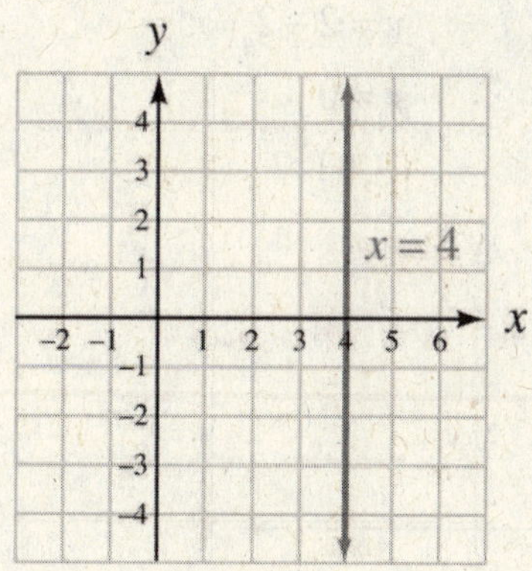

49.

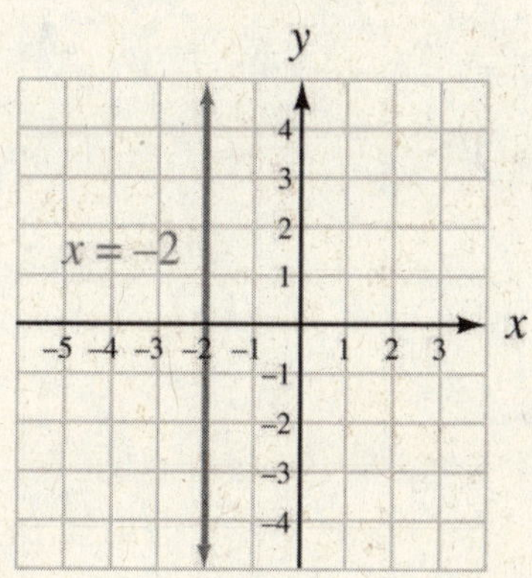

TRY IT YOURSELF

51. y – int:

$x = 0$

$0 + y = 8$

$y = 8$

$(0, 8)$

x – int:

$y = 0$

$x + 0 = 8$

$x = 8$

$(8, 0)$

53. y – int:

$x = 0$

$4(0) + 5y = 100$

$5y = 100$

$y = 20$

$(0, 20)$

x – int:

$y = 0$

$4x + 5(0) = 100$

$4x = 100$

$x = 25$

$(25, 0)$

55.

$x = -3$

$y = \frac{2}{3}(-3) - 2$

$y = -2 - 2$

$y = -4$

$(-3, -4)$

$x = 3$

$y = \frac{2}{3}(3) - 2$

$y = 2 - 2$

$y = 0$

$(3, 0)$

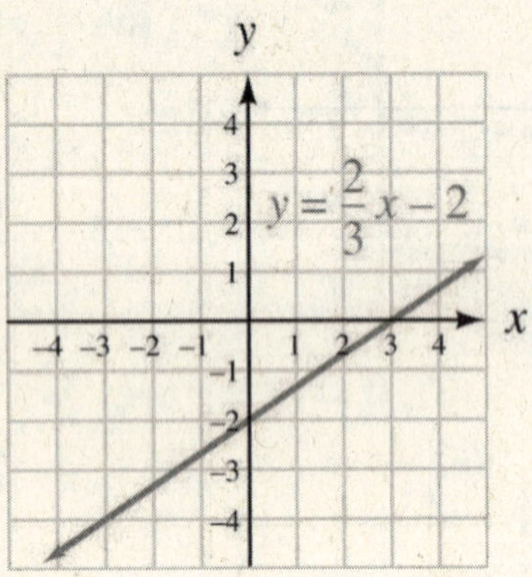

57.

$x = -1$

$-1 + y = 5$

$y = 6$

$(-1, 6)$

$x = 4$

$4 + y = 5$

$y = 1$

$(4, 1)$

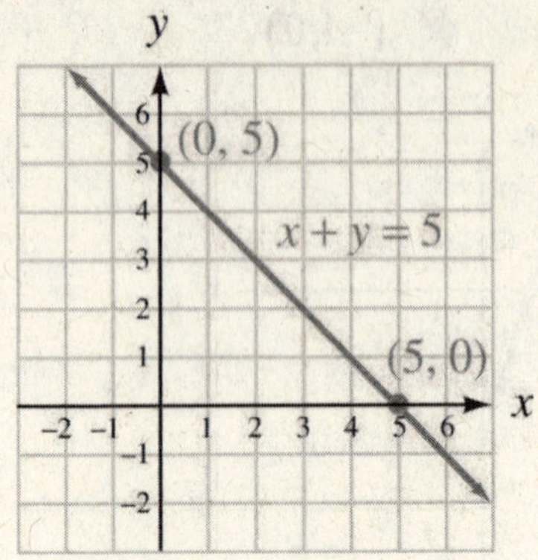

59.

$x = 2$

$y = 2$

$(2, 2)$

$x = -3$

$y = -3$

$(-3, -3)$

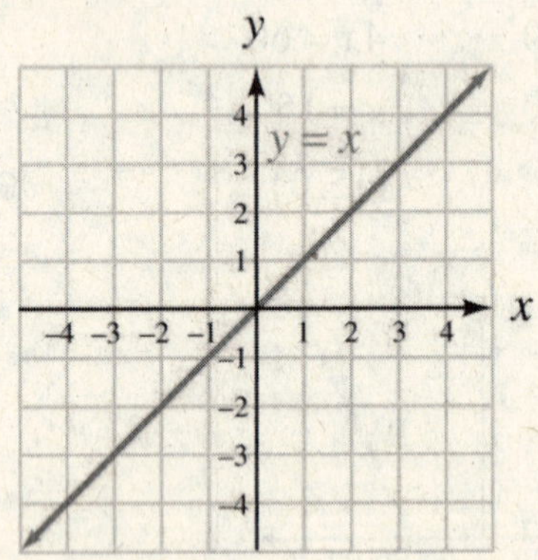

61.

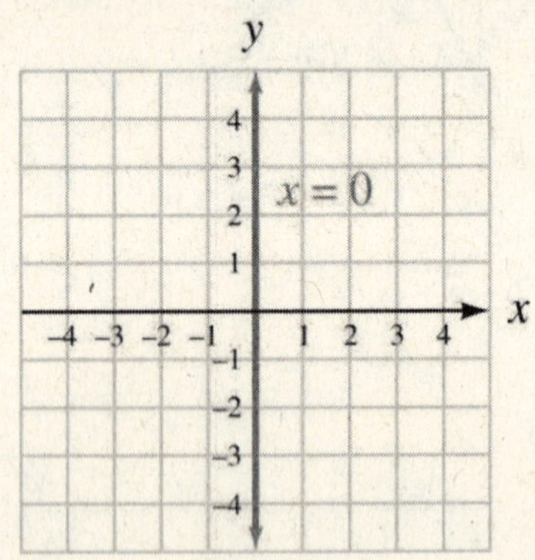

63.

$x=0$	$y=0$
$3(0)+5y=-150$	$3x+5(0)=-150$
$5y=-150$	$3x=-150$
$y=-30$	$x=-50$
$(0,-30)$	$(-50,0)$

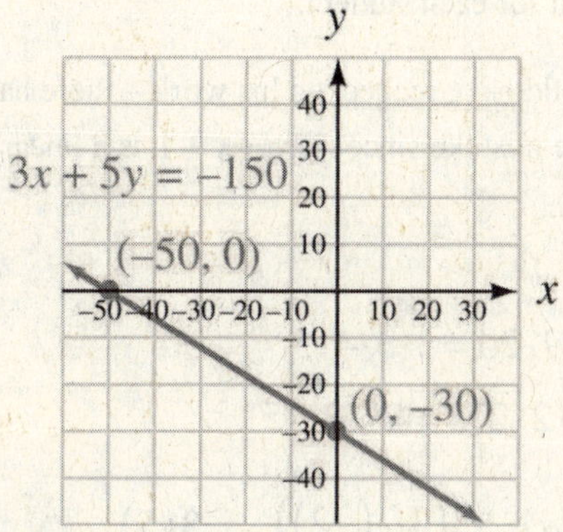

65.

$x=-3$	$x=3$
$y=\frac{-3}{3}$	$y=\frac{3}{3}$
$y=-1$	$y=1$
$(-3,-1)$	$(3,1)$

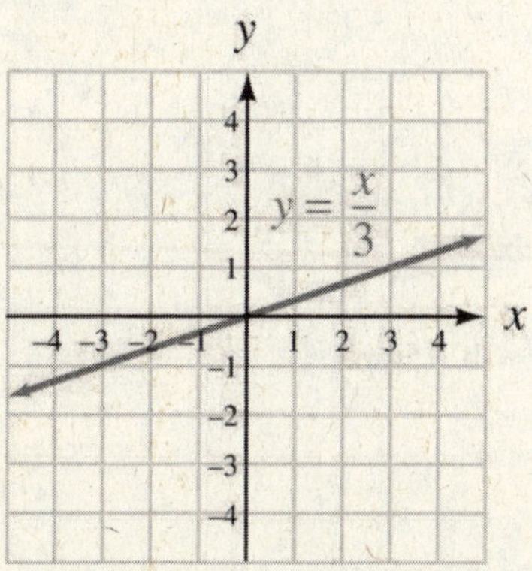

67.

$x=-2$	$x=1$
$y=-50(-2)-25$	$y=-50(1)-25$
$y=100-25$	$y=-50-25$
$y=75$	$y=-75$
$(-2,75)$	$(1,-75)$

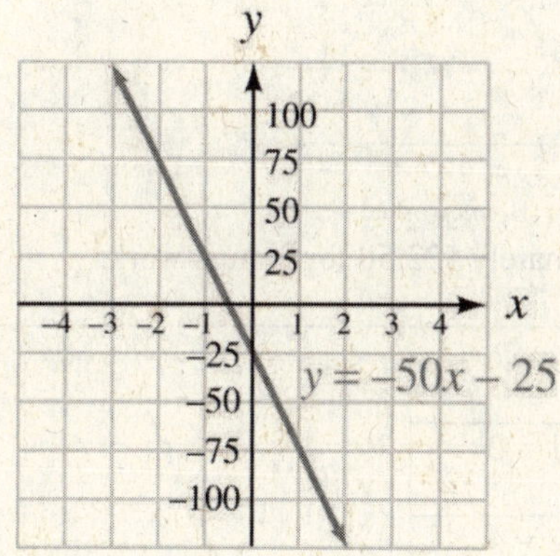

69.

$x=0$	$y=0$
$4(0)-3y=12$	$4x-3(0)=12$
$-3y=12$	$4x=12$
$y=-4$	$x=3$
$(0,-4)$	$(3,0)$

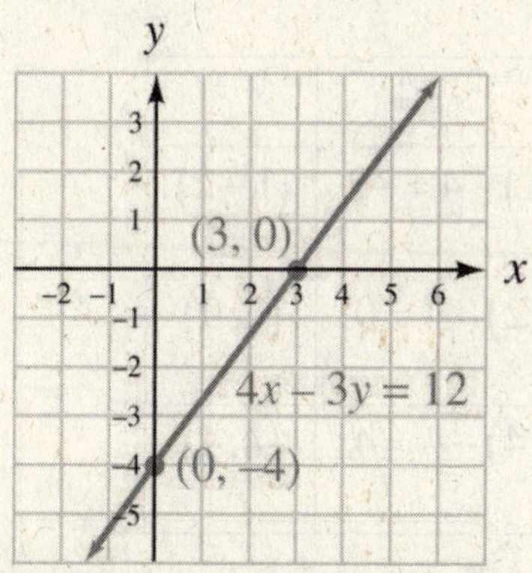

71.

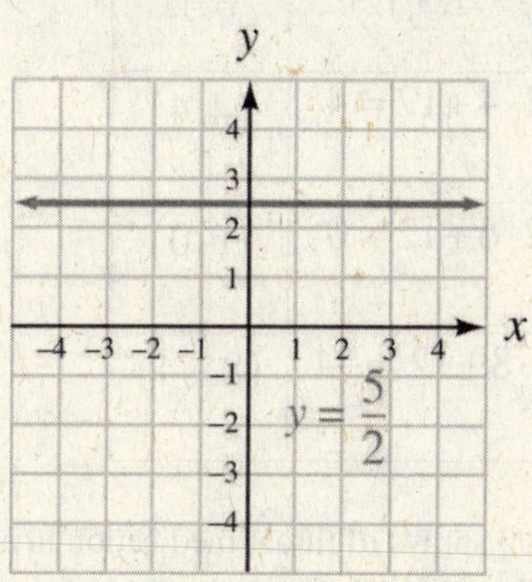

APPLICATIONS

73.

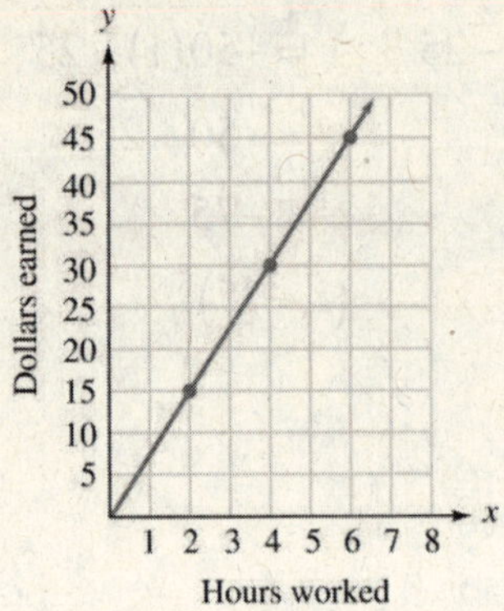

Approximately \$22.50 for 3hours work.

75.

t	$d = 2t$
1	$2\cdot1 = 2$
2	$2\cdot2 = 4$
3	$2\cdot3 = 6$
4	$2\cdot4 = 8$
5	$2\cdot5 = 10$

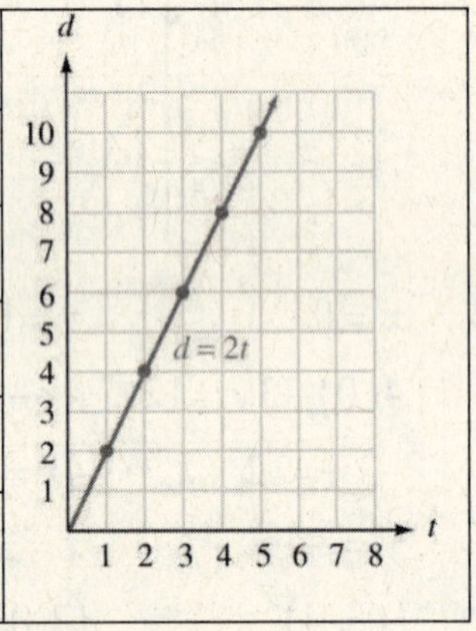

77.

x	$y = 2x - 4$	(x, y)
1	$2(1)-4=-2$	$(1,-2)$
2	$2(2)-4=0$	$(2,0)$
4	$2(4)-4=4$	$(4,4)$

x	$y = -2x + 12$	(x, y)
4	$-2\cdot4+12=4$	$(4,4)$
6	$-2\cdot6+12=0$	$(6,0)$
8	$-2\cdot8+12=-4$	$(8,-4)$

79. There are 300 sprays of throat medication in a bottle. The bottle holds 6 ounces of throat spray.

WRITING

81. We mean that plugging $(-2,-1)$ into $5x-6y=-4$ and simplifying will result in a true statement.

83. They could both be correct since the x values are different for each student.

85. He should have rechecked his work – there has to be some mistake since $y=-x+1$ is a linear equation.

REVIEW

87. $180=9\cdot20=3\cdot3\cdot4\cdot5$
$=3\cdot3\cdot2\cdot2\cdot5=2^2\cdot3^2\cdot5$

89. $\dfrac{3(b-a)}{5a+7}=\dfrac{3(3-(-2))}{5(-2)+7}=\dfrac{3(5)}{-10+7}$
$=\dfrac{15}{-3}=-5$

91. $\dfrac{1}{4}$sec

Chapter 8 Review

1. a. -18° b. -71°

2. a. 30mph b. 15mph

3. 23

4. 58

5. Ukraine, United Kingdom, China, Canada

6. roughly 50 – 19 = 31

7. 29%

8. men; about 43% – 28% = 15% more

9. women

10. They are roughly the same in the “not date a co-worker” category.

11. about 4,100 animals

12. the Columbus zoo; about 6,250 animals

13. roughly 3 elephants worth, or 3,000 animals

14. 4,100 + 6,250 + 1,100 = 11,450 animals

15. oxygen

16. 100% – 3% – 10% – 18% – 65% = 4%

17. $0.10(135) = 13.5lb$

18. $65\% + 18\% = 83\%$

$0.83(200) = 166lb$

19. about 3,000 million eggs

20. about 3,050 million eggs

21. 2007; about 2,975 million eggs

22. about 2,575 + 3,225 = 5,800 million eggs

23. between 2006 and 2007

24. between 2007 and 2008

25. about 3,150 – 2,750 = 400 million eggs

26. about 3,200 – 2,700 = 500 million eggs

27. 60

28. 100 + 80 = 180

29. 80 + 50 + 30 = 160

30.

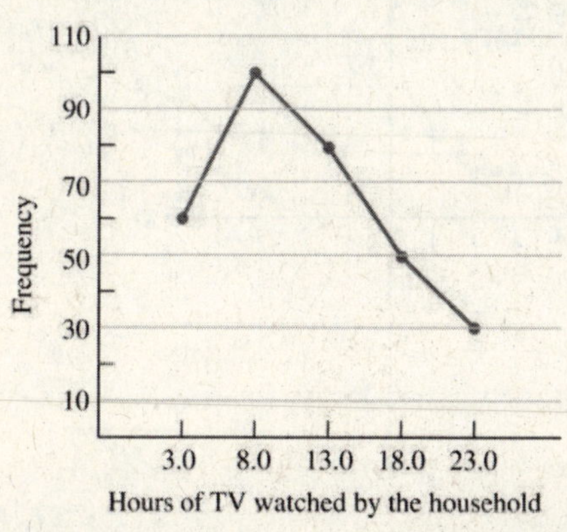

31. $$\frac{87+92+97+100+100+98+90+98}{8}$$

$$= \frac{762}{8} = 95.25$$

He earned an A in the class.

32. Median: 43
Mean: 62
Mode: 40
She was using the median.

33. $1.2oz$ happened most often.

34. $$\frac{0.9(1)+1.0(6)+1.1(18)+1.2(23)+1.3(2)}{1+6+18+23+2}$$

$$= \frac{56.9}{50} \approx 1.138\ oz$$

Range: $1.3 - 0.9 = 0.4oz$

35. mean: $\frac{7.8+\cdots 7.5}{9} = \frac{65.7}{9} = 7.3$ microns

median:

$6.7, 6.8, 6.9, 6.9, 7.2, 7.5, 7.8, 7.9, 8.0$

7.2 microns

mode: 6.9 microns

range: 8.0 – 6.7 = 1.3 microns

36. June has 30 days: $\frac{960}{30} = 32$ pages per day

37. $$\frac{5\cdot 20+10\cdot 65+20\cdot 25+50\cdot 5+100\cdot 10}{20+65+25+5+10}$$

$$= \frac{2,500}{125} = \$20$$

38. $$\frac{4\cdot 5+2\cdot 3+1\cdot 4+4\cdot 1}{5+3+4+1} = \frac{34}{13} \approx 2.62GPA$$

39. $2x + 5y = -11$

$2(2) + 5(-3) \stackrel{?}{=} -11$

$4 - 15 \stackrel{?}{=} -11$

$-11 = -11$

$(2, -3)$ is a solution.

40.
$y=8x+15$
$2\stackrel{?}{=}8(-3)+15$
$2\stackrel{?}{=}-24+15$
$2=-9$
$(-3,2)$ is not a solution.

41.
$x=0$ $\quad$ $y=0$
$3(0)-4y=12$ $\quad$ $3x-4(0)=12$
$-4y=12$ $\quad$ $3x=12$
$y=-3$ $\quad$ $x=4$
$(0,-3)$ $\quad$ $(4,0)$

42.
$x=1$ $\quad$ $y=-11$
$y=-3(1)-2$ $\quad$ $-11=-3x-2$
$y=-3-2$ $\quad$ $-9=-3x$
$y=-5$ $\quad$ $3=x$
$(1,-5)$ $\quad$ $(3,-11)$

$x=-2$
$y=-3(-2)-2$
$y=6-2$
$y=4$
$(-2,4)$

x	y	(x,y)
1	-5	$(1,-5)$
3	-11	$(3,-11)$
-2	4	$(-2,4)$

43.

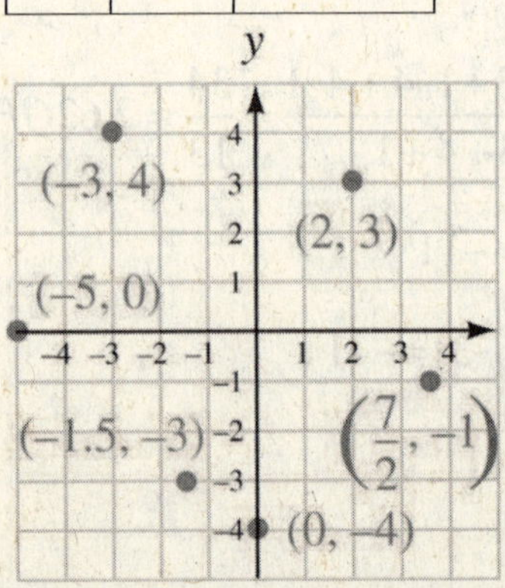

44. A: $(4,3)$ B: $(-3,3)$ C: $(-4,0)$

D: $(-1.5,-3.5)$ E: $(2.5,-1.5)$ F: $(0,0)$

45. Quadrant III

46. The middle seat in the second column is yours.

47.
$x=-1$ $\quad$ $x=2$
$y=2(-1)-3$ $\quad$ $y=2(2)-3$
$y=-2-3$ $\quad$ $y=4-3$
$y=-5$ $\quad$ $y=1$
$(-1,-5)$ $\quad$ $(2,1)$

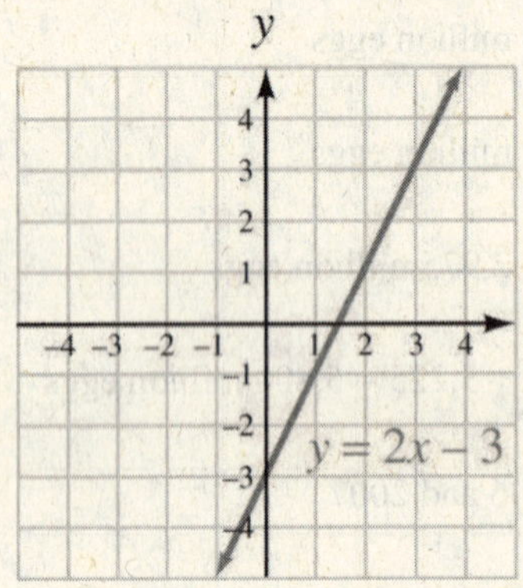

48.
$x=-4$ $\quad$ $x=2$
$y=\frac{1}{2}(-4)-1$ $\quad$ $y=\frac{1}{2}(2)-1$
$y=-2-1$ $\quad$ $y=1-1$
$y=-3$ $\quad$ $y=0$
$(-4,-3)$ $\quad$ $(2,0)$

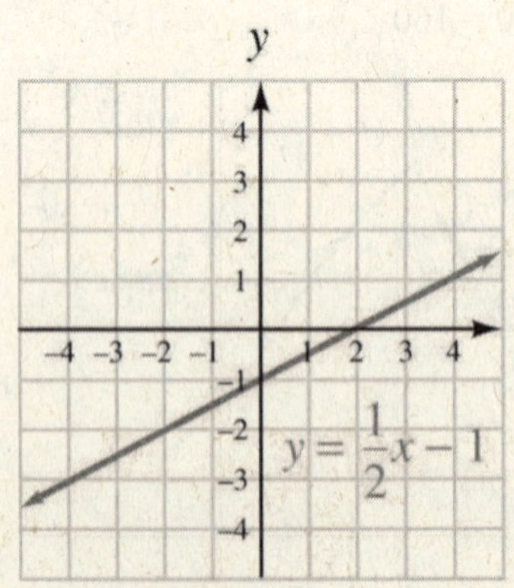

49.

$$\begin{array}{ll} x=0 & x=2 \\ y=-3(0)+2 & y=-3(2)+2 \\ y=0+2 & y=-6+2 \\ y=2 & y=-4 \\ (0,2) & (2,-4) \end{array}$$

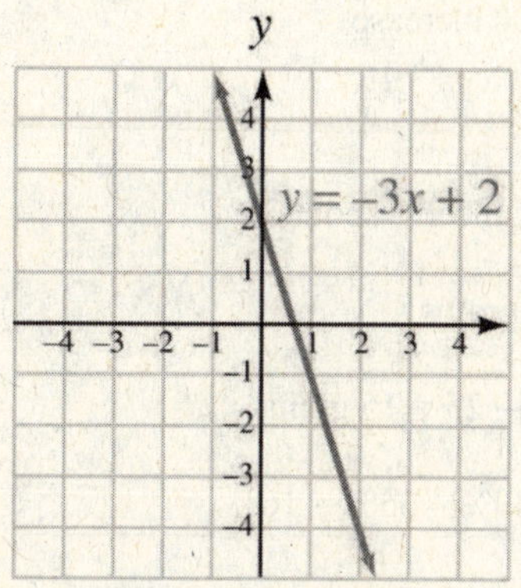

50.

$$\begin{array}{ll} x=-1 & x=1 \\ y=4(-1) & y=4(1) \\ y=-4 & y=4 \\ (-1,-4) & (1,4) \end{array}$$

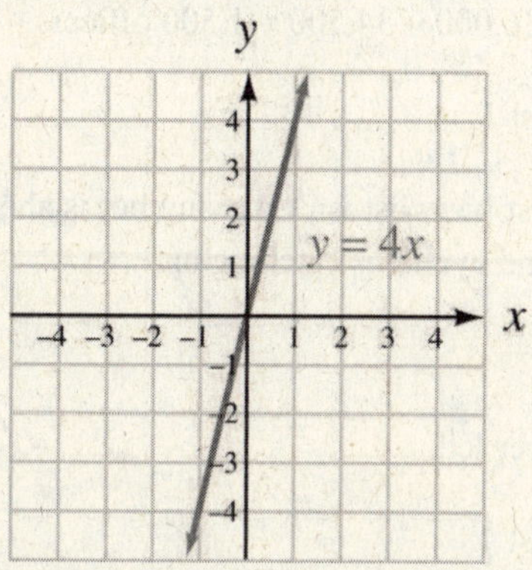

51. $(-4,-3),(-3,-2),(-2,-1)$
$(0,1),(1,2),(2,3)$

52. x-int: $(-3,0)$ y-int: $\left(0,\dfrac{5}{2}\right)$

53.

$$\begin{array}{ll} x=0 & y=0 \\ 8(0)+4y=-24 & 8x+4(0)=-24 \\ 4y=-24 & 8x=-24 \\ y=-6 & x=-3 \\ (0,-6) & (-3,0) \end{array}$$

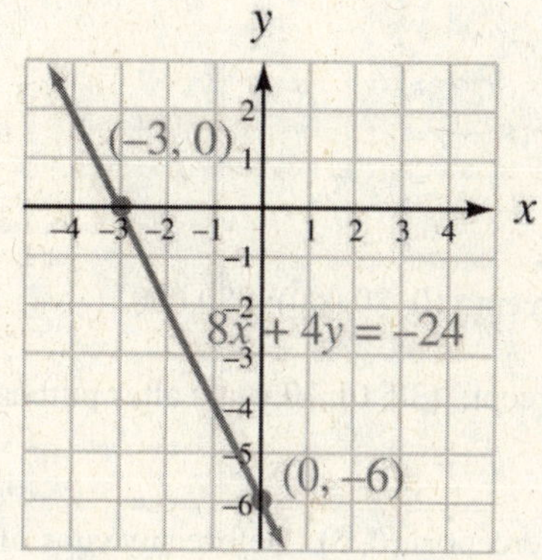

54.

$$\begin{array}{ll} x=0 & y=0 \\ 30(0)-y=30 & 30x-0=30 \\ -y=30 & 30x=30 \\ y=-30 & x=1 \\ (0,-30) & (1,0) \end{array}$$

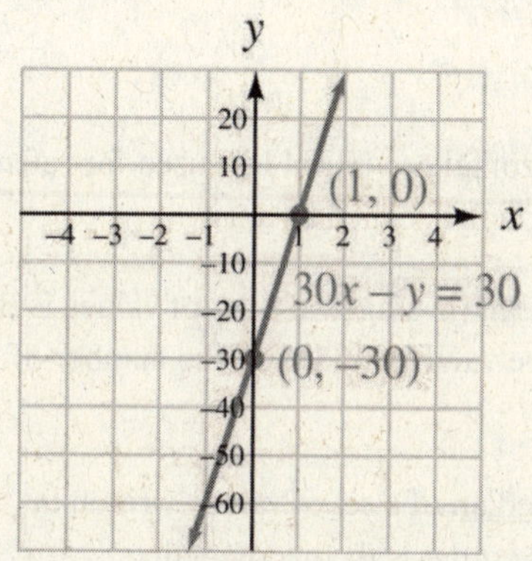

55.

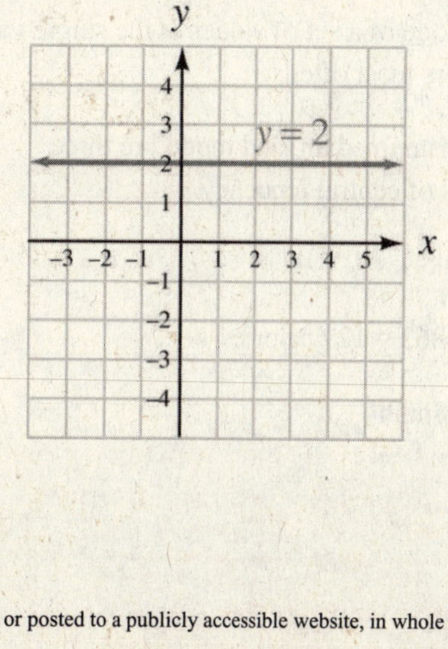

56.

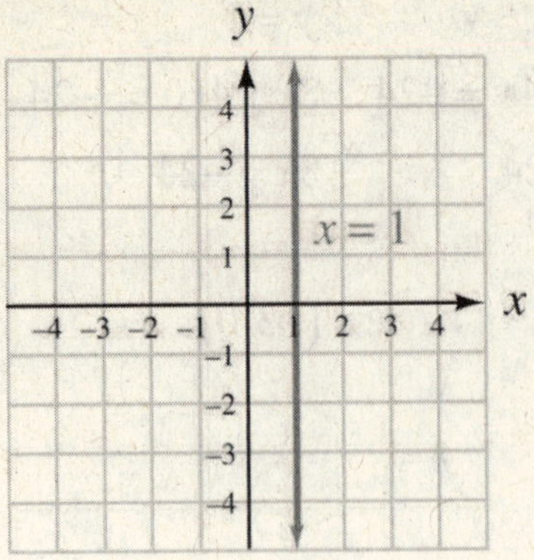

57. a. y-intercept: (0, 80,000); $80,000

b. x-intercept: (30, 0); 30 years after purchase

58. The g-intercept is (0, 5). Before any cups of water have been served from the bottle, it contains 5 gallons of water. The c-intercept is $\left(106\frac{2}{3}, 0\right)$. The bottle will be empty after $106\frac{2}{3}$ six-ounce cups of water have been served from it.

Chapter 8 Test

1. a. A horizontal or vertical line used for reference in a bar graph is called an axis.

b. The mean (average) of a set of values is the sum of the values divided by the number of values in the set.

c. The median of a set of values written in increasing order is the middle value.

d. The mode of a set of values is the single value that occurs most often.

e. The mean, median, and mode are three measures of central tendency.

2. a. 563 calories

b. 992 – 863 = 129 calories

c. about 8mph

3. a. love seat; 150 feet

b. 140 – 90 = 50 feet more

c. 60 + 100 + 90 + 90 =340 feet

4. a. 75%

b. about 15% increase

c. lung cancer

d. prostate cancer – about 35%

5. a. about 38 grams

b. about 40 – 25 = 15 grams

6. a. 100% – 49% – 35% = 16%

b. $0.35(1,389,500) = 486,325$

7. a. about 39,000 officers

b. 2011; about 33,800 officers

c. 2000; about 40,300 officers

d. about 36,000 – 34,500 = 1,500 officers

8. a. bicyclist 1

b. The first bicyclist isn't moving but is ahead. The second cyclist is catching up.

c. time C

d. He never led.

e. bicyclist 1

9. a. 22 employees

b. 8 + 22 = 30 employees

c. 28 + 20 + 9 = 57 employees

10. a. $\frac{4+6+\cdots+1+9}{16} = \frac{120}{16} = 7.5hr$

b. 1,2,4,5,5,5,6,7,8,8,9,9,10,11,12,18: $\frac{7+8}{2} = 7.5hr$

c. $5hr$

d. 18 – 1 = 17hr

11. $$\frac{5\cdot3+4\cdot4+3\cdot5+2\cdot6+1\cdot2}{3+4+5+6+2}$$
$$=\frac{60}{20}=3\,\text{stars}$$

12. $$\frac{2\cdot1+4\cdot3+3\cdot2+4\cdot4+1\cdot1}{11}$$
$$=\frac{37}{11}\approx3.36\,GPA$$

13. mean:
$$\frac{4.70+\cdots+2.57}{8}=\frac{26.19}{8}=3.27\,\text{million}$$

2.57 2.57 2.76 **2.97** **3.01** 3.77 3.84 4.70

median: $$\frac{2.97+3.01}{2}=\frac{5.98}{2}=2.99\,\text{million}$$

mode: 2.57 million

range: $4.70-2.57=3.13$ million

14. Of all the existing single-family homes sold in December, 2012, half of them sold for less than \$180,300 and half of them sold for more than \$180,300.

15. $4x+5y=6$

$4(-1)+5(2)\stackrel{?}{=}6$

$-4+10\stackrel{?}{=}6$

$6=6$

$(-1,2)$ is a solution.

16. $y=12x-34$

$-2\stackrel{?}{=}12(3)-34$

$-2\stackrel{?}{=}36-34$

$-2=2$

$(3,-2)$ is not a solution.

17.

$x=0$	$y=0$	$x=2$
$0-2y=4$	$x-2(0)=4$	$2-2y=4$
$-2y=4$	$x-0=4$	$-2y=2$
$y=-2$	$x=4$	$y=-1$
$(0,-2)$	$(4,0)$	$(2,-1)$

18.

$x=0$	$y=2$	$x=-3$
$0-3y=-3$	$x-3(2)=-3$	$-3-3y=-3$
$-3y=-3$	$x-6=-3$	$-3y=0$
$y=1$	$x=3$	$y=0$
$(0,1)$	$(3,2)$	$(-3,0)$

19. $(30,32)\ (30,34)\ (31,34)\ (38,30)$

20.

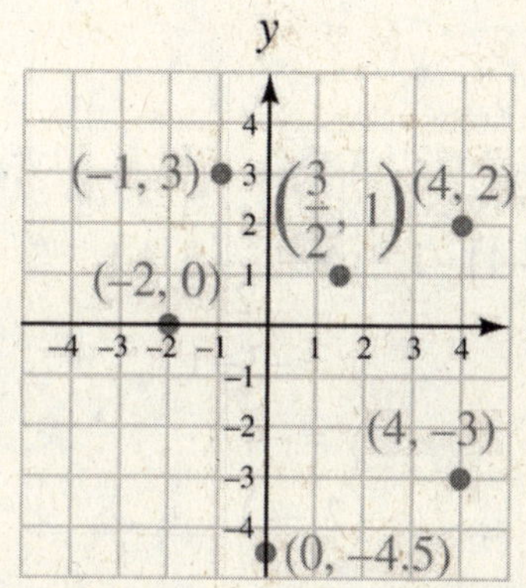

21. A: $(0,0)$ B: $(2.5,3.5)$ C: $(-3,-2)$

D: $(0,-2)$ E: $(4,0)$ F: $(-5,-5)$

22. Quadrant III

23.

$x=0$	$x=1$
$y=4(0)-2$	$y=4(1)-2$
$y=0-2$	$y=4-2$
$y=-2$	$y=2$
$(0,-2)$	$(1,2)$

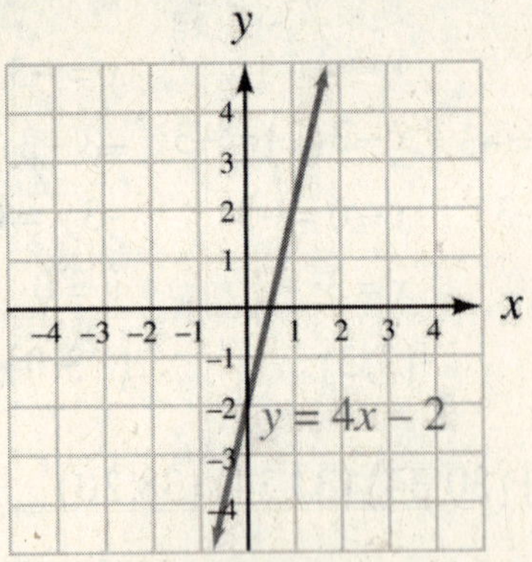

24.

$x=-4$	$x=2$
$y=-\frac{3}{2}(-4)-1$	$y=-\frac{3}{2}(2)-1$
$y=6-1$	$y=-3-1$
$y=5$	$y=-4$
$(-4,5)$	$(2,-4)$

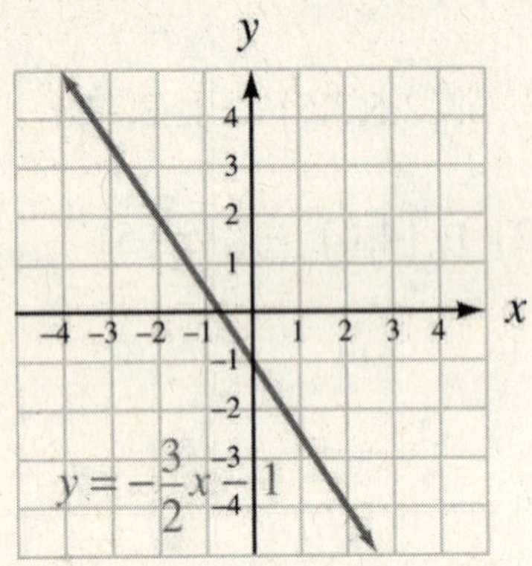

25. a. $(-3,0)$ b. $(0,-1)$

26. $(-3,-1),(0,0),(3,1)$

27.

$x=0$	$y=0$
$3(0)+4y=12$	$3x+4(0)=12$
$4y=12$	$3x=12$
$y=3$	$x=4$
$(0,3)$	$(4,0)$

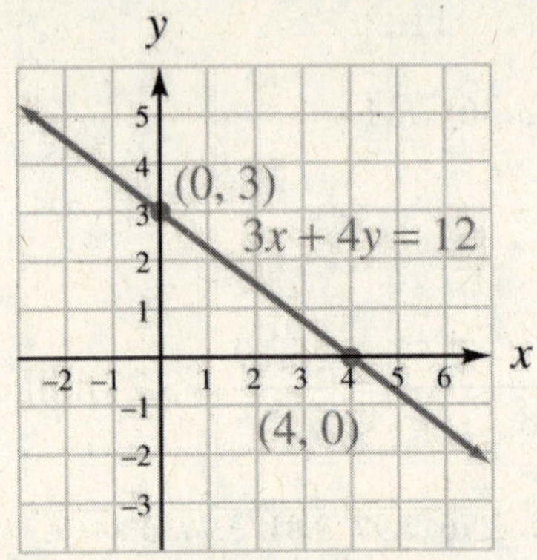

28.

$x=0$	$y=0$
$20(0)+y=-20$	$20x+0=-20$
$0+y=-20$	$20x=-20$
$y=-20$	$x=-1$
$(0,-20)$	$(-1,0)$

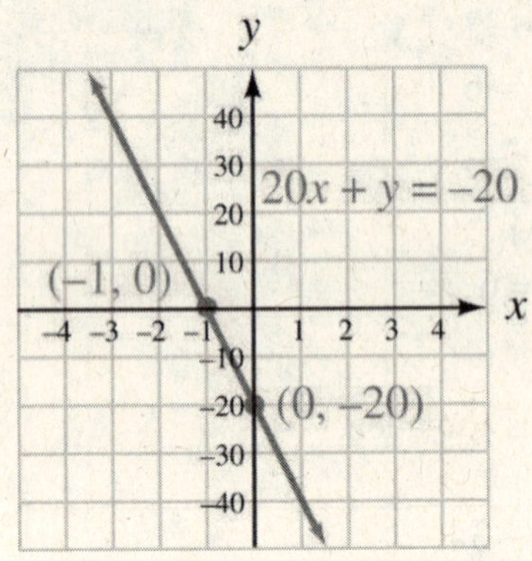

29.

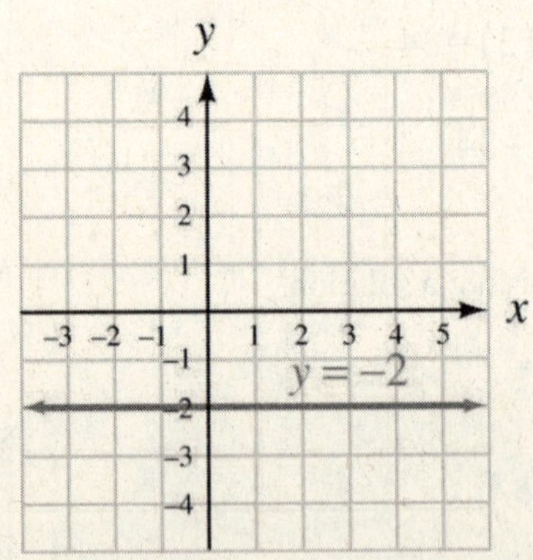

30.

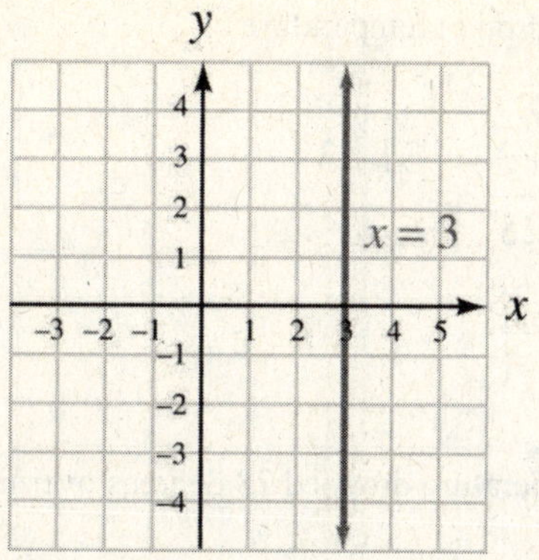

Chapters 1 – 8 Cumulative Review

1. Fifty-nine million, nine hundred twenty-nine thousand, sixteen

50,000,000 + 9,000,000 + 900,000 + 20,000 + 9000+ 10 + 6

2. 60,000

3.

$$\begin{array}{r} \scriptstyle 1\,1\;\,1\,1 \\ 48{,}908 \\ +\ 5{,}696 \\ \hline 54{,}604 \end{array}$$

4.

$$\begin{array}{r} \scriptstyle 6\;\overset{9}{10}\;10 \\ 8{,}\not{7}\,\not{0}\,\not{0} \\ -5{,}4\ 9\ 1 \\ \hline 3{,}2\ 0\ 9 \end{array}$$

5.

$$\begin{array}{r} 408 \\ \times\quad 67 \\ \hline 2{,}856 \\ +24{,}480 \\ \hline 27{,}336 \end{array}$$

6.

$$\begin{array}{r} 23 \\ 87\overline{)2001} \\ \underline{-174} \\ 261 \\ \underline{-261} \\ 0 \end{array}$$

7. Add upwards: $1{,}683+459=2{,}142$

8. $P=2l+2w=2\cdot 6+2\cdot 14=12+28=40in$

$A=lw=6\cdot 14=84in^2$

9. a. 1, 2, 3, 4, 6, 9, 12, 18, 36

b. 2, 3, 5, 7, 11, 13, 17, 19, 23, 29

c. $36=6\cdot 6=2\cdot 3\cdot 2\cdot 3=2^2\cdot 3^2$

10. $8=2\cdot 2\cdot 2$

$12=2\cdot 2\cdot 3$

a. $LCM(8,12)=2\cdot 2\cdot 2\cdot 3=24$

b. $GCF(8,12)=2\cdot 2=4$

11.

$$\begin{aligned} &15+5\left[12-\left(2^2+4\right)\right] \\ &=15+5\left[12-(4+4)\right] \\ &=15+5[12-8]=15+5[4]=15+20 \\ &=35 \end{aligned}$$

12. $\dfrac{12+5\cdot 3}{3^2-2\cdot 3}=\dfrac{12+15}{9-6}=\dfrac{27}{3}=9$

13.

$$\begin{aligned} &96=m-83 \\ &96+83=m-83+83 \\ &179=m \end{aligned}$$

Check: $96=m-83$

$96\overset{?}{=}179-83$

$96=96$

14.

$$\begin{aligned} &73n=219 \\ &\frac{73n}{73}=\frac{219}{73} \\ &n=3 \end{aligned}$$

Check: $73n=219$

$73(3)\overset{?}{=}219$

$219=219$

15.

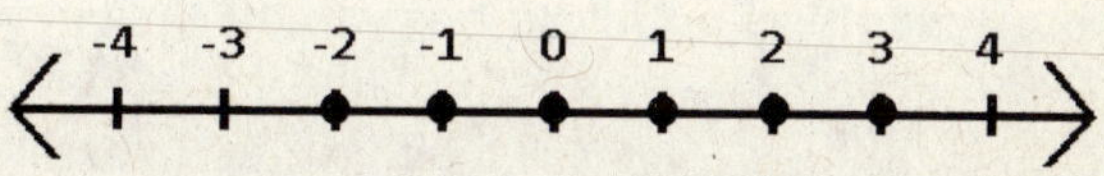

16. a. $-(-6)=6$ b. $|-5|=5$ c. false

17. a. $-35+5=-30$

b. $-35-(-5)=-35+5=-30$

c. $-35(5)=-175$

d. $\frac{-35}{-5}=\frac{35}{5}=7$

18. $810°-(-290°)=810°+290°=1{,}100°F$

19. $\frac{(-6)^2-1^5}{-4-3}=\frac{36-1}{-7}=\frac{35}{-7}=-5$

20. $-3+3(-4-4\cdot 2)^2=-3+3(-4-8)^2$

$=-3+3(-12)^2=-3+3(144)$

$=-3+432=429$

21. $-\left|\frac{45}{-9}-(-9)\right|=-|-5+9|=-|4|=-4$

22. $-10^2-(-10)^2=-100-100=-200$

23. $\frac{x}{-4}+4=-5$

$\frac{x}{-4}+4-4=-5-4$

$\frac{x}{-4}=-9$

$-4\left(\frac{x}{-4}\right)=-4(-9)$

$x=36$

Check: $\frac{x}{-4}+4=-5$

$\frac{36}{-4}+4\overset{?}{=}-5$

$-9+4\overset{?}{=}-5$

$-5=-5$

24. d = the drop in temperature

$-3-d=-31$

$-3-d+3=-31+3$

$-d=-28$

$\frac{-d}{-1}=\frac{-28}{-1}$

$d=28$

The temperature dropped 28 degrees overnight.

25. x^2-4

26. $d=r\cdot t=45\cdot 5=225mi$

27. a. $(-9t)(-8)=((-9)(-8))t=72t$

b. $-5(3x-2y-10)=-15x+10y+50$

28. a. $x+x+x+x=4x$

b. $5(3-2x)+4(2-3x)+19x$

$=15-10x+8-12x+19x$

$=23-3x$

29. $8+4(2a-2)=-16+4a$

$8+8a-8=-16+4a$

$8a=-16+4a$

$8a-4a=-16+4a-4a$

$4a=-16$

$\frac{4a}{4}=\frac{-16}{4}$

$a=-4$

Check: $8+4(2a-2)=-16+4a$

$8+4(2(-4)-2)\overset{?}{=}-16+4(-4)$

$8+4(-8-2)\overset{?}{=}-16-16$

$8+4(-10)\overset{?}{=}-32$

$8-40\overset{?}{=}-32$

$-32=-32$

30. w = the score on the written test

$$w+(w-4)=98$$
$$2w-4=98$$
$$2w-4+4=98+4$$
$$2w=102$$
$$\frac{2w}{2}=\frac{102}{2}$$
$$w=51$$

He scored a 51 on the written portion and 51 – 4 = 47 on the interview portion.

31. a. $\frac{60}{108}=\frac{5\cdot\cancel{12}}{9\cdot\cancel{12}}=\frac{5}{9}$

b. $\frac{24a^3}{16a}=\frac{3\cdot\cancel{8}\cdot\cancel{a}\cdot a\cdot a}{2\cdot\cancel{8}\cdot\cancel{a}}=\frac{3a^2}{2}$

32. a. $\frac{0}{64}=0$ b. $\frac{27}{0}$ is undefined.

33. $\frac{4}{5}\cdot\frac{2}{7}=\frac{4\cdot 2}{5\cdot 7}=\frac{8}{35}$

34. $\frac{8m^2}{63n^5}\div\frac{2m^2}{n^4}=\frac{8m^2}{63n^5}\cdot\frac{n^4}{2m^2}$

$$=\frac{\cancel{2}\cdot 4\cdot\cancel{m^2}\cdot\cancel{n^4}}{63\cdot n\cdot\cancel{n^4}\cdot\cancel{2}\cdot\cancel{m^2}}=\frac{4}{63n}$$

35. $\frac{1}{2}-\frac{2}{3}=\frac{1}{2}\cdot\frac{3}{3}-\frac{2}{3}\cdot\frac{2}{2}=\frac{3}{6}-\frac{4}{6}=-\frac{1}{6}$

36. $\frac{11}{12}+\frac{1}{30}=\frac{11}{12}\cdot\frac{5}{5}+\frac{1}{30}\cdot\frac{2}{2}$

$$=\frac{55}{60}+\frac{2}{60}=\frac{57}{60}=\frac{\cancel{3}\cdot 19}{\cancel{3}\cdot 20}=\frac{19}{20}$$

37. $\frac{x}{9}-\frac{1}{8}=\frac{x}{9}\cdot\frac{8}{8}-\frac{1}{8}\cdot\frac{9}{9}=\frac{8x}{72}-\frac{9}{72}=\frac{8x-9}{72}$

38. $\frac{7}{15}\cdot 300=\frac{7\cdot\cancel{15}\cdot 20}{\cancel{15}}=140$ minutes in the lab per week.

$300-140=160$ minutes in lecture per week.

39. $2\frac{4}{5}\div\left(-2\frac{2}{3}\right)=\frac{14}{5}\div\left(-\frac{8}{3}\right)$

$$=-\frac{14}{5}\cdot\frac{3}{8}=-\frac{\cancel{2}\cdot 7\cdot 3}{5\cdot\cancel{2}\cdot 4}=-\frac{21}{20}=-1\frac{1}{20}$$

40. $26-19\frac{1}{4}=25\frac{4}{4}-19\frac{1}{4}=6\frac{3}{4}\,in$

41.
$$A=\frac{1}{2}h(a+b)$$
$$A=\frac{1}{2}\left(2\frac{1}{8}\right)\left(4\frac{1}{2}+5\frac{1}{2}\right)$$
$$A=\frac{1}{2}\cdot\frac{17}{8}\cdot 10=\frac{170}{16}$$
$$A=10\frac{10}{16}=10\frac{5}{8}$$

42. $\frac{-\frac{1}{5}}{\frac{8}{15}}=-\frac{1}{5}\cdot\frac{15}{8}=-\frac{3\cdot\cancel{5}}{\cancel{5}\cdot 8}=-\frac{3}{8}$

43.
$$-\frac{5}{6}y=-2$$
$$\left(-\frac{6}{5}\right)\left(-\frac{5}{6}y\right)=\left(-\frac{6}{5}\right)(-2)$$
$$y=\frac{12}{5}$$

Check: $-\frac{5}{6}y=-2$

$$-\frac{5}{6}\cdot\frac{12}{5}\overset{?}{=}-2$$
$$-\frac{12}{6}\overset{?}{=}-2$$
$$-2=-2$$

44.

$$\frac{a}{5}-\frac{5}{6}=\frac{a}{3}$$

$$30\cdot\frac{a}{5}-30\cdot\frac{5}{6}=30\cdot\frac{a}{3}$$

$$6a-25=10a$$

$$6a-25-6a=10a-6a$$

$$-25=4a$$

$$\frac{-25}{4}=\frac{4a}{4}$$

$$-\frac{25}{4}=a$$

Check: $\frac{a}{5}-\frac{5}{6}=\frac{a}{3}$

$$\frac{\left(-\frac{25}{4}\right)}{5}-\frac{5}{6}\stackrel{?}{=}\frac{\left(-\frac{25}{4}\right)}{3}$$

$$-\frac{5}{4}-\frac{5}{6}\stackrel{?}{=}-\frac{25}{12}$$

$$-\frac{15}{12}-\frac{10}{12}\stackrel{?}{=}-\frac{25}{12}$$

$$-\frac{25}{12}=-\frac{25}{12}$$

45. t = the number of players on the team

$$\frac{1}{3}t+\frac{1}{5}t+7=t$$

$$15\cdot\frac{1}{3}t+15\cdot\frac{1}{5}t+15\cdot7=15\cdot t$$

$$5t+3t+105=15t$$

$$8t+105=15t$$

$$8t+105-8t=15t-8t$$

$$105=7t$$

$$\frac{105}{7}=\frac{7t}{7}$$

$$15=t$$

There are 15 players on the team.

46. 428.91

47.

$$\begin{array}{r} \scriptstyle 1\,2\,2 \\ 213.16 \\ 1{,}504.80 \\ 89.73 \\ +\quad 7.50 \\ \hline \$1{,}815.19 \end{array}$$

48. a. $3.45\cdot100=345$

b. $3.45\div10{,}000=0.000345$

49.

$$\begin{array}{r} \scriptstyle 9 \\ \scriptstyle 5\;10\;\;12 \\ 7\not{6}\not{0}.\not{2} \\ -614.7 \\ \hline 145.5 \end{array}$$

50. $(-0.31)(2.4)=-0.744$

51. Begin by moving the decimal point 2 units right.

$$\begin{array}{r} 745 \\ 72\overline{)53640} \\ \underline{-504} \\ 324 \\ \underline{-288} \\ 360 \\ \underline{-360} \\ 0 \end{array}$$

52.

$$\begin{array}{r} 0.01825 \\ 4\overline{)0.07300} \\ \underline{-4} \\ 33 \\ \underline{-32} \\ 10 \\ \underline{-8} \\ 20 \\ \underline{-20} \\ 0 \end{array}$$

53.

$$\begin{array}{r} 0.7272 \\ 11\overline{)8.0000} \\ \underline{-77} \\ 30 \\ \underline{-22} \\ 80 \\ \underline{-77} \\ 30 \\ \underline{-22} \\ 8 \end{array}$$

$$\frac{8}{11} = 0.\overline{72}$$

54.

$$15+\sqrt{16}\left[5^2-\left(\sqrt{9}+2\right)\sqrt{4}\right]$$
$$=15+4\left[25-(3+2)2\right]$$
$$=15+4\left[25-5\cdot 2\right]=15+4\left[25-10\right]$$
$$=15+4(15)=15+60=75$$

55.

$$2(3.6t-4.1)+0.9t=16.1$$
$$7.2t-8.2+0.9t=16.1$$
$$8.1t-8.2=16.1$$
$$8.1t-8.2+8.2=16.1+8.2$$
$$8.1t=24.3$$
$$\frac{8.1t}{8.1}=\frac{24.3}{8.1}$$
$$t=3$$

Check: $2(3.6t-4.1)+0.9t=16.1$

$$2(3.6(3)-4.1)+0.9(3)\stackrel{?}{=}16.1$$
$$2(10.8-4.1)+2.7\stackrel{?}{=}16.1$$
$$2(6.7)+2.7\stackrel{?}{=}16.1$$
$$13.4+2.7\stackrel{?}{=}16.1$$
$$16.1=16.1$$

56. c = the number of copies allowed

$$35+0.03c=110$$
$$35+0.03c-35=110-35$$
$$0.03c=75$$
$$\frac{0.03c}{0.03}=\frac{75}{0.03}$$
$$c=2{,}500$$

They can make up to 2,500 copies on their budget.

57.

$$\frac{8\,ft}{4\,yd}=\frac{8\,ft}{12\,ft}=\frac{2\cdot \cancel{4ft}}{3\cdot \cancel{4ft}}=\frac{2}{3}$$

58.

$$\frac{3}{35.97}=\frac{5}{x}$$
$$3\cdot x=5\cdot 35.97$$
$$\frac{3x}{3}=\frac{179.85}{3}$$
$$x=\$59.95$$

59.

$$\frac{\frac{7}{8}}{\frac{1}{2}}=\frac{\frac{1}{4}}{x}$$
$$\frac{7}{8}\cdot x=\frac{1}{2}\cdot\frac{1}{4}$$
$$\frac{7x}{8}=\frac{1}{8}$$
$$7x=1$$
$$\frac{7x}{7}=\frac{1}{7}$$
$$x=\frac{1}{7}$$

60.

$$\frac{8\,pt}{1}\cdot\frac{2c}{1\,pt}\cdot\frac{8\,fl\,oz}{1c}=8\cdot 2\cdot 8\,fl\,oz=128\,fl\,oz$$

61.

$$\frac{640cm}{1}\cdot\frac{1m}{100cm}=\frac{640}{100}m=6.4m$$

62.

$$C=\frac{5}{9}(F-32)$$
$$C=\frac{5}{9}(67.7-32)C=\frac{5}{9}(35.7)\approx 19.8°C$$

63.

Fraction	Decimal	Percent
$\frac{3}{100}$	0.03	3%
$\frac{225}{100}=\frac{9\cdot 25}{4\cdot 25}=\frac{9}{4}$	2.25	225%
$\frac{41}{1,000}$	0.041	4.1%

64. $90 = x\cdot 525$

$\frac{90}{525} = 0.17142\ldots \approx 17\%$

65. $x = 1.05(23.2) = 24.36$

66. $33\frac{1}{3}\% = \frac{1}{3}$

$19.2 = \frac{1}{3}x$

$3(19.2) = \cancel{3}\left(\frac{1}{\cancel{3}}x\right)$

$57.6 = x$

67. $0.08\cdot 98.95 \approx \$7.92$

68. $240 = x\cdot 1,500$

$\frac{240}{1,500} = \frac{\cancel{1,500}x}{\cancel{1,500}}$

$0.16 = x$

The commission rate is 16%.

69. Estimate \$80 for the bill.

10% of 80 is 8 so 5% of 80 is 4.

\$8 + \$4 = \$12

70. $I = P\cdot r\cdot t$

$I = 18,000\cdot 0.092\cdot 2 = 1,656\cdot 2$

$I = \$3,312$

71. $I = P\cdot r\cdot t$

$I = 12,600\cdot 0.18\cdot\frac{90}{365}$

$I = 2,268\cdot\frac{90}{365} \approx \$559.23$

$\$12,600 + \$559.23 = \$13,159.23$

72. a. $100\% - 28\% - 39\% - 15\% - 10\% = 8\%$

b. $0.39\cdot 12,000 = 4,680$

73. a. 2008 and 2010; 36

b. 2007 to 2008; increase of 16 deaths

c. 2010 to 2011; decrease of 11 deaths

74. mean: $\frac{3.04+\cdots+2.75}{12} = \frac{36.24}{12} = 3.02$

median:

2.21 2.25 2.56 2.75 2.75 2.98

3.02 3.04 3.23 3.58 3.87 4.00

$\frac{2.98+3.02}{2} = \frac{6}{2} = 3$

mode: 2.75

range: 4.0 – 2.21 = 1.79

75. $4x - 5y = -23$

$4(-2) - 5(3) \stackrel{?}{=} -23$

$-8 - 15 \stackrel{?}{=} -23$

$-23 = -23$

$(-2,3)$ is a solution.

76. $4x - 5y = -4$

$4x - 5(4) = -4$

$4x - 20 = -4$

$4x - 20 + 20 = -4 + 20$

$4x = 16$

$\frac{4x}{4} = \frac{16}{4}$

$x = 4$

$(4,4)$

77.

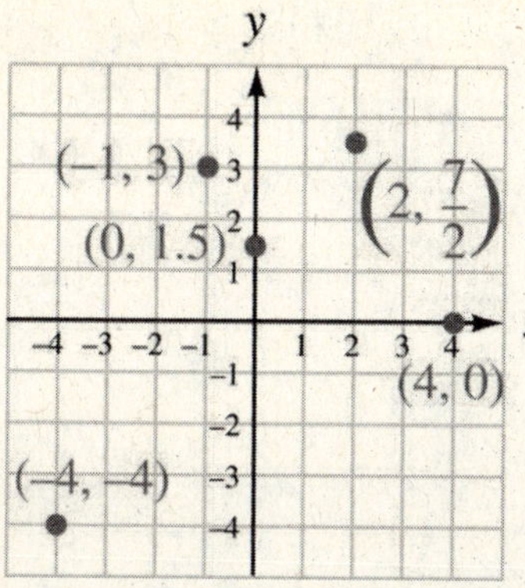

78.

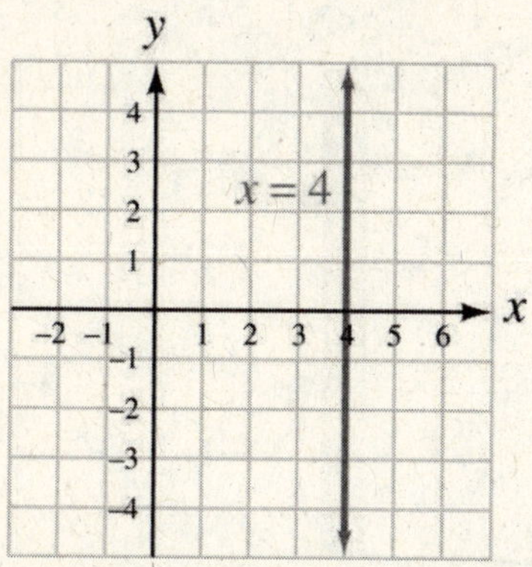

79.

$x = 0$	$y = 0$
$y = -0 - 1$	$0 = -x - 1$
$y = -1$	$x = -1$
$(0, -1)$	$(-1, 0)$

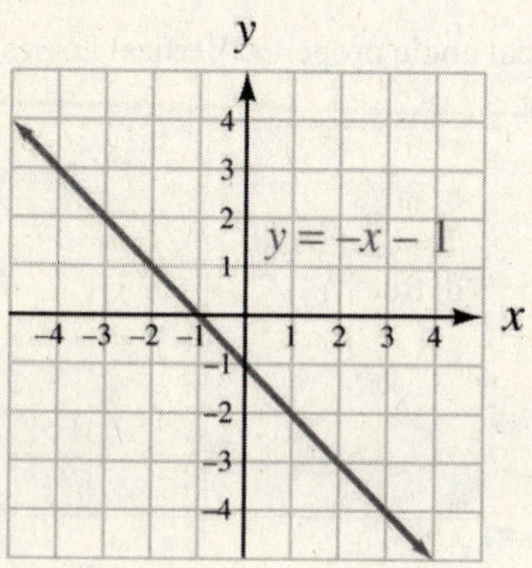

80.

$x = 0$	$y = 0$
$3(0) - 3y = 9$	$3x - 3(0) = 9$
$-3y = 9$	$3x = 9$
$y = -3$	$x = 3$
$(0, -3)$	$(3, 0)$

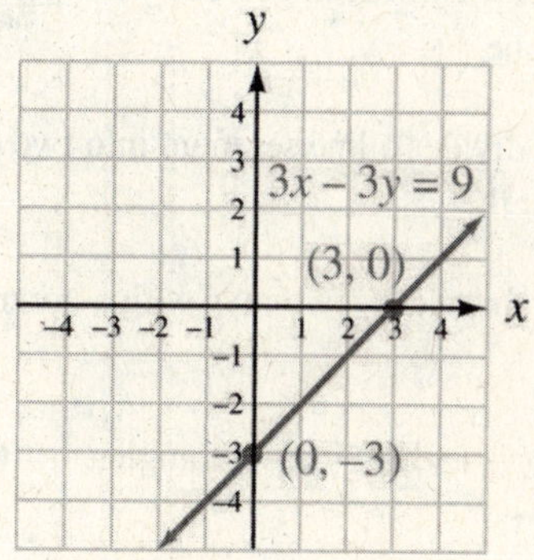

81.

a. (0, 12)

b. When the car has been driven 0 miles, the fuel tank contains 12 gallons of gasoline.

c. (600, 0)

d. After 600 miles of driving, the fuel tank contains 0 gallons of gasoline.

CHAPTER 9 An Introduction to Geometry

Section 9.1: Basic Geometric Figures; Angles

VOCABULARY

1. Three undefined words in geometry are point, line, and plane.

3. A midpoint divides a line segment into two parts of equal length.

5. An angle is formed by two rays with a common endpoint.

7. A protractor is used to measure angles.

9. The measure of a right angle is $90°$.

11. The measure of a straight angle is $180°$.

13. Adjacent angles have the same vertex, are side-by-side, and their interiors do not overlap.

15. When two angles have the same measure, we say that they are congruent.

17. The sum of two complementary angles is $90°$.

CONCEPTS

19. a. one

b. In general, two different points determine exactly one line.

21. a. $\overrightarrow{ST}$, $\overrightarrow{SR}$

b. S

c. $\angle RST$, $\angle TSR$, $\angle S$, $\angle 1$

23. a. 

b. 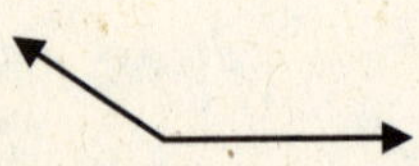

c. 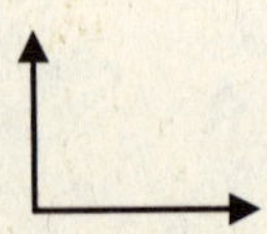

d. 

25. a.

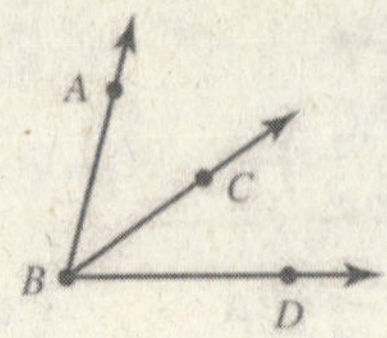

b.

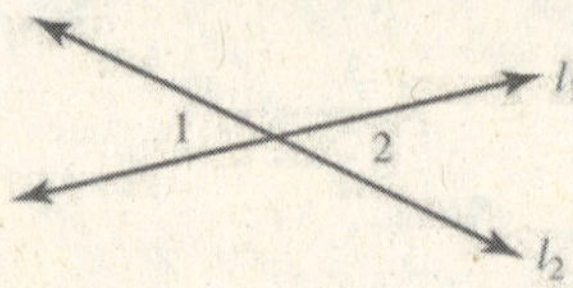

c.

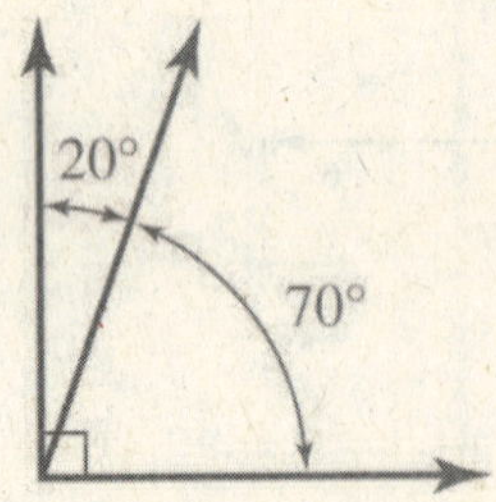

d.

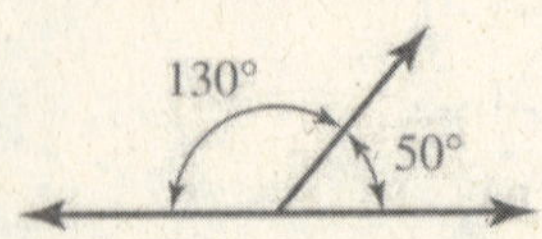

27. The vertical angle property: Vertical angles are congruent.

29. a. false b. false

c. false d. true

31. true

33. false

NOTATION

35. The symbol $\overleftrightarrow{AB}$ is read as "line AB."

37. The symbol $\overrightarrow{AB}$ is read as "ray AB."

39. We read $\angle ABC$ as "angle ABC."

41. The symbol for degree is a small raised circle, $°$.

43. The symbol $\cong$ is read as "is congruent to."

GUIDED PRACTICE

45. a.

b.

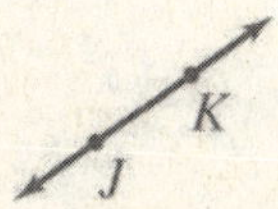

c.

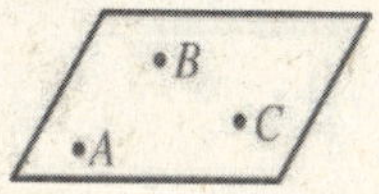

47. a. 2 b. 3 c. 1 d. 6

49. $50°$

51. $25°$

53. $180° - 105° = 75°$

55. $180° - 50° = 130°$

57. right

59. acute

61. straight

63. obtuse

65. $x = 55° - 45° = 10°$

67. $x = 50° - 22.5° = 27.5°$

69. Vertical angles with $\angle OYZ$: 70°

71. Vertical angles with $\angle MYX$: 65°

73.
$$2x = x + 30°$$
$$2x - x = x + 30° - x$$
$$x = 30°$$
$$m(\angle ABD) = 2 \cdot 30° = 60°$$
$$m(\angle DBE) = 180° - 60° = 120°$$

75.
$$4x + 15° = 7x - 60°$$
$$4x + 15° + 60° = 7x - 60° + 60°$$
$$4x + 75° = 7x$$
$$4x + 75° - 4x = 7x - 4x$$
$$75° = 3x$$
$$\frac{75°}{3} = \frac{3x}{3}$$
$$25° = x$$
$$m(\angle ZYQ) = 7 \cdot 25° - 60° = 175° - 60°$$
$$= 115°$$
$$m(\angle PYQ) = 180° - 115° = 65°$$

77.
$$x + 30° = 90°$$
$$x + 30° - 30° = 90° - 30°$$
$$x = 60°$$

79.
$$x + 105° = 180°$$
$$x + 105° - 105° = 180° - 105°$$
$$x = 75°$$

TRY IT YOURSELF

81. a. true

b. False. It is a segment so it has two endpoints.

c. False. It is a line so it has no endpoints.

d. False. G is the vertex of that angle.

e. true

f. true

83. $40°$

85. $135°$

87. a. vertical angles: $m(\angle 3)=50°$

b. supplementary angles:

$m(\angle 1)+m(\angle 4)=180°$

$50°+m(\angle 4)=180°$

$50°+m(\angle 4)-50°=180°-50°$

$m(\angle 4)=130°$

c. $m(\angle 2)=m(\angle 4)=130°$

$m(\angle 1)+m(\angle 2)+m(\angle 3)$

$=50°+130°+50°=230°$

d. $m(\angle 2)+m(\angle 4)=130°+130°=260°$

89. a. $m(\angle BAC)=24°$

$90°-24°=66°$

b. $m(\angle BAC)=24°$

$180°-24°=156°$

91. complement: $90°-51°=39°$

supplement of that: $180°-39°=141°$

93. complement: $90°-1°=89°$

complement of that: $90°-89°=1°$

APPLICATIONS

95. a. about $80°$

b. about $30°$

c. about $65°$

97. a. $90-63=27°$ from horizontal

b. $180-150=30°$ with the ground

WRITING

99. a. It means that he reversed his direction.

b. She flipped completely over and landed facing in the same direction as she started.

101. Yes. If two angles each measure $45°$ they are complementary and equal.

REVIEW

103. $x+3x+20=180$

$4x+20=180$

$4x+20-20=180-20$

$4x=160$

$x=40$

105. $5x=6x-15$

$5x-6x=6x-15-6x$

$-x=-15$

$x=15$

Section 9.2: Parallel and Perpendicular Lines

VOCABULARY

1. Two lines that lie in the same plane are called <u>coplanar</u>. Two lines that lie in different planes are called <u>noncoplanar</u>.

3. <u>Perpendicular</u> lines are lines that intersect and form right angles.

5. $\angle 4$ and $\angle 6$ are <u>alternate</u> interior angles.

CONCEPTS

7. a.

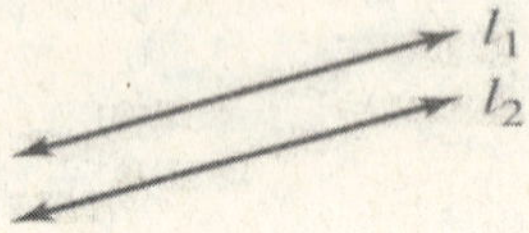

b.

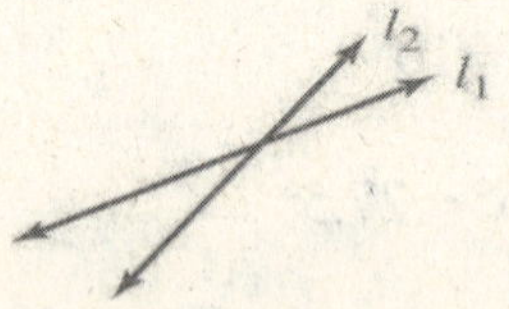

9. a.

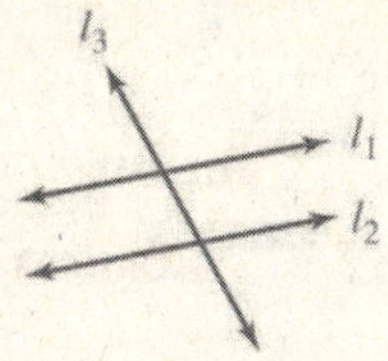

b.

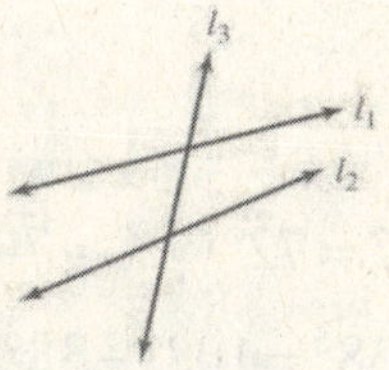

11. When two parallel lines are cut by a transversal, corresponding angles are congruent.

13. When two parallel lines are cut by a transversal, interior angles on the same side of the transversal are supplementary.

15. They are perpendicular.

NOTATION

17. The symbol ⌉ indicates a right angle.

19. The symbol ⊥ is read "is perpendicular to."

GUIDED PRACTICE

21. a. $\angle 1$ & $\angle 5$; $\angle 4$ & $\angle 8$;
$\angle 2$ & $\angle 6$; $\angle 3$ & $\angle 7$

b. $\angle 3$, $\angle 4$, $\angle 5$, $\angle 6$

c. $\angle 3$ & $\angle 5$; $\angle 4$ & $\angle 6$

23. By corresponding angles and vertical angles:
$m(\angle 1) = m(\angle 4) = m(\angle 5) = m(\angle 8) = 130°$

By supplementary angles, since $180° - 130° = 50°$,
$m(\angle 2) = m(\angle 3) = m(\angle 6) = m(\angle 7) = 50°$

25. $\angle 1 \cong \angle X$; $\angle 2 \cong \angle N$

27.
$$4x - 8° = 2x + 16°$$
$$4x - 8° + 8° = 2x + 16° + 8°$$
$$4x = 2x + 24°$$
$$4x - 2x = 2x + 24° - 2x$$
$$2x = 24°$$
$$\frac{2x}{2} = \frac{24°}{2}$$
$$x = 12°$$

$$4x - 8 = 4(12) - 8 = 48 - 8 = 40°$$
$$2x + 16 = 2(12) + 16 = 24 + 16 = 40°$$

29.
$$5x + 6x + 70° = 180°$$
$$11x + 70° = 180°$$
$$11x + 70° - 70° = 180° - 70°$$
$$11x = 110°$$
$$\frac{11x}{11} = \frac{110°}{11}$$
$$x = 10°$$

$$5(10°) = 50°$$
$$6(10°) + 70° = 60° + 70° = 130°$$

TRY IT YOURSELF

31. a. alternate interior: $m(\angle 1) = 50°$

alternate interior: $m(\angle 3) = 45°$

supplementary: $m(\angle 2) + 45° = 180°$

$m(\angle 2) = 135°$

$50° + m(\angle 4) + m(\angle 3) = 180°$

$50° + m(\angle 4) + 45° = 180°$

$m(\angle 4) + 95° = 180°$

$m(\angle 4) = 85°$

b. They form a straight angle, so the sum is $180°$.

c. $m(\angle 1) = 50°$

$m(\angle \text{ABC}) = 45°$

$m(\angle 4) = 85°$

$m(\angle 1) + m(\angle \text{ABC}) + m(\angle 5)$

$= 50° + 45° + 85° = 180°$

33. vertical angles: $\angle 1 \cong \angle 2$

alternate interior: $\angle A \cong \angle E$; $\angle B \cong \angle D$

35. $(3x + 20°) + (x) = 180°$

$4x + 20° = 180°$

$4x + 20° - 20° = 180° - 20°$

$4x = 160°$

$\frac{4x}{4} = \frac{160°}{4}$

$x = 40°$

$3x + 20° = 3(40°) + 20° = 120° + 20° = 140°$

37. $6x - 2° = 9x - 38°$

$6x - 2° + 38° = 9x - 38° + 38°$

$6x + 36° = 9x$

$6x + 36° + 6x = 9x - 6x$

$36° = 3x$

$\frac{36°}{3} = \frac{3x}{3}$

$12° = x$

$6x - 2° = 6(12°) - 2° = 72° - 2° = 70°$

$9x - 38° = 9(12°) - 38° = 108° - 38° = 70°$

APPLICATIONS

39. The plummet string should hang perpendicular to the top of the stones.

41. $\angle DCF \cong \angle ABC$

$m(\angle ABE) + m(\angle ABC) = 180°$

$m(\angle ABE) + 130° = 180°$

$m(\angle ABE) = 50°$

43. Wallpaper strips need to be perpendicular to the ceiling (and floor), and must also be parallel to one another to preserve the pattern.

45. $m(\angle 2) = 105°$ (corresponding angles)

$105° + m(\angle 1) = 180°$

$m(\angle 1) = 75°$

$m(\angle 3) = 75°$ (corresponding angles with $\angle 1$)

WRITING

47. On the North side of the street cars will park parallel to the West – East line. On the South side of the street East of the planter cars will park perpendicular to the West – East line. On the South side of the street West of the planter the cars will park at slant to the West – East line.

49. Since they are corresponding angles, $\angle CBE \cong \angle FEH$.

Since they are vertical angles, $\angle CBE \cong \angle ABD$.

Therefore, they all have the same measurement of $100°$

51. No. Pairs of alternate interior angles are congruent only when the transversal crosses parallel lines.

REVIEW

53. $\dfrac{60}{100}=\dfrac{x}{120}$

$60\cdot 120 = x\cdot 100$

$\dfrac{7200}{100}=\dfrac{x\cdot 100}{100}$

$72 = x$

55. $\dfrac{x}{100}=\dfrac{225}{500}$

$500\cdot x = 225\cdot 100$

$\dfrac{500\cdot x}{500}=\dfrac{22{,}500}{500}$

$x = 45$

225 is 45% of 500.

57. yes

59. $\dfrac{4oz}{12oz}=\dfrac{\cancel{4}\,\cancel{oz}}{3\cdot\cancel{4}\,\cancel{oz}}=\dfrac{1}{3}$

Section 9.3: Triangles

VOCABULARY

1. A polygon is a closed geometric figure with at least three line segments for its sides.

3. A point where two sides of a polygon intersect is called a vertex of the polygon.

5. A triangles with three sides of equal length is called an equilateral triangle. An isosceles triangle has at least two sides of equal length. A scalene triangle has no sides of equal length.

7. The longest side of a right triangle is called the hypotenuse. The other two sides of a right triangle are called legs.

9. In this section we discussed the sum of the measures of the angles of a triangle. The word *sum* indicates the operation of addition.

CONCEPTS

11. a. b. c.

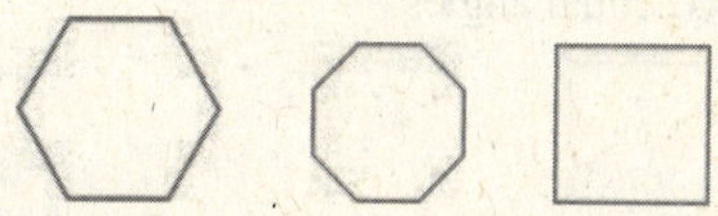

d. e. f.

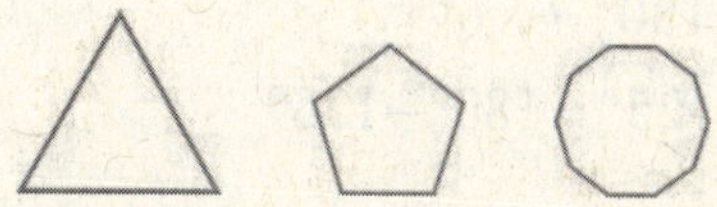

13. a. b. c.

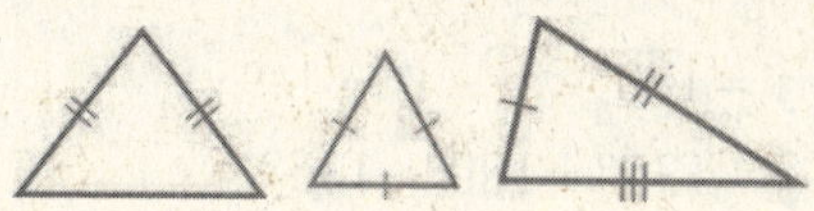

d. e. f.

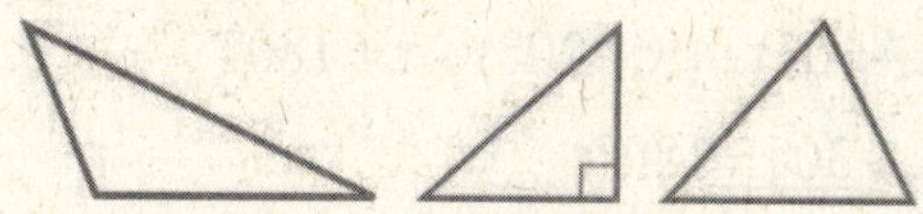

15. a. 90° b. right

c. $\overline{AB}$, $\overline{BC}$ d. $\overline{AC}$

e. $\overline{AC}$ f. $\overline{AC}$

17. a. The isosceles triangle theorem states that if two sides of a triangle are congruent, then the angles opposite those sides are congruent.

b. The converse of the isosceles triangle theorem states that if two angles of a triangle are congruent, then the sides opposite the angles have the same length, and the triangle is isosceles.

19. a. $\overline{EF}\cong\overline{GF}$ b. isosceles

NOTATION

21. The symbol Δ means triangle.

23. $\overline{AB}\cong\overline{CB}$

GUIDED PRACTICE

25. a. 4, quadrilateral, 4 b. 6, hexagon, 6

27. a. 7, heptagon, 7 b. 9, nonagon, 9

29. a. scalene b. isosceles

31. a. equilateral b. scalene

33. Yes, it has two equal angles.

35. no

37. $35° + 90° + y = 180°$
$125° + y = 180°$
$125° + y - 125° = 180° - 125°$
$y = 55°$

39. $45° + 90° + y = 180°$
$135° + y = 180°$
$135° + y - 135° = 180° - 135°$
$y = 45°$

41. $(x + 10°) + (x + 20°) + x = 180°$
$3x + 30° = 180°$
$3x + 30° - 30° = 180° - 30°$
$3x = 150°$
$\frac{3x}{3} = \frac{150°}{3}$
$x = 50°$
$50° + 10° = 60°$
$50° + 20° = 70°$

43. $4x + 4x + x = 180°$
$9x = 180°$
$\frac{9x}{9} = \frac{180°}{9}$
$x = 20°$
$4x = 4(20°) = 80°$
The angles measure $20°, 20°,$ and $80°$.

45. $56° + 56° + x = 180°$
$112° + x = 180°$
$112° + x - 112° = 180° - 112°$
$x = 68°$

47. $85.5° + 85.5° + x = 180°$
$171° + x = 180°$
$171° + x - 171° = 180° - 171°$
$x = 9°$

49. $x + x + 102° = 180°$
$2x + 102° = 180°$
$2x + 102° - 102° = 180° - 102°$
$2x = 78°$
$\frac{2x}{2} = \frac{78°}{2}$
$x = 39°$

51. $x + x + 90.5° = 180°$
$2x + 90.5° = 180°$
$2x + 90.5° - 90.5° = 180° - 90.5°$
$2x = 89.5°$
$\frac{2x}{2} = \frac{89.5°}{2}$
$x = 44.75°$

TRY IT YOURSELF

53. $x + 76° + 76° = 180°$
$x + 152° = 180°$
$x + 152° - 152° = 180° - 152°$
$x = 28°$

55. $x + 53.5° + 53.5° = 180°$
$x + 107° = 180°$
$x + 107° - 107° = 180° - 107°$
$x = 73°$

57. $30° + 60° + x = 180°$
$90° + x = 180°$
$90° + x - 90° = 180° - 90°$
$x = 90°$

59. $100° + 35° + x = 180°$
$135° + x = 180°$
$135° + x - 135° = 180° - 135°$
$x = 45°$

61. $25.5° + 63.8° + x = 180°$
$89.3° + x = 180°$
$89.3° + x - 89.3° = 180° - 89.3°$
$x = 90.7°$

63. $29° + 89.5° + x = 180°$

$118.5° + x = 180°$

$118.5° + x - 118.5° = 180° - 118.5°$

$x = 61.5°$

65. $x + x + 156° = 180°$

$2x + 156° = 180°$

$2x + 156° - 156° = 180° - 156°$

$2x = 24°$

$\frac{2x}{2} = \frac{24°}{2}$

$x = 12°$

67. $x + x + 75° = 180°$

$2x + 75° = 180°$

$2x + 75° - 75° = 180° - 75°$

$2x = 105°$

$\frac{2x}{2} = \frac{105°}{2}$

$x = 52.5°$

69. If it is a base angle:

$39° + 39° + x = 180°$

$78° + x = 180°$

$x = 102°$

$39°, 39°, 102°$

If it is the vertex angle:

$x + x + 39° = 180°$

$2x + 39° = 180°$

$2x = 141°$

$x = 70.5°$

$39°, 70.5°, 70.5°$

71. $m(\angle DBA) = 180° - 73° - 61° = 46°$

$46° + 49° + m(\angle EBC) = 180°$

$95° + m(\angle EBC) = 180°$

$m(\angle EBC) = 85°$

$85° + 22° + m(\angle C) = 180°$

$107° + m(\angle C) = 180°$

$m(\angle C) = 73°$

73. $m(\angle ONM) = 180° - 79° = 101°$

$m(\angle NOM) = 180° - 64° - 101° = 15°$

$m(\angle NOM) + m(\angle NOQ) + 90° = 180°$

$15° + m(\angle NOQ) + 90° = 180°$

$105° + m(\angle NOQ) = 180°$

$m(\angle NOQ) = 75°$

APPLICATIONS

75. a. octagon b. triangle c. pentagon

77. As the jack is raised, the two sides remain the same length.

79. equilateral

WRITING

81. The Pentagon is the headquarters of the U.S. Military. It is large office building in the shape of a regular pentagon.

83. A triangle can not have two right angles because that would equal the full $180°$ and not leave anything for the third required angle.

REVIEW

85. $\frac{20}{100} = \frac{x}{110}$

$20 \cdot 110 = x \cdot 100$

$\frac{2,200}{100} = \frac{x \cdot 100}{100}$

$22 = x$

87. $\frac{x}{100}=\frac{80}{200}$

$x\cdot 200=80\cdot 100$

$\frac{x\cdot 200}{200}=\frac{8,000}{200}$

$x=40$

80 is 40% of 200.

89. $0.85\div 2(0.25)$

$=0.425(0.25)$

$=0.10625$

Section 9.4: The Pythagorean Theorem

VOCABULARY

1. In a right triangle, the side opposite the $90°$ angle is called the hypotenuse. The other two sides are called the legs.

3. The Pythagorean theorem states that in any right triangle, the square of the length of the hypotenuse is equal to the sum of the squares of the lengths of the two legs.

CONCEPTS

5. If a and b are the lengths of two legs of a right triangle and c is the hypotenuse, then $a^2+b^2=c^2$.

7. The converse of the Pythagorean theorem: If a triangle has three sides of lengths $a, b,$ and c, such that $a^2+b^2=c^2$, then the triangle is a right triangle.

9. a. $\overline{BC}$ b. $\overline{AB}$ c. $\overline{AC}$

NOTATION

11. $8^2+6^2=c^2$

$64+36=c^2$

$100=c^2$

$\sqrt{100}=c$

$10=c$

GUIDED PRACTICE

13. $6^2+8^2=c^2$

$36+64=c^2$

$100=c^2$

$\sqrt{100}=c$

$10ft=c$

15. $5^2+12^2=c^2$

$25+144=c^2$

$169=c^2$

$\sqrt{169}=c$

$13m=c$

17. $48^2+55^2=c^2$

$2,304+3,025=c^2$

$5,329=c^2$

$\sqrt{5,329}=c$

$73mi=c$

19. $88^2+105^2=c^2$

$7,744+11,025=c^2$

$18,769=c^2$

$\sqrt{18,769}=c$

$137cm=c$

21. $10^2+b^2=26^2$

$100+b^2=676$

$100+b^2-100=676-100$

$b^2=576$

$b=\sqrt{576}$

$b=24cm$

23. $a^2+18^2=82^2$

$a^2+324=6,724$

$a^2+324-324=6,724-324$

$a^2=6,400$

$a=\sqrt{6,400}$

$a=80m$

25.
$$a^2+21^2=29^2$$
$$a^2+441=841$$
$$a^2+441-441=841-441$$
$$a^2=400$$
$$a=\sqrt{400}$$
$$a=20m$$

27.
$$180^2+b^2=181^2$$
$$32{,}400+b^2=32{,}761$$
$$32{,}400+b^2-32{,}400=32{,}761-32{,}400$$
$$b^2=361$$
$$b=19m$$

29.
$$5^2+b^2=6^2$$
$$25+b^2=36$$
$$25+b^2-25=36-25$$
$$b^2=11$$
$$b=\sqrt{11}cm\approx 3.32cm$$

31.
$$12^2+8^2=c^2$$
$$144+64=c^2$$
$$208=c^2$$
$$\sqrt{208}m=c\approx 14.42m$$

33.
$$9^2+3^2=c^2$$
$$81+9=c^2$$
$$90=c^2$$
$$\sqrt{90}in=c\approx 9.49in$$

35.
$$a^2+4^2=6^2$$
$$a^2+16=36$$
$$a^2+16-16=36-16$$
$$a^2=20$$
$$a=\sqrt{20}in\approx 4.47in$$

37.
$$12^2+14^2\overset{?}{=}15^2$$
$$144+196\overset{?}{=}225$$
$$340=225$$
This is not a right triangle.

39.
$$33^2+56^2\overset{?}{=}65^2$$
$$1{,}089+3{,}136\overset{?}{=}4{,}225$$
$$4{,}225=4{,}225$$
This is a right triangle.

APPLICATIONS

41.
$$20^2=16^2+x^2$$
$$400=256+x^2$$
$$400-256=256+x^2-256$$
$$144=x^2$$
$$\sqrt{144}=x$$
$$12\,ft=x$$

43.
$$20^2+15^2=x^2$$
$$400+225=x^2$$
$$625=x^2$$
$$\sqrt{625}=x$$
$$25in=x$$

45.
$$90^2+90^2=x^2$$
$$8{,}100+8{,}100=x^2$$
$$16{,}200=x^2$$
$$\sqrt{16{,}200}=x$$
$$127.28\,ft\approx x$$

47.
$$9^2+h^2=37^2$$
$$81+h^2=1{,}369$$
$$81+h^2-81=1{,}369-81$$
$$h^2=1{,}288$$
$$h=\sqrt{1{,}288}\approx 35.89\,ft$$
Applying the Pythagorean Theorem with the length of the ladder as the hypotenuse and the distance from the wall as a leg, the other leg of the triangle (and the maximum distance up the wall that the ladder will reach) is about 35.89 feet. Since the window is 35 feet from the ground, the ladder will reach the window.

WRITING

49. If you add the squares of the shorter sides of a right triangle, you get the square of the longest side.

51. If you count the total number of little squares in the two smaller squares you get the same number as the number of little squares in the large square.

REVIEW

53. $2b+3=-15$

$2(-8)+3\stackrel{?}{=}-15$

$-16+3\stackrel{?}{=}-15$

$-13=-15$

-8 is not a solution.

55. $0.5x=2.9$

$0.5(5)\stackrel{?}{=}2.9$

$2.5=2.9$

5 is not a solution.

57. $33-\frac{x}{2}=30$

$33-\frac{(-6)}{2}\stackrel{?}{=}30$

$33-(-3)\stackrel{?}{=}30$

$36=30$

-6 is not a solution.

59. $3x-2=4x-5$

$3(12)-2\stackrel{?}{=}4(12)-5$

$36-2\stackrel{?}{=}48-5$

$33=43$

12 is not a solution.

Section 9.5: Congruent and Similar Triangles

VOCABULARY

1. Congruent triangles are the same size and the same shape.

3. Two angles or two line segments with the same measure are said to be congruent.

5. If two triangles are similar, they have the same shape but not necessarily the same size.

CONCEPTS

7. a. No, they are different sizes.

b. Yes, they appear to have the same shape.

9. $\Delta XYZ \cong \Delta PRQ$

11. $\Delta RST \sim \Delta MNO$

13. $\angle A \cong \angle B$, $\angle Y \cong \angle T$, $\angle Z \cong \angle R$

$\overline{YZ} \cong \overline{TR}$, $\overline{AZ} \cong \overline{BR}$, $\overline{AY} \cong \overline{BT}$

15. Two triangles are congruent if and only if their vertices can be matched so that the corresponding sides and the corresponding angles are congruent.

17. SAS property: If two sides and the angle between them in one triangle are congruent, respectively, to two sides and the angle between them in a second triangle, the triangles are congruent.

19. $\frac{x}{15}=\frac{20}{3}$

$x\cdot 3=20\cdot 15$

$\frac{x\cdot 3}{3}=\frac{300}{3}$

$x=100$

21. $\frac{h}{2.6}=\frac{27}{13}$

$h\cdot 13=27\cdot 2.6$

$\frac{h\cdot 13}{13}=\frac{70.2}{13}$

$h=5.4$

23. Two triangles are similar if and only if their vertices can be matched so that corresponding angles are congruent and the lengths of corresponding sides are proportional.

25. Congruent triangles are always similar, but similar triangles are not always <u>congruent</u>.

NOTATION

27. The symbol $\cong$ is read as "is congruent to."

29.

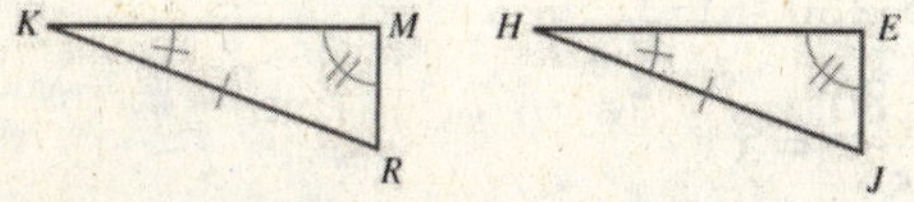

GUIDED PRACTICE

31.
$\overline{AC} \cong \overline{DF}$
$\overline{DE} \cong \overline{AB}$
$\overline{BC} \cong \overline{EF}$
$\angle A \cong \angle D$
$\angle E \cong \angle B$
$\angle F \cong \angle C$

33. a. $\overline{BC} \cong \overline{MN}$
$\overline{CD} \cong \overline{NO}$
$\overline{BD} \cong \overline{MO}$
$\angle B \cong \angle M$
$\angle C \cong \angle N$
$\angle D \cong \angle O$

b. $72°$ c. $10\,ft$ d. $9\,ft$

35. yes: SSS

37. Not necessarily. Despite appearances the other sides for SAS (or angle for ASA) may not be of equal size.

39. a. $\angle L \cong \angle H$
$\angle M \cong \angle J$
$\angle R \cong \angle E$

b. $\dfrac{LM}{HJ} = \dfrac{MR}{JE}, \ \dfrac{MR}{JE} = \dfrac{LR}{HE}, \ \dfrac{LM}{HJ} = \dfrac{LR}{HE}$

c. $\dfrac{LM}{HJ} = \dfrac{MR}{JE} = \dfrac{LR}{HE}$

41. Yes. vertical angle, right angle, third angle

43. Not necessarily. Two angles are needed.

45. Yes

47. Not necessarily. They may not be right triangles.

49. Yes

51. Not necessarily

52. Not necessarily

53.

$$\frac{x}{28} = \frac{6}{21} \qquad x \cdot 21 = 6 \cdot 28 \qquad \frac{x \cdot 21}{21} = \frac{168}{21} \qquad x = 8$$

$$\frac{y}{10} = \frac{21}{6} \qquad y \cdot 6 = 21 \cdot 10 \qquad \frac{y \cdot 6}{6} = \frac{210}{6} \qquad y = 35$$

55.

$$\frac{x}{40} = \frac{75}{50} \qquad x \cdot 50 = 75 \cdot 40 \qquad \frac{x \cdot 50}{50} = \frac{3{,}000}{50} \qquad x = 60$$

$$\frac{y}{57} = \frac{50}{75} \qquad y \cdot 75 = 50 \cdot 57 \qquad \frac{y \cdot 75}{75} = \frac{2{,}850}{75} \qquad y = 38$$

TRY IT YOURSELF

57. True. SSS

59. False. The angle needs to be the included angle.

61. Yes. SSS

63. Yes. SAS

65. Yes. ASA (alternate interior angles)

67. Not necessarily

69. $x = m(\angle F) = m(\angle C)80°$
$y = m(\overline{BC}) = m(\overline{EF}) = 2\,yd$

71. $x = m(\angle D) = m(\angle C) = 19°$
$y = m(\overline{AC}) = m(\overline{AD}) = 14m$

73. $x = 6mm$

75. $x = 50°$

77.
$$\frac{x}{10}=\frac{5}{12}$$
$$x\cdot 12=5\cdot 10$$
$$\frac{x\cdot 12}{12}=\frac{50}{12}$$
$$x=\frac{25}{6}=4\frac{1}{6}$$

79.
$$\frac{x}{12}=\frac{x+4}{15}$$
$$x\cdot 15=(x+4)12$$
$$15x=12x+48$$
$$15x-12x=12x+48-12x$$
$$3x=48$$
$$\frac{3x}{3}=\frac{48}{3}$$
$$x=16$$

APPLICATIONS

81.
$$m(\overline{AO})=m(\overline{CO})=9.5cm$$
$$9.5+8=17.5cm$$

83.
$$\frac{25}{74}=\frac{20}{w}$$
$$25\cdot w=20\cdot 74$$
$$\frac{25\cdot w}{25}=\frac{1480}{25}$$
$$w=59.2\,ft$$

85.
$$\frac{6}{4}=\frac{t}{24}$$
$$6\cdot 24=t\cdot 4$$
$$\frac{144}{4}=\frac{t\cdot 4}{4}$$
$$36\,ft=t$$

87.
$$\frac{3}{2.5}=\frac{h}{29}$$
$$3\cdot 29=h\cdot 2.5$$
$$\frac{87}{2.5}=\frac{h\cdot 2.5}{2.5}$$
$$34.8\,ft=h$$

89.
$$\frac{200}{1{,}000}=\frac{x}{5{,}280}$$
$$\frac{1}{5}=\frac{x}{5{,}280}$$
$$1\cdot 5{,}280=x\cdot 5$$
$$\frac{5{,}280}{5}=x$$
$$1{,}056\,ft=x$$

WRITING

91. There is no SSA property for triangles because the angle could be acute in the first triangle and obtuse in the second triangle – the included angle would be different in each case.

REVIEW

93.
$$21=3\cdot 7$$
$$27=3\cdot 3\cdot 3$$
$$LCM(21,27)=3\cdot 3\cdot 3\cdot 7=189$$

95.
$$63=3\cdot 3\cdot 7$$
$$84=2\cdot 2\cdot 3\cdot 7$$
$$GCF(63,84)=3\cdot 7=21$$

Section 9.6: Quadrilaterals and Other Polygons

VOCABULARY

1. A quadrilateral is a polygon with four sides.

3. A rectangle is a quadrilateral with four right angles.

5. A rhombus is a parallelogram with four sides of equal length.

7. A trapezoid has two sides that are parallel and two sides that are not parallel. The parallel sides are called bases. The legs of an isosceles trapezoid have the same length.

CONCEPTS

9. a. four; A, B, C, D

b. four; $\overline{AB}, \overline{BC}, \overline{CD}, \overline{DA}$

c. two; $\overline{AC}, \overline{BD}$

d. yes, no, no, yes (The vertices must be consecutive.)

11. a. $\overline{ST} \parallel \overline{VU}$ b. $\overline{SV} \parallel \overline{TU}$

13. a. All four angles are right angles.

b. Opposite sides are parallel.

c. Opposite sides have equal length.

d. The diagonals have equal length.

e. The diagonals intersect at their midpoint.

15. It is a rectangle.

17. a. no b. yes c. no

d. yes e. no f. yes

19. a. isosceles b. $\angle J, \angle M$

c. $\angle K, \angle L$ d. $m(\angle J) = m(\angle M)$

$m(\angle K) = m(\angle L)$

$m(\overline{JK}) = m(\overline{ML})$

NOTATION

21. The four sides of the quadrilateral have the same length.

23. S represents the sum of the angles in the polygon.
n represents the number of sides of the polygon.

GUIDED PRACTICE

25. a. square b. rhombus

c. trapezoid d. square

27. a. $90°$

b. 9 (diagonals cross at the midpoint)

c. $9+9=18$

d. 18 (diagonals are the same length)

29. Interior angles are supplementary:

a. $138° + x = 180°$

$138° + x - 138° = 180° - 138°$

$x = 42°$

b. $y + 85° = 180°$

$y + 85° - 85° = 180° - 85°$

$y = 95°$

31. a. 9 b. $70°$

c. $m(\angle x) + m(\angle y) = 180°$

$70° + m(\angle y) = 180°$

$70° + m(\angle y) - 70° = 180° - 70°$

$m(\angle y) = 110°$

d. $110°$

33. $S = (n-2)180° = (14-2)180°$
$= (12)180° = 2{,}160°$

35. $S = (n-2)180° = (20-2)180°$
$= (18)180° = 3{,}240°$

37. $S = (n-2)180° = (8-2)180°$
$= (6)180° = 1{,}080°$

39. $S = (n-2)180° = (12-2)180°$
$= (10)180° = 1{,}800°$

41. $540° = (n-2)180°$

$540° = 180°n - 360°$

$540° + 360° = 180°n - 360° + 360°$

$900° = 180°n$

$\dfrac{900°}{180°} = \dfrac{180°n}{180°}$

$5 = n$

43.

$$900° = (n-2)180°$$
$$900° = 180°n - 360°$$
$$900° + 360° = 180°n - 360° + 360°$$
$$1,260° = 180°n$$
$$\frac{1,260°}{180°} = \frac{180°n}{180°}$$
$$7 = n$$

45.

$$1,980° = (n-2)180°$$
$$1,980° = 180°n - 360°$$
$$1,980° + 360° = 180°n - 360° + 360°$$
$$2,340° = 180°n$$
$$\frac{2,340°}{180°} = \frac{180°n}{180°}$$
$$13 = n$$

47.

$$2,160° = (n-2)180°$$
$$2,160° = 180°n - 360°$$
$$2,160° + 360° = 180°n - 360° + 360°$$
$$2,520° = 180°n$$
$$\frac{2,520°}{180°} = \frac{180°n}{180°}$$
$$14 = n$$

TRY IT YOURSELF

49.

a. $m(\angle 1) + 60° = 90°$

$m(\angle 1) + 60° - 60° = 90° - 60°$

$m(\angle 1) = 30°$

b. alternate interior: $m(\angle 3) = m(\angle 1) = 30°$

c. alternate interior: $m(\angle 2) = 60°$

d. diagonals are of equal length: 8cm

e. $\frac{1}{2}(8) = 4cm$

51.

$$S = (n-2)180°$$
$$S = (4-2)180° = (2)180° = 360°$$

$$2x + (2x + 10°) + (3x + 30)° + x = 360°$$
$$8x + 40° = 360°$$
$$8x + 40° - 40° = 360° - 40°$$
$$8x = 320°$$
$$\frac{8x}{8} = \frac{320°}{8}$$
$$x = 40°$$

$$m(\angle C) = x = 40°$$
$$m(\angle D) = 2x = 2 \cdot 40° = 80°$$
$$m(\angle A) = 2x + 10° = 2 \cdot 40° + 10° = 90°$$
$$m(\angle B) = 3x + 30° = 3 \cdot 40° + 30° = 150°$$

APPLICATIONS

53. a. trapezoid b. square c. rectangle

d. trapezoid e. parallelogram

55.

$$S = (n-2)180° = (5-2)180°$$
$$= 3(180°) = 540°$$

WRITING

57. A square is a rectangle because its opposite sides are parallel and all the angles are right angles.

59. She can check that the diagonals are of the same length.

REVIEW

61. two hundred fifty-four thousand, three hundred nine

63. eighty-two million, four hundred fifteen

Section 9.7: Perimeters and Areas of Polygons

VOCABULARY

1. The distance around a polygon is called the perimeter.

3. The measure of the surface enclosed by a polygon is called its area.

5. The area of a polygon is measured in square units.

CONCEPTS

7. $8ft \cdot 16ft = 128ft^2$

9. a. $P = 4s$ b. $P = 2l + 2w$

11.

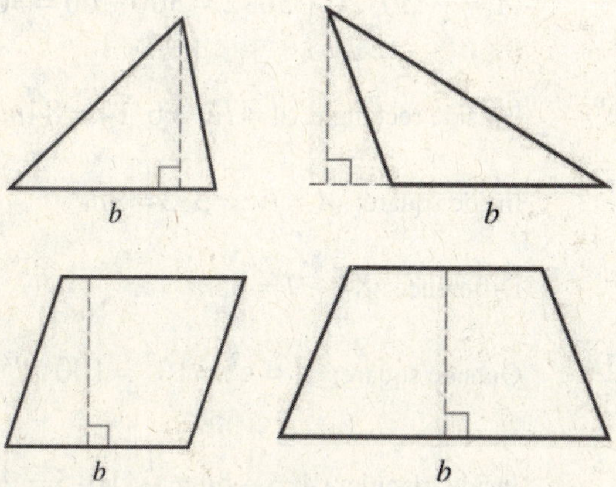

13. a rectangle and a triangle

NOTATION

15. a. The symbol $1\ in^2$ means one square inch.

b. One square meter is expressed as $1m^2$.

GUIDED PRACTICE

17. $P = 4s = 4 \cdot 8 = 32in$

19. $P = 4s = 4 \cdot 5.75 = 23mi$

21. First convert: $2ft = 24in$

$P = 2l + 2w = 2 \cdot 24 + 2 \cdot 7$

$= 48 + 14 = 62in$

23. First convert: $3ft = 36in$

$P = 2l + 2w = 2 \cdot 11 + 2 \cdot 36$

$= 22 + 72 = 94in$

x = the length of the third side

25.

$10 + 10 + x = 35$

$20 + x = 35$

$20 + x - 20 = 35 - 20$

$x = 15ft$

27. x = the length of the legs

$x + 10 + x + 15 = 35$

$2x + 25 = 35$

$2x + 25 - 25 = 35 - 25$

$2x = 10$

$\frac{2x}{2} = \frac{10}{2}$

$x = 5m$

29. $A = s^2 = 4^2 = 16cm^2$

31. $A = s^2 = 2.5^2 = 6.25m^2$

33. $1ft^2 = 1ft \cdot 1ft = 12in \cdot 12in = 144in^2$

35. $1m^2 = 1m \cdot 1m = 1,000mm \cdot 1,000mm$

$= 1,000,000mm^2$

37. $1mi^2 = 1mi \cdot 1mi = 5,280ft \cdot 5,280ft$

$= 27,878,400ft^2$

39. $1km^2 = 1km \cdot 1km = 1,000m \cdot 1,000m$

$= 1,000,000m^2$

41. First convert to feet.
The dimensions are 9ft by 15ft.

$A = lw = 9 \cdot 15 = 135ft^2$

43. First convert to feet.
The dimensions are 60ft by 186ft.

$A = lw = 60 \cdot 186 = 11,160ft^2$

45. $A = \frac{1}{2}bh = \frac{1}{2} \cdot 10 \cdot 5 = 25in^2$

47. $A = \frac{1}{2}bh = \frac{1}{2} \cdot 9 \cdot 6 = 27cm^2$

49. $A = \frac{1}{2}bh = \frac{1}{2} \cdot 5 \cdot 3 = 7.5in^2$

51. $A=\frac{1}{2}bh=\frac{1}{2}\cdot 7\cdot 3=10.5mi^2$

53. $A=\frac{1}{2}h(b_1+b_2)=\frac{1}{2}\cdot 4(8+12)$

$=2(20)=40ft^2$

55. $A=\frac{1}{2}h(b_1+b_2)=\frac{1}{2}\cdot 7(10+16)$

$=\frac{7}{2}(26)=91cm^2$

57. $A=bh$

$60=b\cdot 15$

$\frac{60}{15}=\frac{b\cdot 15}{15}$

$4m=b$

59. $A=lw$

$36=3\cdot w$

$\frac{36}{3}=\frac{3\cdot w}{3}$

$12cm=w$

61. $A=\frac{1}{2}bh$

$54=\frac{1}{2}\cdot 3\cdot h$

$54=\frac{3}{2}\cdot h$

$\frac{2}{3}\cdot 54=\frac{2}{3}\cdot\frac{3}{2}\cdot h$

$36m=h$

63. $P=2l+2w$

$64=2\cdot 21+2w$

$64=42+2w$

$64-42=42+2w-42$

$22=2w$

$\frac{22}{2}=\frac{2w}{2}$

$11mi=w$

65. Break down the figure into a rectangle of dimensions 12in by 6in and a triangle with base 12in and height 5in.

$A=\frac{1}{2}bh+lw$

$A=\frac{1}{2}\cdot 12\cdot 5+12\cdot 6=30+72=102in^2$

67. Break down the figure into a rectangle of dimensions 2 ft by 30 ft and a triangle with a base of 30 ft and a height of 20 ft.

$A=\frac{1}{2}bh+lw$

$A=\frac{1}{2}\cdot 30\cdot 20+30\cdot 2=300+60=360ft^2$

69. Outside rectangle: $A=lw=6\cdot 14=84m^2$

Inside square: $A=lw=3\cdot 3=9m^2$

Difference: $84-9=75m^2$

71. Outside square: $A=s^2=10^2=100yd^2$

Inside triangle: $A=\frac{1}{2}bh=\frac{1}{2}\cdot 10\cdot 5=25yd^2$

Difference: $100-25=75yd^2$

73. Area of the floor: $A=lw=8\cdot 5=40yd^2$

$40\cdot \$30=\$1,200$ for all the flooring.

75. Perimeter of the yard:

$2l+2w=2\cdot 110+2\cdot 85=390ft.$

$390\cdot \$12.50=\$4,875$ to enclose the yard.

TRY IT YOURSELF

77.

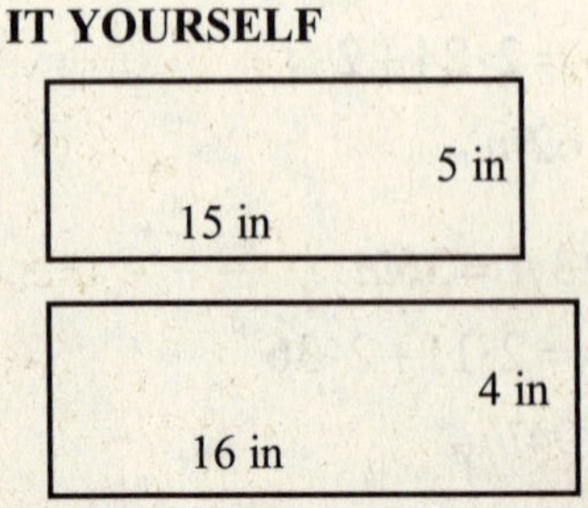

79.

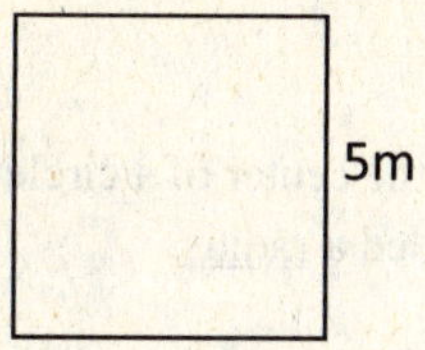

81.

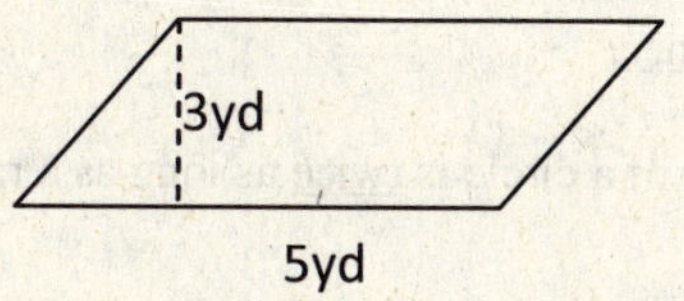

83.

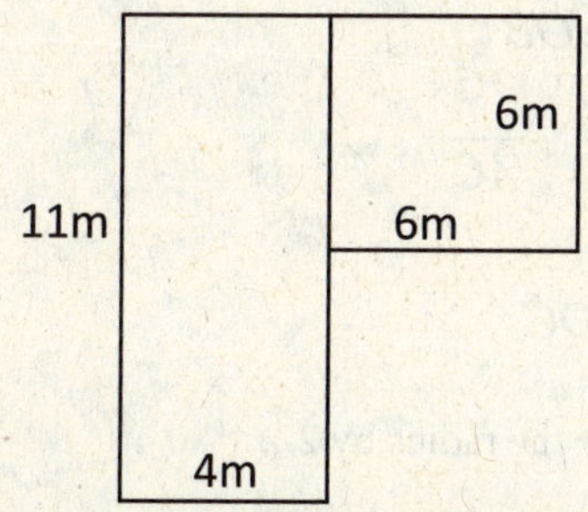

85. $A = bh = 15 \cdot 4 = 60cm^2$

87. x = the length of the third side

$22 + 22 + x = 80$

$44 + x = 80$

$44 + x - 44 = 80 - 44$

$x = 36m$

89. x = the length of a side

$x + x + x = 85$

$3x = 85$

$\frac{3x}{3} = \frac{85}{3}$

$x = \frac{85}{3} ft = 28\frac{1}{3} ft$

91. The missing segment has length $10 - 4 - 4 = 2m$.

$P = 10 + 6 + 4 + 2 + 2 + 2 + 4 + 6 = 36m$

93. $x = 9.1 - 5.4 = 3.7 ft$

$y = 16.3 - 6.2 = 10.1 ft$

$P = 2 \cdot 9.1 + 2 \cdot 16.3 = 18.2 + 32.6 = 50.8 ft$

APPLICATIONS

95. The 'upper' and 'lower' sides have $\frac{60}{3} = 20$ trees. The side opposite the house has $\frac{120}{3} = 40$ trees. With the addition of the first tree (as noted in the figure) there are $1 + 20 + 40 + 20 = 81$ trees.

97. $1\ yd^2 = 3 \cdot 3 = 9 ft^2$

Ceramic: $\$3.75 / ft^2$

Vinyl: $\frac{\$34.95}{9 ft^2} \approx \$3.88 / ft^2$

The vinyl is more expensive.

99. $A = 14 ft \cdot 20 ft = 280 ft^2$

$280 \cdot \$1.29 = \$361.20$

101. Area of the sail is $A = \frac{1}{2} \cdot 12 ft \cdot 24 ft = 144 ft^2$

$1\ yd^2 = 3 \cdot 3 = 9 ft^2$

Area of the sail is $\frac{144}{9} = 16 yd^2$

$16 \cdot 12 = \$192$ for the whole sail.

103. $A = \frac{1}{2} h(b_1 + b_2) = \frac{1}{2} \cdot 315mi(205mi + 505mi)$

$= \frac{315mi}{2}(710mi) = 111{,}825mi^2$

105. Area of the sides:

$$48\,ft\cdot 12\,ft+48\,ft\cdot 12\,ft$$
$$=576\,ft^2+576\,ft^2=1{,}152\,ft^2$$

Area of the front and back:

$$2(12\,ft\cdot 20\,ft)=2(240\,ft^2)=480\,ft^2$$

Total area = $1{,}152\,ft^2+480\,ft^2=1{,}632\,ft^2$

Each piece of sheetrock covers

$$4\,ft\cdot 8\,ft=32\,ft^2$$

$\frac{1{,}632}{32}=51$ sheets of sheetrock

WRITING

107. Perimeter is the distance around an object while area is what is enclosed in the object.

109. He put the ^{2} in the wrong place. It should be $25\,ft^2$.

REVIEW

111. $8\left(\frac{3}{4}t\right)=\frac{2\cdot \cancel{4}}{1}\cdot\frac{3}{\cancel{4}}t=6t$

113. $-\frac{2}{3}(3w-6)=\left(-\frac{2}{3}\right)3w-\left(-\frac{2}{3}\right)6$

$$=-\frac{6w}{3}+\frac{12}{3}=-2w+4$$

115. $-\frac{7}{16}x-\frac{3}{16}x=-\frac{10}{16}x$

$$=-\frac{\cancel{2}\cdot 5}{\cancel{2}\cdot 8}x=-\frac{5}{8}x$$

117. $60\left(\frac{3}{20}r-\frac{4}{15}\right)=60\left(\frac{3}{20}r\right)-60\left(\frac{4}{15}\right)$

$$=\overset{3}{\cancel{60}}\left(\frac{3}{\cancel{20}}r\right)-\overset{4}{\cancel{60}}\left(\frac{4}{\cancel{15}}\right)=9r-16$$

Section 9.8: Circles

VOCABULARY

1. A segment drawn from the center of a circle to a point on the circle is called a radius.

3. A diameter is a chord that passes through the center of a circle.

5. The distance around a circle is called its circumference.

7. A diameter of a circle is twice as long as a radius.

CONCEPTS

9. $\overline{OA},\overline{OC},\overline{OB}$

11. $\overline{DA}$, $\overline{DC}$, $\overline{AC}$

13. ABC, ADC

15. a. Multiply the radius by 2.

b. Divide the diameter by 2.

17. If C is the circumference of a circle and D is its diameter, then $\frac{C}{D}=\pi$.

19. Square the 6.

NOTATION

21. The symbol AB is read as "arc AB."

23. a. multiplication: $2\cdot\pi\cdot r$

b. raising to a power and multiplication: $\pi\cdot r^2$

GUIDED PRACTICE

25. $C=2\pi r=2\cdot\pi\cdot 4=8\pi\,ft\approx 25.1\,ft$

27. $C=2\pi r=2\cdot\pi\cdot 6=12\pi\,m\approx 37.7m$

29. The upper portion is a semicircle with radius 6cm.

$$P=\frac{1}{2}(2\cdot\pi\cdot 6)+10+12+10$$
$$=6\pi+32\approx 18.85+32=50.85cm$$

31. The upper portion is a semicircle with radius 3m.

$$P=\frac{1}{2}(2\cdot\pi\cdot3)+8+8+6$$
$$=3\pi+22\approx9.42+22=31.42m$$

33. $D=6in\rightarrow r=3in$

$$A=\pi r^2=\pi\cdot3^2=9\pi\,in^2\approx28.3in^2$$

35. $D=18in\rightarrow r=9in$

$$A=\pi r^2=\pi\cdot9^2=81\pi\,in^2\approx254.5in^2$$

37. The radius of the semicircle is 6cm.

$$A=\frac{1}{2}(\pi\cdot6^2)+\frac{1}{2}\cdot12\cdot12=18\pi+72$$
$$\approx56.5+72=128.5cm^2$$

39. The radius of the semicircle is 4cm.

$$A=\frac{1}{2}(\pi\cdot4^2)+4\cdot8=8\pi+32$$
$$\approx25.1+32=57.1cm^2$$

TRY IT YOURSELF

41. Area of rectangle: $4\cdot10=40in^2$

Area of circle: $\pi\cdot2^2=4\pi\,in^2\approx12.6in^2$

Difference: $40-12.6=27.4in^2$

43. Area of parallelogram: $13\cdot9=117in^2$

Area of circle: $\pi\cdot4^2=16\pi\,in^2\approx50.3in^2$

Difference: $117-50.3=66.7in^2$

45. $C=\pi D=$

$$C\pi\cdot50=50\pi\,yd\approx157.08yd$$

47. If the square has side length 6 inches, the diameter of the circle is 6 inches.

$$C=\pi D=6\pi\,in\approx18.8in$$

49. If the square has side length 9mm, the radius of the circle is 4.5mm.

$$A=\pi r^2=\pi\cdot4.5^2=20.25\pi\,mm^2\approx63.6mm^2$$

APPLICATIONS

51. a. $1\,in$ b. $2\cdot1=2in$

c. $2\cdot\pi\cdot1=2\pi in\approx2(3.14)=6.28in$

d. $\pi\cdot1^2=\pi in^2\approx3.14in^2$

53. $D=2mi\rightarrow r=1mi$

$$A=\pi\cdot1^2=\pi\,mi^2\approx3.14mi^2$$

55. $C=\pi D=32.66\pi\,ft\approx32.66(3.14)$

$$=102.60\,ft$$

57. Each lap is $\pi\cdot D=\pi\cdot\frac{1}{4}\approx0.79mi$.

$\frac{10}{0.79}\approx12.65$ laps.

She needs to run 13 times around the track.

59. Area of the target: $\pi\cdot2^2=4\pi\,ft^2\approx12.57\,ft^2$

Area of the bull's eye:

$$\pi\cdot\left(\frac{1}{2}\right)^2=\frac{\pi}{4}\,ft^2\approx0.79\,ft^2$$

$$\frac{\frac{\pi}{4}}{4\pi}=\frac{\pi}{4}\cdot\frac{1}{4\pi}=\frac{1}{16}=0.0625=6.25\%$$

WRITING

61. The circumference of a circle is the distance around the outside.

63. π is the ratio of a circle's circumference to its diameter.

REVIEW

65. $\frac{9}{10}=0.9=90\%$

67. $0.827=82.7\%$

69. $\frac{\$1.29}{24oz}=\$0.05375/oz=5.375¢/oz$

71. five

Section 9.9: Volume

VOCABULARY

1. The volume of a three-dimensional figure is a measure of its capacity.

3. cone

5. cylinder

7. pyramid

CONCEPTS

9.

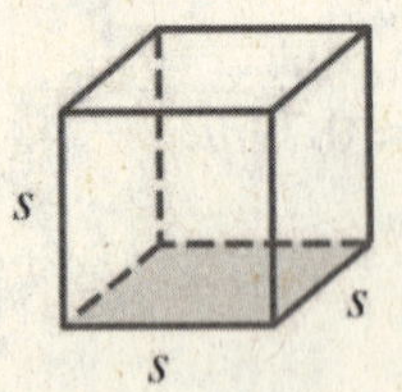

11.

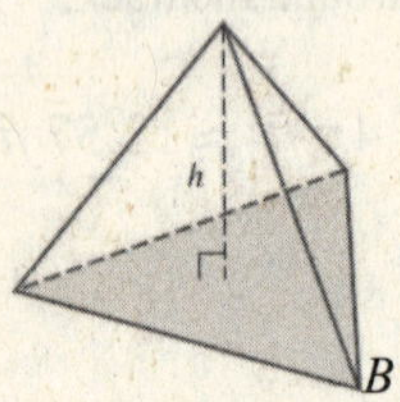

13.

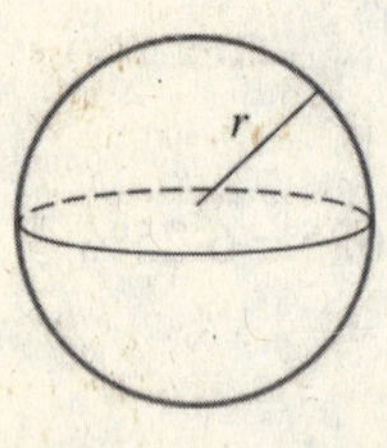

15. cubic inches, mi^3, m^3

17. a. perimeter b. volume c. area

d. volume e. area f. circumference

19. a. $\frac{1}{3}\pi(25)(6) = \frac{150}{3}\pi = 50\pi$

b. $\frac{4}{3}\pi(125) = \frac{500}{3}\pi$

NOTATION

21. a. cubic inch b. $1cm^3$

23. a right angle

GUIDED PRACTICE

25. $1yd^3 = 1yd \cdot 1yd \cdot 1yd = 3ft \cdot 3ft \cdot 3ft = 27ft^3$

27. $1km^3 = 1km \cdot 1km \cdot 1km$

$= 1{,}000m \cdot 1{,}000m \cdot 1{,}000m$

$= 1{,}000{,}000{,}000m^3$

29. $V = lwh = 4 \cdot 2 \cdot 7 = 56ft^3$

31. $V = lwh = 5 \cdot 5 \cdot 5 = 125in^3$

33. First convert: $0.2m = 20cm$

$B = \frac{1}{2}bh = \frac{1}{2} \cdot 3 \cdot 4 = 6cm^2$

$V = Bh = 6 \cdot 20 = 120cm^3$

35. First convert: $2ft = 24in$

$B = \frac{1}{2}bh = \frac{1}{2} \cdot 9 \cdot 12 = 54in^2$

$V = Bh = 54 \cdot 24 = 1{,}296in^3$

37. $B = 10 \cdot 10 = 100yd^2$

$V = \frac{1}{3}Bh = \frac{1}{3} \cdot 100 \cdot 21 = 700yd^3$

39. $B = 8 \cdot 2 = 16ft^2$

$V = \frac{1}{3}Bh = \frac{1}{3} \cdot 16 \cdot 6 = 32ft^3$

41. $B = \frac{1}{2} \cdot 7.2 \cdot 8.3 = 29.88ft^2$

$V = \frac{1}{3}Bh = \frac{1}{3} \cdot 29.88 \cdot 7 = 69.72ft^3$

43. $B = 9yd^2$

$V = \frac{1}{3}Bh = \frac{1}{3} \cdot 9 \cdot 2 = 6yd^3$

45. $V = \pi r^2 h = \pi \cdot 4^2 \cdot 12$

$= 192\pi \, ft^3 \approx 603.19ft^3$

47. $D = 30cm \rightarrow r = 15cm$

$V = \pi r^2 h = \pi \cdot 15^2 \cdot 14$

$= 3{,}150\pi \, cm^3 \approx 9{,}896.02cm^3$

49. $D = 6m \rightarrow r = 3m$

$$V = \frac{1}{3}\pi r^2 h = \frac{1}{3}\cdot\pi\cdot 3^2\cdot 13$$
$$= 39\pi\, m^3 \approx 122.52m^3$$

51. $$V = \frac{1}{3}\pi r^2 h = \frac{1}{3}\cdot\pi\cdot 9^2\cdot 7$$
$$= 189\pi\, yd^3 \approx 593.76yd^3$$

53. $$V = \frac{4}{3}\pi r^3 = \frac{4}{3}\cdot\pi\cdot 6^3$$
$$= 288\pi\, in^3 \approx 904.8in^3$$

55. $D = 4cm \rightarrow r = 2cm$

$$V = \frac{4}{3}\pi r^3 = \frac{4}{3}\cdot\pi\cdot 2^3$$
$$= \frac{32}{3}\pi\, cm^3 \approx 33.5cm^3$$

TRY IT YOURSELF

57. A sphere would be

$V = \frac{4}{3}\pi r^3 = \frac{4}{3}\cdot\pi\cdot 9^3 = 972\pi\, ft^3$ so a hemisphere would have volume

$\frac{972\pi}{2} = 486\pi\, in^3 \approx 1{,}526.81in^3$.

59. $$V = \pi r^2 h = \pi\cdot 6^2\cdot 12$$
$$= 432\pi\, m^3 \approx 1{,}357.17m^3$$

61. $V = lwh = 3\cdot 4\cdot 5 = 60cm^3$

63. $D = 10cm \rightarrow r = 5cm$

$$V = \frac{1}{3}\pi r^2 h = \frac{1}{3}\cdot\pi\cdot 5^2\cdot 12$$
$$= 100\pi\, cm^3 \approx 314.16cm^3$$

65. $B = 10\cdot 10 = 100m^2$

$$V = \frac{1}{3}Bh = \frac{1}{3}\cdot 100\cdot 12 = 400m^3$$

67. $$B = \frac{1}{2}bh = \frac{1}{2}\cdot 3\cdot 4 = 6m^2$$
$$V = Bh = 6\cdot 8 = 48m^3$$

69. Break the figure down into a rectangular solid and a pyramid.

Rectangular solid: $V = 8\cdot 8\cdot 8 = 512cm^3$

Pyramid: $V = \frac{1}{3}Bh = \frac{1}{3}(8\cdot 8)3 = 64cm^3$

Total volume: $512 + 64 = 576cm^3$

71. Note that the outsides of the cylinder form a sphere.

Cylinder: $V = \pi r^2 h = \pi\cdot 3^2\cdot 16 = 144\pi cm^3$

Sphere: $V = \frac{4}{3}\pi r^3 = \frac{4}{3}\cdot\pi\cdot 3^3 = 36\pi$

Total volume: $144\pi + 36\pi = 180\pi\, cm^3 \approx 565.49cm^3$

APPLICATIONS

73. $$V = s^3 = \left(\frac{1}{2}\right)^3 = \frac{1}{8}in^3 = 0.125in^3$$

75. Begin by converting everything to feet.

$$V = (27")(17")(8") = \frac{27}{12}\cdot\frac{17}{12}\cdot\frac{8}{12}$$
$$= \frac{9}{4}\cdot\frac{17}{12}\cdot\frac{2}{3} = \frac{306}{144} = 2.125 \text{ cubic feet}$$

77. $D = 6ft \rightarrow r = 3ft$

$$V = \pi r^2 h = \pi\cdot 3^2\cdot 7 = \pi\cdot 9\cdot 7 = 63\pi\, ft^3$$
$$63\pi\, ft^3 \approx 197.92ft^3$$

79. $D = 40ft \rightarrow r = 20ft$

$$V = \frac{4}{3}\pi r^3 = \frac{4}{3}\pi(20)^3 = \frac{32{,}000}{3}\pi\, ft^3$$
$$\frac{32{,}000}{3}\pi\, ft^3 \approx 33{,}510.32ft^3$$

81. $$\frac{30.4}{3.8} = \frac{304}{38} = \frac{\cancel{2}\cdot 2\cdot 2\cdot 2\cdot \cancel{19}}{\cancel{2}\cdot\cancel{19}} = \frac{8}{1} \text{ or } 8:1$$

83. a. the radius of the hemisphere is 15 inches.

$$V=\frac{1}{2}\left(\frac{4}{3}\pi r^3\right)=\frac{2}{3}\pi\left(15^3\right)=2,250\pi in^3$$

$$2,250\pi in^3 \approx 7,068.58in^3$$

b. $\frac{7,068.58}{231}\approx 30.6gal$

WRITING

85. Volume is the measure of capacity of a figure.

87. Yes. Area is a two dimensional measurement while volume is a three dimensional measurement.

REVIEW

89. $-5(5-2)^2+3=-5(3)^2+3=-5\cdot 9+3$

$=-45+3=-42$

91. $-x=4$

$\frac{-x}{-1}=\frac{4}{-1}$

$x=-4$

93. $\frac{3in}{15in}=\frac{\cancel{3}\,\cancel{in}}{\cancel{3}\cdot 5\cancel{in}}=\frac{1}{5}$ or 1:5

95. $\frac{2.4m}{1}\cdot\frac{1,000mm}{1m}=2,400mm$

Chapter 9 Review

1. point: C or D

line: CD

plane: GHI

2. a. 6 units b. *E* c. yes

3. $\angle ABC$; $\angle CBA$; $\angle B$; $\angle 1$

4. a. acute b. *B* c. $\overrightarrow{BA}$, $\overrightarrow{BC}$ d. 48°

5. acute: $\angle 1$, $\angle 2$

right: $\angle ABD$, $\angle CBD$

obtuse: $\angle CBE$

straight: $\angle ABC$

6. Yes. They have the same degree measure.

7. Yes. They have the same origin and direction.

8. a. obtuse angle b. right angle

c. straight angle d. acute angle

9. $x+35°=50°$

$x+35°-35°=50°-35°$

$x=15°$

10. $y+30°=180°$

$y+30°-30°=180°-30°$

$y=150°$

11. a. vertical angles: $m(\angle 1)=65°$

b. supplementary angles:

$m(\angle 2)+65°=180°$

$m(\angle 2)+65°-65°=180°-65°$

$m(\angle 2)=115°$

12. a. vertical angles: $m(\angle ABG)=39°$

b. right angle: $m(\angle FBE)=90°$

c. complementary angles:

$$m(\angle CBD)+39°=90°$$
$$m(\angle CBD)+39°-39°=90°-39°$$
$$m(\angle CBD)=51°$$

d. vertical angles:

$$m(\angle FBG)=m(\angle CBD)=51°$$

e. yes

13. a. vertical angles: $5x+25°=6x+5°$

$$5x+25°-5°=6x+5°-5°$$
$$5x+20°=6x$$
$$5x+20°-5x=6x-5x$$
$$20°=x$$

b. substitute x: $m(\angle HFI)=6(20°)+5°$

$$=120°+5°=125°$$

c. supplementary angles:

$$m(\angle GFH)+125°=180°$$
$$m(\angle GFH)+125°-125°=180°-125°$$
$$m(\angle GFH)=55°$$

14. $90°-71°=19°$

15. $180°-143°=37°$

16. No. Only two angles can be supplementary.

17. a. parallel b. transversal c. perpendicular

18. $\angle 4 \& \angle 6$; $\angle 3 \& \angle 5$

19. $\angle 1 \& \angle 5$; $\angle 4 \& \angle 8$

$\angle 2 \& \angle 6$; $\angle 3 \& \angle 7$

20. $\angle 1 \& \angle 3$; $\angle 2 \& \angle 4$

$\angle 5 \& \angle 7$; $\angle 6 \& \angle 8$

21.

$$m(\angle 2)=m(\angle 4)=m(\angle 6)=110°$$
$$m(\angle 1)+110°=180°$$
$$m(\angle 1)+110°-110°=180°-110°$$
$$m(\angle 1)=70°$$
$$m(\angle 1)=m(\angle 3)=m(\angle 5)=m(\angle 7)=70°$$

22.

$$m(\angle 4)=180°-70°-60°=50°$$
$$m(\angle 3)=180°-50°=130°$$
$$m(\angle 2)=180°-60°=120°$$
$$m(\angle 1)=180°-120°=60°$$

23. a. alternate interior angles:

$$2x-30°=x+10°$$
$$2x-30°+30°=x+10°+30°$$
$$2x=x+40°$$
$$2x-x=x+40°-x$$
$$x=40°$$

b. both are $40°+10°=50°$

24. a. The angles sum to $180°$.

$$3x+50°+4x-10°=180°$$
$$7x+40°=180°$$
$$7x+40°-40°=180°-40°$$
$$7x=140°$$
$$\frac{7x}{7}=\frac{140°}{7}$$
$$x=20°$$

b. $3(20°)+50°=60°+50°=110°$

$4(20°)-10°=80°-10°=70°$

25. a. The angles are congruent.

$$7x-46°=2x+9°$$
$$7x-46°+46°=2x+9°+46°$$
$$7x=2x+55°$$
$$7x-2x=2x+55°-2x$$
$$5x=55°$$
$$\frac{5x}{5}=\frac{55°}{5}$$
$$x=11°$$

b. both are $2(11°)+9°=22°+9°=31°$

26. a. alternate interior angles:

$$3x+13°=5x-33°$$
$$3x+13°-13°=5x-33°-13°$$
$$3x=5x-46°$$
$$3x-5x=-46°$$
$$-2x=-46°$$
$$x=23°$$

b. both are $3(23°)+13°=69°+13°=82°$

27. a. 8, octagon, 8 b. 5, pentagon, 5

c. 3, triangle, 3 d. 6, hexagon, 6

e. 4, quadrilateral, 4 f. 10, decagon, 10

28. a. isosceles b. scalene

c. equilateral d. isosceles

29. a. acute b. right

c. obtuse d. acute

30. a. $90°$ b. right c. $\overline{XY}$, $\overline{XZ}$

d. $\overline{YZ}$ e. $\overline{YZ}$ f. $\overline{YZ}$

31.

$$70°+20°+x=180°$$
$$90°+x=180°$$
$$90°+x-90°=180°-90°$$
$$x=90°$$

32.

$$70°+60°+x=180°$$
$$130°+x=180°$$
$$130°+x-130°=180°-130°$$
$$x=50°$$

33.

$$m(\angle A)+32°+77°=180°$$
$$m(\angle A)+109°=180°$$
$$m(\angle A)+109°-109°=180°-109°$$
$$m(\angle A)=71°$$

34.

$$2x+5x+26°+x+10°=180°$$
$$8x+36°=180°$$
$$8x+36°-36°=180°-36°$$
$$8x=144°$$
$$\frac{8x}{8}=\frac{144°}{8}$$
$$x=18°$$
$$2x=2(18°)=36°$$
$$5x+26°=5(18°)+26°=116°$$
$$x+10°=18°+10°=28°$$

35.

$$2(65°)+x=180°$$
$$130°+x=180°$$
$$130°+x-130°=180°-130°$$
$$x=50°$$

36.

$$2x+68°=180°$$
$$2x+68°-68°=180°-68°$$
$$2x=112°$$
$$\frac{2x}{2}=\frac{112°}{2}$$
$$x=56°$$

37.

$$2(56.5°)+m(\angle C)=180°$$
$$113°+m(\angle C)=180°$$
$$113°+m(\angle C)-113°=180°-113°$$
$$m(\angle C)=67°$$

38. $m(\angle DBA) = 180° - 81° - 47° = 52°$

$m(\angle EBC) = 180° - 52° - 50° = 78°$

$m(\angle ECB) = 180° - 78° - 19° = 83°$

39. $a^2 + b^2 = c^2$

$5^2 + 12^2 = c^2$

$25 + 144 = c^2$

$169 = c^2$

$\sqrt{169} = c$

$13cm = c$

40. $a^2 + b^2 = c^2$

$8^2 + 15^2 = c^2$

$64 + 225 = c^2$

$289 = c^2$

$\sqrt{289} = c$

$17ft = c$

41. $a^2 + b^2 = c^2$

$a^2 + 77^2 = 85^2$

$a^2 + 5{,}929 = 7{,}225$

$a^2 + 5{,}929 - 5{,}929 = 7{,}225 - 5{,}929$

$a^2 = 1{,}296$

$a = \sqrt{1{,}296}$

$a = 36in$

42. $a^2 + b^2 = c^2$

$21^2 + b^2 = 29^2$

$441 + b^2 = 841$

$441 + b^2 - 441 = 841 - 441$

$b^2 = 400$

$b = \sqrt{400}$

$b = 20ft$

43. $a^2 + 5^2 = 16^2$

$a^2 + 25 = 256$

$a^2 + 25 - 25 = 256 - 25$

$a^2 = 231$

$a = \sqrt{231}m \approx 15.20m$

44. $20^2 + 30^2 = c^2$

$400 + 900 = c^2$

$1300 = c^2$

$\sqrt{1300}in = c \approx 36.06in$

45. $48^2 + 55^2 = c^2$

$2{,}304 + 3{,}025 = c^2$

$5{,}329 = c^2$

$\sqrt{5{,}329} = c$

$73in = c$

46. $a^2 + 41^2 = 52^2$

$a^2 + 1{,}681 = 2{,}704$

$a^2 + 1{,}681 - 1{,}681 = 2{,}704 - 1{,}681$

$a^2 = 1{,}023$

$a = \sqrt{1{,}023}in \approx 32in$

47. $8^2 + 11^2 \stackrel{?}{=} 15^2$

$64 + 121 \stackrel{?}{=} 225$

$185 = 225$

This is not a right triangle.

48. $2^2 + 7^2 \stackrel{?}{=} 9^2$

$4 + 49 \stackrel{?}{=} 81$

$53 = 81$

This is not a right triangle.

49. a. $\angle D$ b. $\angle E$ c. $\angle F$

d. $\overline{DF}$ e. $\overline{DE}$ f. $\overline{EF}$

50. a. $m(\angle X)=m(\angle A)=32°$

b. $m(\angle B)=m(\angle Y)=61°$

c. $m(\overline{YZ})=m(\overline{BC})=6in$

d. $m(\overline{AB})=m(\overline{XY})=9in$

51. congruent: SSS

52. congruent: SAS

53. not necessarily congruent

54. congruent: ASA

55. yes

56. yes

57.

$$\frac{x}{16}=\frac{8}{32} \qquad \frac{y}{7}=\frac{32}{8}$$
$$32\cdot x=8\cdot 16 \qquad 8\cdot y=32\cdot 7$$
$$\frac{32x}{32}=\frac{128}{32} \qquad \frac{8y}{8}=\frac{224}{8}$$
$$x=4 \qquad y=28$$

58. t = the height of the tree

$$\frac{t}{26}=\frac{5}{2}$$
$$t\cdot 2=5\cdot 26$$
$$\frac{2t}{2}=\frac{130}{2}$$
$$t=65$$

The tree is 65 feet tall.

59. a. trapezoid b. square c. parallelogram

d. rectangle e. rhombus f. rectangle

60. We know that all 4 triangles are isosceles.

a. Since it is a rectangle, the diagonals are equal.

$$m(\overline{BD})=m(\overline{AC})=15cm$$

b. alternate interior angles:

$$m(\angle 1)=m(\angle BDC)=40°$$

c. sum of a triangle's angles:

$$m(\angle 2)+2(40°)=180°$$
$$m(\angle 2)+80°=180°$$
$$m(\angle 2)+80°-80°=180°-80°$$
$$m(\angle 2)=100°$$

d. The diagonals bisect.

$$m(\overline{EC})=\frac{1}{2}m(\overline{AC})=\frac{1}{2}(15)=7.5cm$$

e. It is a rectangle.

$$m(\overline{AB})=m(\overline{DC})=14cm$$

61. a. true b. true c. true d. false

62. a. isosceles trapezoid: $m(\angle B)=m(\angle A)=65°$

b. isosceles trapezoid: $m(\angle C)=m(\angle D)=115°$

c. isosceles trapezoid: $m(\overline{CB})=m(\overline{AD})=4yd$

63.

$$S=(n-2)180°$$
$$S=(8-2)180°$$
$$S=6\cdot 180°$$
$$S=1{,}080°$$

64.

$$S=(n-2)180°$$
$$3{,}240°=(n-2)180°$$
$$\frac{3{,}240°}{180°}=\frac{(n-2)180°}{180°}$$
$$18=n-2$$
$$18+2=n-2+2$$
$$20=n$$

65. $P=4l=4\cdot 18=72in$

66. $3ft = 36in$

$P = 2l + 2w$

$P = 2\cdot 36 + 2\cdot 7 = 72 + 14 = 86in$

67. $8+6+8+4+4 = 30m$

68. The upper missing side is $6+4=10m$

The right missing side is $8-4=4m$

$P = 8+10+4+4+4+6 = 36m$

69. $2(34)+b = 107$

$68 + b = 107$

$68 + b - 68 = 107 - 68$

$b = 39ft$

70. a. $1yd = 3ft \rightarrow 1yd^2 = (3ft)^2 = 3^2 ft^2 = 9ft^2$

b. $1ft = 12in \rightarrow 1ft^2 = (12in)^2$

$= 12^2 in^2 = 144in^2$

71. $A = s^2 = 3.1^2 = 9.61cm^2$

72. $A = l\cdot w = 50\cdot 150 = 7{,}500ft^2$

73. $A = b\cdot h = 30\cdot 15 = 450ft^2$

74. $A = \frac{1}{2}bh = \frac{1}{2}\cdot 40\cdot 10 = 20\cdot 10 = 200in^2$

75. $A = \frac{1}{2}h(b_1+b_2) = \frac{1}{2}\cdot 8(12+18)$

$= 4(30) = 120cm^2$

76. The missing vertical side is $14-8=6ft$

The area of the lower rectangle is $20\cdot 8 = 160ft^2$

The area of the upper rectangle is $12\cdot 6 = 72ft^2$

The area of the figure is $160+72 = 232ft^2$

77. Trapezoid: $\frac{1}{2}\cdot 8(12+20) = 4(32) = 128ft^2$

Triangle: $\frac{1}{2}\cdot 12\cdot 4 = 6\cdot 4 = 24ft^2$

Area of figure: $128+24 = 152ft^2$

78. The area of the parallelogram (if the missing triangle were solid) is $A = bh = 15\cdot 10 = 150m^2$. The area of the triangle is $A = \frac{1}{2}bh = \frac{1}{2}\cdot 15\cdot 4 = 30m^2$.

The shaded region has area $150-30 = 120m^2$.

79. $A = bh$

$240 = 30\cdot h$

$\frac{240}{30} = \frac{30h}{30}$

$8ft = h$

80. $P = 2l + 2w$

$48 = 2l + 2\cdot 6$

$48 = 2l + 12$

$48 - 12 = 2l + 12 - 12$

$36 = 2l$

$\frac{36}{2} = \frac{2l}{2}$

$18mm = l$

81. $P = 2l + 2w = 2\cdot 115 + 2\cdot 78$

$= 230 + 156 = 386ft$

$386(8.5) = \$3{,}281$

82. First convert to yards. 12 yards by 8 yards.

$A = lw = 12\cdot 8 = 96$

$96(48) = \$4{,}608$

83. a. $\overline{CD}$, $\overline{AB}$ b. $\overline{AB}$

c. $\overline{OA}$, $\overline{OC}$, $\overline{OD}$, $\overline{OB}$ d. O

84. $C = \pi d = \pi\cdot 21 = 21\pi ft \approx 65.97ft$

85. Outer circle: $C = \pi d = 8\pi cm \approx 25.1cm$

Total perimeter: $25.1 + 10 + 10 = 45.1cm$.

86. $d = 18 \rightarrow r = 9$

$A = \pi r^2 = \pi \cdot 9^2 = 81\pi in^2 \approx 254.47in^2$

87. The area of the circle (which has radius 4 centimeters) is $A = \pi r^2 = \pi \cdot 4^2 = 16\pi cm^2 \approx 50.3cm^2$. The remaining rectangle has area $l \cdot w = 8 \cdot 10 = 80cm^2$. The total area of the figure is approximately

$80 + 50.3 = 130.3cm^2$

88. Since the square is 100 inches wide, the radius of the semicircle is 50 inches, so its area is

$\frac{1}{2}\pi r^2 = \frac{1}{2}\pi \cdot 50^2 = 1,250\pi in^2 \approx 3,927in^2$.

The area of the square is $s^2 = 100^2 = 10,000in^2$, so the shaded area is

$10,000 - 3,927 = 6,073.0in^2$.

89. $V = s^3 = 5^3 = 125cm^3$

90. $V = lwh = 8 \cdot 10 \cdot 6 = 480m^3$

91. $B = \frac{1}{2} \cdot 12 \cdot 16 = 96mm^2$

$V = Bh = 96 \cdot 18 = 1,728mm^2$

92. $V = \frac{4}{3}\pi r^3 = \frac{4}{3}\pi \cdot 5^3 = \frac{4}{3}\pi \cdot 125$

$= \frac{500}{3}\pi in^3 \approx 523.60in^3$

93. $V = \frac{1}{3}\pi r^2 h = \frac{1}{3}\pi \cdot 5^2 \cdot 30$

$= 250\pi in^3 \approx 785.40in^3$

94. $B = 20 \cdot 20 = 400yd^2$

$V = \frac{1}{3}Bh = \frac{1}{3} \cdot 400 \cdot 15 = 2,000yd^3$

95. $B = \frac{1}{2} \cdot 12 \cdot 35 = 210m^2$

$A = \frac{1}{3}Bh = \frac{1}{3} \cdot 210 \cdot 42 = 2,940m^3$

96. The radius of the sphere is 8 inches.

$V = \frac{4}{3}\pi r^3 = \frac{4}{3}\pi \cdot 8^3 = \frac{4}{3}\pi \cdot 512 = \frac{2,048}{3}\pi in^3$.

Since we only have half the sphere,

$V = \frac{1}{2} \cdot \frac{2,048}{3}\pi = \frac{1,024}{3}\pi in^3 \approx 1,072.33in^3$

97. Combine half a sphere (of radius 5 feet) with a cylinder.

$V = \frac{1}{2} \cdot \frac{4}{3}\pi \cdot 5^3 + \pi \cdot 5^2 \cdot 16$

$= \frac{250}{3}\pi + 400\pi$

$\approx 1,518 ft^3$

98. Note that the radius of the waffle cone is 1.25 inches.

$V = \frac{1}{3}\pi r^2 h = \frac{1}{3}\pi (1.25)^2 \cdot 6$

$= 3.125\pi in^3 \approx 9.8in^3$

99. $1ft = 12in$

$1ft^3 = (12in)^3 = 12^3 in^3 = 1,728in^3$

100. $1yd = 3ft$

$1yd^3 = (3ft)^3 = 3^3 ft^3 = 27ft^3$

$2yd^3 = 2 \cdot 27ft^3 = 54ft^3$

Chapter 9 Test

1. a. $135°$; obtuse b. $90°$; right

c. $40°$; acute d. $180°$; straight

2. a. If $\angle ABC \cong \angle DEF$, then the angles have the same measure.

b. Two congruent segments have the same length.

c. Two different points determine one line.

d. Two angles are called complementary if the sum of their measures is $90°$.

3. *D*

4. a. false b. true c. true

d. true e. false

5. Vertical angles have the same measure.

$3x = 2x + 20°$

$3x - 2x = 2x + 20° - 2x$

$x = 20°$

$m(\angle ABD) = m(\angle CBE) = 3x = 3(20°) = 60°$

6. $x + 47° = 180°$

$x + 47° - 47 = 180° - 47° = 133°$

7. a. l_1 intersects two coplanar lines. It is called a transversal.

b. $\angle 4$ and $\angle 6$ are alternate interior angles.

c. $\angle 3$ and $\angle 7$ are corresponding angles.

8. $m(\angle 2) = m(\angle 4) = m(\angle 5) = m(\angle 7) = 25°$

$m(\angle 1) + 25° = 180°$

$m(\angle 1) + 25° - 25° = 180° - 25°$

$m(\angle 1) = 155°$

$m(\angle 1) = m(\angle 3) = m(\angle 6) = m(\angle 8) = 155°$

9. The two angles sum to $180°$.

$x + 20° + 2x + 10° = 180°$

$3x + 30° = 180°$

$3x + 30° - 30° = 180° - 30°$

$3x = 150°$

$\frac{3x}{3} = \frac{150°}{3}$

$x = 50°$

$x + 20° = 50° + 20° = 70°$

$2x + 10° = 2(50°) + 10° = 110°$

10. a. 8 octagon, 8 b. 5, pentagon, 5

c. 6, hexagon, 6 d. 4, quadrilateral, 4

11. a. isosceles b. scalene

c. equilateral d. isosceles

12. $90° + 20° + x = 180°$

$110° + x = 180°$

$110° + x - 110° = 180° - 110°$

$x = 70°$

13. $2b + 12° = 180°$

$2b + 12° - 12° = 180° - 12°$

$2b = 168°$

$b = 84°$

14. a. $m(\overline{HG}) = m(\overline{EF}) = 12$

b. $m(FH) = 2 \cdot m(Ex) = 2 \cdot 6.5 = 13$

c. $m(\angle FGH) = 90°$

d. $m(EH) = m(FG) = 5$

15. a. $m(\overline{RS}) = m(\overline{QT}) = 10$

b. $x = m(\angle T) = 65°$

c. $y = 180° - x = 180° - 65° = 115°$

d. $z = y = 115°$

16. $S=(n-2)180°$

$S=(10-2)180°=8\cdot 180°=1{,}440°$

17. $25+36+37+48+42=188in$

18. $3x=45.6$

$\frac{3x}{3}=\frac{45.6}{3}$

$x=15.2m$

19. Area of rectangle: $lw=25\cdot 16=400cm^2$

Area of triangle: $\frac{1}{2}bh=\frac{1}{2}\cdot 10\cdot 8=40cm^2$

Area shaded: $400-40=360cm^2$

20. Convert to yards: top = $\frac{20}{3}yd$, bottom = $4yd$, side = $9yd$

$A=\frac{1}{2}h(b_1+b_2)=\frac{1}{2}\cdot 9\left(4+\frac{20}{3}\right)$

$=\frac{9}{2}\left(\frac{12}{3}+\frac{20}{3}\right)=\frac{9}{2}\cdot\frac{32}{3}=48yd^2$

$48\cdot 18=\$864$

21. $1ft=12in\rightarrow 1ft^2=(12in)^2=12^2in^2=144in^2$

22. Convert from feet to inches first.

10 ft = 10·12=120 in

$A=l\cdot w=120\cdot 1=120in^2$

23. a. $\overline{RS}$, $\overline{XY}$ b. $\overline{XY}$

c. $\overline{OX}$, $\overline{OR}$, $\overline{OS}$, $\overline{OY}$

24. If C is the circumference of a circle and D is the length of its diameter, then $\frac{C}{D}=\pi$.

25. $C=\pi d=\pi\cdot 21=21\pi ft\approx 66.0ft$

26. The outer circle has circumference $C=\pi d=12\pi cm$. The total perimeter, then, is $12\pi+20+20=(40+12\pi)ft\approx 77.7ft$

27. Since the diameter is 30 meters, the radius is 15 meters.

$A=\pi r^2=\pi\cdot 15^2=225\pi m^2\approx 706.9m^2$.

28. $\angle M\cong\angle R$ $\overline{MO}\cong\overline{RT}$

$\angle N\cong\angle S$ $\overline{MN}\cong\overline{RS}$

$\angle O\cong\angle T$ $\overline{NO}\cong\overline{ST}$

29. a. congruent: SSS b. congruent: ASA

c. not necessarily congruent d. congruent: SAS

30. a. $m(\overline{DE})=m(\overline{BA})=8in$

b. $m(\angle E)=m(\angle B)=50°$

31. a. yes b. yes

32. $\frac{x}{9}=\frac{4}{6}$ $\frac{y}{8}=\frac{6}{4}$

$x\cdot 6=4\cdot 9$ $y\cdot 4=6\cdot 8$

$\frac{6x}{6}=\frac{36}{6}$ $\frac{4y}{4}=\frac{48}{4}$

$x=6m$ $y=12m$

33. $\frac{t}{7}=\frac{6}{2}$

$t\cdot 2=6\cdot 7$

$\frac{2t}{2}=\frac{42}{2}$

$t=21ft$

34. a. $a^2+b^2=c^2$

$10^2+24^2=c^2$

$100+576=c^2$

$676=c^2$

$\sqrt{676}=c$

$26cm=c$

b. $a^2+b^2=c^2$

$6^2+b^2=8^2$

$36+b^2=64$

$36+b^2-36=64-36$

$b^2=28$

$b=\sqrt{28}in\approx 5.3in$

35. $25^2+19^2=d^2$

$625+361=d^2$

$986=d^2$

$\sqrt{986}in=d\approx 31.4in$

36. $1ft=12in\rightarrow 1ft^3=(12in)^3=12^3in^3=1,728in^3$

37. $V=s^3=6^3=216m^3$

38. $V=lwh=10\cdot 8\cdot 6=480m^3$

39. $V=\frac{1}{3}\pi r^2h=\frac{1}{3}\pi\cdot 12^2\cdot 27$

$=1,296\pi\, in^3\approx 4,071.50in^3$

40. $B=30in^2$

$V=Bh=30\cdot 20=600in^3$

41. $B=\frac{1}{2}\cdot 20\cdot 21=210ft^2$

$V=\frac{1}{3}Bh=\frac{1}{3}\cdot 210\cdot 27=1,890ft^3$

42. $V=\pi r^2h=\pi\cdot 3^2\cdot 7=63\pi\, yd^3\approx 197.92yd^3$

43. $B=10\cdot 10=100yd^2$

$V=\frac{1}{3}Bh=\frac{1}{3}\cdot 100\cdot 12=400mi^3$

44. $V=\frac{4}{3}\pi r^3=\frac{4}{3}\pi\cdot 4^3=\frac{4}{3}\pi\cdot 64$

$=\frac{256}{3}\pi\, in^3\approx 268.08in^3$

45. Combine half a sphere (of radius 15 feet) with a cylinder.

$V=\frac{1}{2}\cdot\frac{4}{3}\pi\cdot 15^3+\pi\cdot 15^2\cdot 40$

$=2,250\pi+9,000\pi$

$=11,250\pi\, ft^3\approx 35,343ft^3$

46. The length of a fence enclosing a lawn is the perimeter of the lawn and is measured in feet, yards, meters, etc.

The amount of carpet needed to cover a floor is the area of the floor and is measured in square feet, square meters, etc.

The amount of space in the refrigerator is its volume and is measured in cubic feet, cubic yards, etc.

Chapters 1 – 9 Cumulative Review

1. one hundred four million, fifty-two thousand, five

2.
$$\begin{array}{r} {}^{1}\ \ {}^{1}{}^{1}\ \ \\ 99,085 \\ 4,101 \\ 257 \\ +\quad 33 \\ \hline 103,476 \end{array}$$

3.
$$\begin{array}{r} 196 \\ \times\quad 78 \\ \hline 1,568 \\ +13,720 \\ \hline 15,288 \end{array}$$

4.
$$\begin{array}{r} 59 \\ 34\overline{)2006} \\ -170 \\ \hline 306 \\ -306 \\ \hline 0 \end{array}$$

5. a. 1, 2, 4, 8, 16, 32

b. $140 = 14 \cdot 10 = 2 \cdot 7 \cdot 2 \cdot 5 = 2^2 \cdot 5 \cdot 7$

6.
$$35 = 5 \cdot 7$$
$$45 = 3 \cdot 3 \cdot 5$$
$$LCM(35,45) = 3 \cdot 3 \cdot 5 \cdot 7 = 315$$
$$GCF(35,45) = 5$$

7.
$$46 + 3[5^2 - 4(9-5)]$$
$$= 46 + 3[25 - 4(4)] = 46 + 3[25 - 16]$$
$$= 46 + 3(9) = 46 + 27 = 73$$

8. a. $x + 33 = 91$
$$x + 33 - 33 = 91 - 33$$
$$x = 58$$
Check: $x + 33 = 91$
$$58 + 33 \stackrel{?}{=} 91$$
$$91 = 91$$

b. $24m = 312$
$$\frac{24m}{24} = \frac{312}{24}$$
$$m = 13$$
Check: $24m = 312$
$$24(13) \stackrel{?}{=} 312$$
$$312 = 312$$

9. true

10. a. $-17 + 8 = -9$

b. $-14 - (-2) = -14 + 2 = -12$

c. $-3(72) = -216$

d. $\frac{-60}{0}$ is undefined.

11.
$$\frac{2+3[5-(1-10)]}{\left|2[(-2)^3+2]+10\right|} = \frac{2+3[5-(-9)]}{\left|2[-8+2]+10\right|}$$
$$= \frac{2+3[5+9]}{\left|2[-6]+10\right|} = \frac{2+3[14]}{\left|-12+10\right|} = \frac{2+42}{\left|-2\right|}$$
$$= \frac{44}{2} = 22$$

12.
$$-2(5) = \frac{y}{-3} + 3$$
$$-10 = \frac{y}{-3} + 3$$
$$-10 - 3 = \frac{y}{-3} + 3 - 3$$
$$-13 = \frac{y}{-3}$$
$$-3(-13) = -3 \cdot \left(\frac{y}{-3}\right)$$
$$39 = y$$
Check: $-2(5) = \frac{y}{-3} + 3$
$$-2(5) \stackrel{?}{=} \frac{39}{-3} + 3$$
$$-10 \stackrel{?}{=} -13 + 3$$
$$-10 = -10$$

13. b = the original balance
$$b + 55 + 80 = -82$$
$$b + 135 = -82$$
$$b + 135 - 135 = -82 - 135$$
$$b = -217$$
Before the deposit, her balance was -$217.

14. $2w - 18$

15.
$$b^2 - 4ac = (-5)^2 - 4(-1)(-2)$$
$$= 25 - 8 = 17$$

16. $-4(6u)(-2) = [(-4)6(-2)u] = 48u$

17.
$$2x + 2(x+4) = 2x + 2x + 8$$
$$= (4x + 8)\ ft$$

18.
$$6(2j+6)+4j=4(j-30)$$
$$12j+36+4j=4j-120$$
$$16j+36=4j-120$$
$$16j+36-36=4j-120-36$$
$$16j=4j-156$$
$$16j-4j=4j-156-4j$$
$$12j=-156$$
$$\frac{12j}{12}=\frac{-156}{12}$$
$$j=-13$$

19. r = the amount of the refund
$$\frac{1}{2}r-250=395$$
$$\frac{1}{2}r-250+250=395+250$$
$$\frac{1}{2}r=645$$
$$2\cdot\frac{1}{2}r=2\cdot645$$
$$r=1,290$$
The refund was \$1,290.

20.
$$\frac{60x^5}{108x^3}=\frac{5\cdot12\cdot x\cdot x\cdot x\cdot x\cdot x}{9\cdot12\cdot x\cdot x\cdot x}$$
$$=\frac{5\cdot\cancel{12}\cdot\cancel{x}\cdot\cancel{x}\cdot\cancel{x}\cdot x\cdot x}{9\cdot\cancel{12}\cdot\cancel{x}\cdot\cancel{x}\cdot\cancel{x}}=\frac{5x^2}{9}$$

21.
$$\frac{4}{5}\cdot\frac{15}{16}=\cdot\frac{\cancel{4}\cdot3\cdot\cancel{5}}{\cancel{5}\cdot\cancel{4}\cdot4}=\frac{3}{4}$$

22.
$$-\frac{7a}{4b^2}\div\left(-\frac{21a}{8b}\right)=\frac{7a}{4b^2}\cdot\frac{8b}{21a}$$
$$=\frac{\cancel{7}\cdot\cancel{a}\cdot2\cdot\cancel{4}\cdot\cancel{b}}{\cancel{4}\cdot\cancel{b}\cdot b\cdot3\cdot\cancel{7}\cdot\cancel{a}}=\frac{2}{3b}$$

23.
$$\frac{4}{n}-\frac{3}{4}=\frac{4}{n}\cdot\frac{4}{4}-\frac{3}{4}\cdot\frac{n}{n}$$
$$=\frac{16}{4n}-\frac{3n}{4n}=\frac{16-3n}{4n}$$

24.
$$\frac{7}{10}+\frac{3}{14}=\frac{7}{10}\cdot\frac{7}{7}+\frac{3}{14}\cdot\frac{5}{5}=\frac{49}{70}+\frac{15}{70}$$
$$=\frac{64}{70}=\frac{\cancel{2}\cdot32}{\cancel{2}\cdot35}=\frac{32}{35}$$

25.
$$A=lw=6\left(7\frac{1}{2}\right)=6\left(\frac{15}{2}\right)=3(15)=45in^2$$

26.
$$\begin{array}{r} 140\frac{5}{6} \\ -129\frac{4}{5} \\ \hline \end{array}\qquad \begin{array}{r} 140\frac{5}{6}\cdot\frac{5}{5} \\ -129\frac{4}{5}\cdot\frac{6}{6} \\ \hline \end{array}\qquad \begin{array}{r} 140\frac{25}{30} \\ -129\frac{24}{30} \\ \hline 11\frac{1}{30} \end{array}$$

27.
$$\frac{-\frac{1}{2}+\frac{7}{8}}{\frac{3}{4}-\frac{1}{2}}=\frac{-\frac{1}{2}\cdot8+\frac{7}{8}\cdot8}{\frac{3}{4}\cdot8-\frac{1}{2}\cdot8}$$
$$=\frac{-4+7}{6-4}=\frac{3}{2}$$

28. x = the number of cars serviced
$$\frac{2}{5}x=140$$
$$\frac{5}{2}\cdot\frac{2}{5}x=\frac{5}{2}\cdot140$$
$$x=5(70)=350$$
The shop serviced 350 cars last year.

29.
$$-\frac{5}{6}x=10$$
$$-\frac{6}{5}\left(-\frac{5}{6}x\right)=-\frac{6}{5}(10)$$
$$x=-6(2)$$
$$x=-12$$

Check: $-\frac{5}{6}x=10$
$$-\frac{5}{6}(-12)\stackrel{?}{=}10$$
$$-5(-2)\stackrel{?}{=}10$$
$$10=10$$

30.
$$\frac{12}{5}+\frac{n}{2}=\frac{n}{10}$$
$$10\cdot\frac{12}{5}+10\cdot\frac{n}{2}=10\cdot\frac{n}{10}$$
$$24+5n=n$$
$$24+5n-5n=n-5n$$
$$24=-4n$$
$$\frac{24}{-4}=\frac{-4n}{-4}$$
$$-6=n$$

Check: $\frac{12}{5}+\frac{n}{2}=\frac{n}{10}$
$$\frac{12}{5}+\frac{-6}{2}\stackrel{?}{=}\frac{-6}{10}$$
$$\frac{12}{5}-3\stackrel{?}{=}-\frac{3}{5}$$
$$\frac{12}{5}-\frac{15}{5}\stackrel{?}{=}-\frac{3}{5}$$
$$-\frac{3}{5}=-\frac{3}{5}$$

31. 9.510

32. $361.5+89.8=451.3$ The answer is correct.

33. $A=lw=(2.72)(3.81)=10.3632\,ft^2$

34. $\frac{14.637}{4.1}=3.57$ so $\frac{14.637}{-4.1}=-3.57$

35.
$$\begin{array}{r} 0.9166 \\ 12\overline{)11.0000} \\ \underline{-108} \\ 20 \\ \underline{-12} \\ 80 \\ \underline{-72} \\ 80 \end{array}$$
$$\frac{11}{12}=0.91\overline{6}$$

36.
$$-4\sqrt{81}+2\sqrt{4}=-4\cdot 9+2\cdot 2$$
$$=-36+4=-32$$

37.
$$3(y-1.1)+3.2=2.3$$
$$3y-3.3+3.2=2.3$$
$$3y-0.1=2.3$$
$$3y-0.1+0.1=2.3+0.1$$
$$3y=2.4$$
$$\frac{3y}{3}=\frac{2.4}{3}$$
$$y=0.8$$

Check: $3(y-1.1)+3.2=2.3$
$$3(0.8-1.1)+3.2\stackrel{?}{=}2.3$$
$$3(-0.3)+3.2\stackrel{?}{=}2.3$$
$$-0.9+3.2\stackrel{?}{=}2.3$$
$$2.3=2.3$$

38.
$$\frac{21in}{3ft}=\frac{21in}{36in}=\frac{3\cdot 7in}{3\cdot 12in}=\frac{\cancel{3}\cdot 7\cancel{in}}{\cancel{3}\cdot 12\cancel{in}}=\frac{7}{12}$$

39.
$$\frac{1\frac{1}{2}}{15}=\frac{x}{260}$$
$$\frac{3}{2}(260)=15x$$
$$390=15x$$
$$\frac{390}{15}=\frac{15x}{15}$$
$$26=x$$

They need 26 cups of flour.

40.

a. $\frac{50lb}{1}\cdot\frac{16oz}{1lb}=50\cdot 16oz=800oz$

b. $\frac{500mL}{1}\frac{1L}{1000mL}=\frac{500}{1000}L=0.5L$

c. $\frac{35cm}{1}\cdot\frac{1in}{2.54cm}=\frac{35}{2.54}in\approx 13.8in$

41.
$$\frac{27}{50}=\frac{27}{50}\cdot\frac{2}{2}=\frac{54}{100}=0.54=54\%$$

42. $36 = 0.24x$

$\dfrac{36}{0.24} = \dfrac{0.24x}{0.24}$

$150 = x$

43. $0.043(196.56) \approx \$8.45$

The total price of the purchase was $\$196.56 + \$8.45 = \$205.01$.

44. 10% of 7,000 is 700. Twice that gives 20%, which is 1,400.

45. $I = P \cdot r \cdot t = 1{,}250 \cdot (0.10) \cdot (0.25) = \$31.25$

46. mean: $\dfrac{73 + 5 + \cdots + 45}{9} = \dfrac{315}{9} = 35$

median: 5 11 27 35 **36** 38 45 45 73 = 36

mode: 45

range: $73 - 5 = 68$

47. A: $(2,3)$ B: $(-3,4)$ C: $(-3,-4)$

D: $(4,-4)$ E: $(0,0)$ F: $(4,0)$

G: $(-4,0)$ H: $(1.5,-2.5)$

48. $x + 5y = -7$

$9 + 5(-3) \stackrel{?}{=} -7$

$9 - 15 \stackrel{?}{=} -7$

$-6 = -7$

$(9,-3)$ is not a solution.

49.

$x = 0$	$x = -2$
$y = 3 \cdot 0 + 1$	$y = 3(-2) + 1$
$y = 0 + 1$	$y = -6 + 1$
$y = 1$	$y = -5$
$(0,1)$	$(-2,-5)$

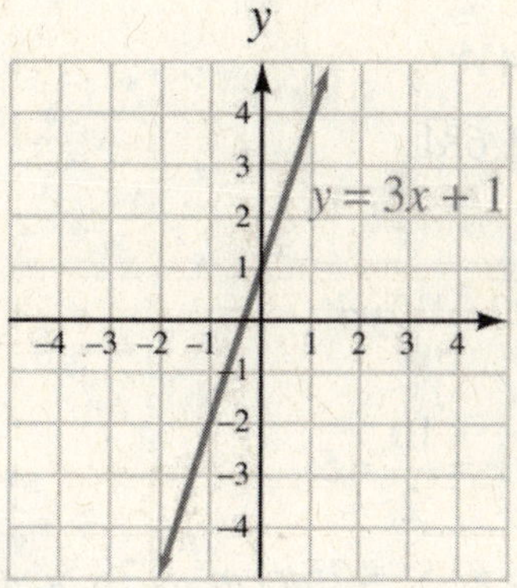

50.

$x = 0$	$y = 0$
$5(0) + 15y = -15$	$5x + 15(0) = -15$
$15y = -15$	$5x = -15$
$y = -1$	$x = -3$
$(0,-1)$	$(-3,0)$

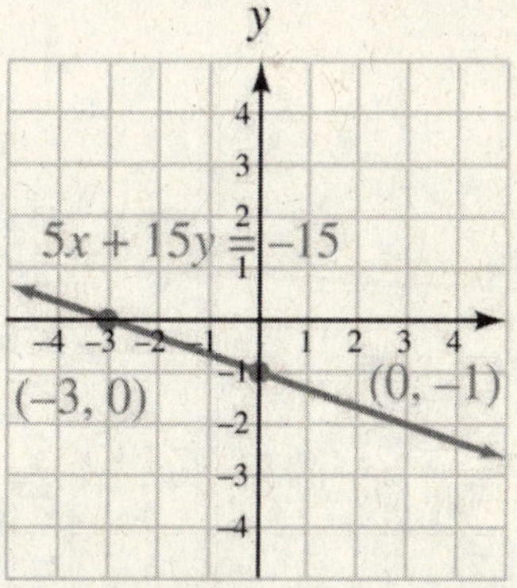

51. Vertical angles:

$6x + 8° = 4x + 32°$

$6x + 8° - 4x = 4x + 32° - 4x$

$2x + 8° = 32°$

$2x + 8° - 8° = 32° - 8°$

$2x = 24°$

$\dfrac{2x}{2} = \dfrac{24°}{2}$

$x = 12°$

$m(\angle ABD) = 6(12°) + 8° = 80°$

$m(\angle DBE) = 180° - m(\angle ABD)$

$= 180° - 80° = 100°$

52. $2(76°)+x=180°$

$152°+x=180°$

$152°+x-152°=180°-152°$

$x=28°$

53. $a^2+b^2=c^2$

$9^2+b^2=41^2$

$81+b^2=1{,}681$

$b^2=1{,}600$

$b=\sqrt{1{,}600}=40\,yd$

54. $d=10 \rightarrow r=\frac{10}{2}=5$

$A=\pi r^2=\pi\cdot 5^2=25\pi\ cm^2\approx 78.5cm^2$

CHAPTER 10 Exponents and Polynomials

10.1: Multiplication Rules for Exponents

VOCABULARY

1. Expressions such as $x^4, 10^3, \text{and} (5t)^2$ are called <u>exponential</u> expressions.

CONCEPTS

3. a. $(3x)^4 = 3x \cdot 3x \cdot 3x \cdot 3x$

   b. $(-5y)(-5y)(-5y) = (-5y)^3$

5. a. add b. multiply c. multiply

7. a. $x^2 + x^2 = 2x^2$ b. $x^2 \cdot x^2 = x^{2+2} = x^4$

9. a. doesn't simplify b. $x^3 \cdot x^2 = x^{3+2} = x^5$

NOTATION

11. $(x^4x^2)^3 = (x^6)^3 = x^{18}$

GUIDED PRACTICE

13. base: 4 ; exponent 3

15. base x ; exponent 5

17. base $-3x$; exponent 2

19. base y ; exponent 6

21. base m ; exponent 12

23. base $y+9$; exponent 4

25. m^5

27. $(4t)^4$

29. $4t^5$

31. a^2b^3

33. $5^3 \cdot 5^4 = 5^{3+4} = 5^7$

35. $a^3 \cdot a^3 = a^{3+3} = a^6$

37. $bb^2b^3 = b^{1+2+3} = b^6$

39. $(c^5)(c^8) = c^{5+8} = c^{13}$

41. $(a^2b^3)(a^3b^3) = a^2a^3b^3b^3$
$= a^{2+3}b^{3+3} = a^5b^6$

43. $cd^4 \cdot cd = ccd^4d = c^{1+1}d^{4+1} = c^2d^5$

45. $x^2 \cdot y \cdot x \cdot y^{10} = x^{2+1}y^{1+10} = x^3y^{11}$

47. $m^{100} \cdot m^{100} = m^{100+100} = m^{200}$

49. $(3^2)^4 = 3^{2\cdot 4} = 3^8$

51. $\left[(-4.3)^3\right]^8 = (-4.3)^{3\cdot 8} = (-4.3)^{24}$

53. $(m^{50})^{10} = m^{50\cdot 10} = m^{500}$

55. $(y^5)^3 = y^{5\cdot 3} = y^{15}$

57. $(x^2x^3)^5 = (x^5)^5 = x^{25}$

59. $(p^2p^3)^5 = (p^5)^5 = p^{25}$

61. $(t^3)^4(t^2)^3 = t^{12}t^6 = t^{18}$

63. $(u^4)^2(u^3)^2 = u^8u^6 = u^{14}$

65. $(6a)^2 = 6^2a^2 = 36a^2$

67. $(5y)^4 = 5^4y^4 = 625y^4$

69. $(3a^4b^7)^3 = 3^3(a^4)^3(b^7)^3 = 27a^{12}b^{21}$

71. $(-2r^2s^3)^3 = (-2)^3(r^2)^3(s^3)^3 = -8r^6s^9$

73. $(2c^3)^3(3c^4)^2 = 2^3(c^3)^3 \cdot 3^2(c^4)^2$
$= 8c^9 \cdot 9c^8 = 8 \cdot 9 \cdot c^9 \cdot c^8 = 72c^{17}$

75. $\left(10d^7\right)^2\left(4d^9\right)^3=10^2\left(d^7\right)^2\cdot 4^3\left(d^9\right)^3$
$=100d^{14}\cdot 64d^{27}=100\cdot 64\cdot d^{14}\cdot d^{27}$
$=6,400d^{41}$

TRY IT YOURSELF

77. $\left(7a^9\right)^2=7^2\left(a^9\right)^2=49a^{18}$

79. $t^4\cdot t^5\cdot t=t^{4+5+1}=t^{10}$

81. $y^3y^2y^4=y^{3+2+4}=y^9$

83. $\left(-6a^3b^2\right)^3=\left(-6\right)^3\left(a^3\right)^3\left(b^2\right)^3=-216a^9b^6$

85. $\left(n^4n\right)^3\left(n^3\right)^6=\left(n^5\right)^3\left(n^3\right)^6=n^{15}n^{18}=n^{33}$

87. $\left(b^2b^3\right)^{12}=\left(b^5\right)^{12}=b^{60}$

89. $\left(2b^4b\right)^5\left(3b\right)^2=\left(2b^5\right)^5\left(3b\right)^2$
$=2^5\left(b^5\right)^5 3^2b^2=32b^{25}\cdot 9b^2$
$=32\cdot 9\cdot b^{25}\cdot b^2=288b^{27}$

91. $\left(c^2\right)^3\left(c^4\right)^2=c^6\cdot c^8=c^{14}$

93. $\left(3s^4t^3\right)^3\left(2st\right)^4=3^3\left(s^4\right)^3\left(t^3\right)^3\cdot 2^4s^4t^4$
$=27s^{12}t^9\cdot 16s^4t^4=27\cdot 16\cdot s^{12}\cdot s^4\cdot t^9\cdot t^4$
$=432s^{16}t^{13}$

95. $x\cdot x^2\cdot x^3\cdot x^4\cdot x^5=x^{1+2+3+4+5}=x^{15}$

APPLICATIONS

97. $\left(5x\right)^2=5^2x^2=25x^2\ ft^2$

WRITING

99. When multiplying two factors with the same base, keep the original base.

REVIEW

101. $\dfrac{18}{24}=\dfrac{3\cdot \not{6}}{4\cdot \not{6}}=\dfrac{3}{4}$

103. $\dfrac{-25}{-5}=\dfrac{5\cdot \not{5}}{\not{5}}=5$

105. $2\left(\dfrac{12}{-3}\right)+3\left(5\right)=2\left(-4\right)+15=-8+15=7$

107. $-x=-12$
$\dfrac{-x}{-1}=\dfrac{-12}{-1}$
$x=12$

10.2 Introduction to Polynomials

VOCABULARY

1. A <u>polynomial</u> is a single term or a sum of terms in which all variables have whole number exponents and no variable appears in the denominator.

3. A polynomial with three terms is called a <u>trinomial</u>.

CONCEPTS

5. a. 2 b. 1 c. 3

7. The degree of a polynomial is equal to the <u>highest</u> degree of any term of the polynomial.

NOTATION

9. $3a^2+2a-7=3\left(2\right)^2+2\left(2\right)-7$
$=3\left(4\right)+4-7$
$=12+4-7$
$=16-7$
$=9$

GUIDED PRACTICE

11. binomial

13. monomial

15. monomial

17. trinomial

19. binomial

21. trinomial

23. 3

25. 2

27. 1

29. 7

31. 2

33. 4

35. $3x+4=3(3)+4=9+4=13$

37. $2x^2+42(-1)^2+4=2(1)+4=2+4=6$

39. $\frac{1}{2}x-3=\frac{1}{2}(-6)-3=-3-3=-6$

41. $0.5t^3-1=0.5\cdot4^3-1=0.5\cdot64-1$
$=32-1=31$

43. $\frac{2}{3}b^2-b+1=\frac{2}{3}\cdot3^2-3+1$
$=6-3+1=4$

45. $-2s^2-2s+1=-2(-1)^2-2(-1)+1$
$=-2\cdot1+2+1=-2+2+1=0+1=1$

APPLICATIONS

47. $h=-16t^2+64t=-16\cdot0^2+64\cdot0$
$=-0+0=0\,ft$

49. $h=-16t^2+64t=-16\cdot2^2+64\cdot2$
$=-64+128=64\,ft$

51. $d=0.04v^2+0.9v=0.04\cdot30^2+0.9\cdot30$
$=36+27=63\,ft$

53. $d=0.04v^2+0.9v=0.04\cdot60^2+0.9\cdot60$
$=144+54=198\,ft$

WRITING

55. To find the degree of the polynomial look at the degrees of the individual terms and then take the largest of those numbers.

REVIEW

57. $\frac{2}{3}+\frac{4}{3}=\frac{6}{3}=2$

59. $\frac{5}{12}\cdot\frac{18}{5}=\frac{\cancel{5}\cdot3\cdot\cancel{6}}{2\cdot\cancel{6}\cdot\cancel{5}}=\frac{3}{2}=1\frac{1}{2}$

61. $x-4=12$
$x-4+4=12+4$
$x=16$

63. $2(x-3)=6$
$2x-6=6$
$2x-6+6=6+6$
$2x=12$
$\frac{2x}{2}=\frac{12}{2}$
$x=6$

10.3 Adding and Subtracting Polynomials

VOCABULARY

1. If two algebraic terms have exactly the same variables and exponents, they are called <u>like</u> terms.

CONCEPTS

3. To add two monomials, we add the <u>coefficients</u> and keep the same <u>variables</u> and exponents.

5. yes, $3y+4y=7y$

7. no

9. yes, $3x^3+4x^3+6x^3=13x^3$

11. yes, $-5x^2+13x^2+7x^2=15x^2$

NOTATION

13. $(3x^2+2x-5)+(2x^2-7x)$
$=(3x^2+2x^2)+(2x-7x)+(5)$
$=5x^2+(-5x)+5=5x^2-5x+5$

GUIDED PRACTICE

15. $4y+5y=9y$

17. $8t^2+4t^2=12t^2$

19. $\frac{1}{8}a+\frac{3}{8}a+\frac{5}{8}a=\frac{1+3+5}{8}a=\frac{9}{8}a$

21. $\frac{2}{3}c^2 + \frac{1}{3}c^2 + \frac{2}{3}c^2 = \frac{2+1+2}{3}c^2 = \frac{5}{3}c^2$

23. $(3x+7)+(4x-3) = (3x+4x)+(7-3)$
$= 7x+4$

25. $(2x^2+3)+(5x^2-10)$
$= (2x^2+5x^2)+(3-10) = 7x^2-7$

27. $(5x^3-42x)+(7x^3-107x)$
$= (5x^3+7x^3)+(-42x-107x)$
$= 12x^3-149x$

29. $(3x^2+2x-4)+(5x^2-17)$
$= (3x^2+5x^2)+2x+(-4-17)$
$= 8x^2+2x-21$

31. $(2.5a^2+3a-9)+(3.6a^2+7a-10)$
$= (2.5a^2+3.6a^2)+(3a+7a)+(-9-10)$
$= 6.1a^2+10a-19$

33. $(3n^2-5.8n+7)+(-n^2+5.8n-2)$
$= (3n^2-n^2)+(-5.8n+5.8n)+(7-2)$
$= 2n^2+5$

35.
$$\begin{array}{r} 3x^2+4x+\ 5 \\ +2x^2-3x+\ 6 \\ \hline 5x^2+\ x+11 \end{array}$$

37.
$$\begin{array}{r} -3x^2 \qquad -7 \\ +\ -4x^2-5x+6 \\ \hline -7x^2-5x-1 \end{array}$$

39. $32u^3-16u^3 = 16u^3$

41. $18x^5-11x^5 = 7x^5$

43. $(4.5a+3.7)-(2.9a-4.3)$
$= 4.5a+3.7-2.9a+4.3 = 1.6a+8$

45. $(7.2x^2-3.1x)-(9.4x^2+6.8x)$
$= 7.2x^2-3.1x-9.4x^2-6.8x = -2.2x^2-9.9x$

47. $(2b^2+3b-5)-(2b^2-4b-9)$
$= 2b^2+3b-5-2b^2+4b+9 = 7b+4$

49. $(5p^2-p+71)-(4p^2+p+71)$
$= 5p^2-p+71-4p^2-p-71 = p^2-2p$

51. $(3x^2+4x-5)-(-2x^2-2x+3)$
$= 3x^2+4x-5+2x^2+2x-3 = 5x^2+6x-8$

53. $(-2x^2-4x+12)-(10x^2+9x-24)$
$= -2x^2-4x+12-10x^2-9x+24$
$= -12x^2-13x+36$

TRY IT YOURSELF

55. $(30x^2-4)-(11x^2+1)$
$= 30x^2-4-11x^2-1 = 19x^2-5$

57. $(7y^2+5y)+(y^2-y-2)$
$= 8y^2+4y-2$

59. $(3x^2-3x-2)+(3x^2+4x-3)$
$= 6x^2+x-5$

61. $(m^2-m-5)-(m^2+5.5m-75)$
$= m^2-m-5-m^2-5.5m+75$
$= -6.5m+70$

63. $(t^2-4.5t+5)-(2t^2-3.1t-1)$
$= t^2-4.5t+5-2t^2+3.1t+1$
$= -t^2-1.4t+6$

65.
$$\begin{array}{r} -3x^2+4x+25.4 \\ +\quad 5x^2-3x-12.5 \\ \hline 2x^2+\ x+12.9 \end{array}$$

67. $3s^2+4s^2+7s^2 = 14s^2$

68. $-2a^3+7a^3-3a^3 = 2a^3$

69. $\frac{1}{3}b^4+\frac{2}{3}b^4-\frac{5}{3}b^4=\frac{1+2-5}{3}b^4=-\frac{2}{3}b^4$

71.
$$\begin{array}{r} z^3+6z^2-7z+16 \\ +\ \underline{9z^3-6z^2+8z-18} \\ 10z^3 \qquad\quad +\ z-\ 2 \end{array}$$

73. $(-4h^3+5h^2+15)-(h^3-15)$
$=-4h^3+5h^2+15-h^3+15$
$=-5h^3+5h^2+30$

75. $0.6x^3+0.8x^4+0.7x^3+(-0.8x^4)=1.3x^3$

77. $(12.1h^3-9.9h^2+9.5)+(7.3h^3-1.2h^2-10.1)$
$=12.1h^3-9.9h^2+9.5+7.3h^3-1.2h^2-10.1$
$=19.4h^3-11.1h^2-0.6$

79.
$$\begin{array}{r} 4x^3-3x+10 \\ \underline{-(5x^3-4x-4)} \\ -x^3 \quad +x+14 \end{array}$$

APPLICATIONS

81. a. $2x$ *ft*

b. $2x+2(2x)=2x+4x=6x$ *ft*

83. a. Length $=(x^2-x+6)+(4x+3)$
$=(x^2+3x+9)$ *ft*

Width $=(x^2-6x+3)+(3x+1)$
$=(x^2-3x+4)$ *ft*

Difference =
$(x^2+3x+9)-(x^2-3x+4)$
$=x^2+3x+9-x^2+3x-4$
$=(6x+5)$ *ft*

b. Perimeter =
$2(x^2+3x+9)+2(x^2-3x+4)$
$=2x^2+6x+18+2x^2-6x+8$
$=(4x^2+26)$ *ft*

85. $(x^2-3x+2)-(5x-10)=x^2-3x+2-5x+10$
$=(x^2-8x+12)$ *ft*

WRITING

87. Like terms are terms with exactly the same bases and the same powers.

89. To subtract two polynomials, add the opposite of the coefficients of any like terms from the subtracted polynomial.

REVIEW

91. $16.8-9.6=7.2oz$

10.4 Multiplying Polynomials

VOCABULARY

1. $(2x^3)(3x^4)$ is the product of two <u>monomials</u>.

3. In the acronym FOIL, F stands for <u>first</u> terms, O for <u>outer</u> terms, I for <u>inner</u> terms, and L for <u>last</u> terms.

CONCEPTS

5. To multiply two polynomials, multiply <u>each</u> term of one polynomial by <u>each</u> term of the other polynomial, and then combine like terms.

7. a. $6x^2+x-12$

b. $5x^4+8x^2+3$

NOTATION

9. $(9n^3)(8n^2)=(9\cdot 8)(n^3\cdot n^2)=72n^5$

11. $(2x+5)(3x-2)$
$=2x(3x)-2x(2)+5(3x)-5(2)$
$=6x^2-4x+15x-10=6x^2+11x-10$

GUIDED PRACTICE

13. $(3x^2)(4x^3)=12x^5$

15. $(3b^2)(-2b)=-6b^3$

17. $2x^2(3x^2+x)=2x^2\cdot 3x^2+2x^2\cdot x$
$=6x^4+2x^3$

19. $2x(3x^2+4x-7)=2x\cdot 3x^2+2x\cdot 4x-2x\cdot 7$
$=6x^3+8x^2-14x$

21. $(a+4)(a+5)=a^2+5a+4a+20$
$=a^2+9a+20$

23. $(3x-2)(x+4)=3x^2+12x-2x-8$
$=3x^2+10x-8$

25. $(2x+3)^2=(2x+3)(2x+3)$
$=4x^2+6x+6x+9=4x^2+12x+9$

27. $(9b-2)^2=(9b-2)(9b-2)$
$=81b^2-18b-18b+4=81b^2-36b+4$

29. $(2x+1)(3x^2-2x+1)$
$=2x(3x^2-2x+1)+1(3x^2-2x+1)$
$=6x^3-4x^2+2x+3x^2-2x+1$
$=6x^3-x^2+1$

31. $(x-1)(x^2+x+1)$
$=x(x^2+x+1)-1(x^2+x+1)$
$=x^3+x^2+x-x^2-x-1$
$=x^3-1$

33.
$$\begin{array}{r} x^2-x+1 \\ \times\ x+1 \\ \hline x^2-x+1 \\ +x^3-x^2+x \\ \hline x^3+1 \end{array}$$

35.
$$\begin{array}{r} 4x^2+3x-4 \\ \times\ 3x+2 \\ \hline 8x^2+6x-8 \\ +12x^3+9x^2-12x \\ \hline 12x^3+17x^2-6x-8 \end{array}$$

TRY IT YOURSELF

37. $(2a+4)(3a-5)=6a^2-10a+12a-20$
$=6a^2+2a-20$

39. $-p(2p^2-3p+2)$
$=-p(2p^2)-(-p)3p+(-p)2$
$=-2p^3+3p^2-2p$

41. $(-2x^2)(3x^3)=-6x^5$

43.
$$\begin{array}{r} 4x+3 \\ \times\ x+2 \\ \hline 8x+6 \\ +4x^2+3x \\ \hline 4x^2+11x+6 \end{array}$$

45. $(2x-3)^2=(2x-3)(2x-3)$
$=4x^2-6x-6x+9=4x^2-12x+9$

47. $3q^2(q^2-2q+7)$
$=3q^2\cdot q^2-3q^2\cdot 2q+3q^2\cdot 7$
$=3q^4-6q^3+21q^2$

49. $\left(-\frac{2}{3}y^5\right)\left(\frac{3}{4}y^2\right)=-\frac{6}{12}y^7=-\frac{1}{2}y^7$

51. $(x+2)(x^2-3x+1)$
$=x(x^2-3x+1)+2(x^2-3x+1)$
$=x^3-3x^2+x+2x^2-6x+2$
$=x^3-x^2-5x+2$

53.
$$\begin{array}{r} 2a^2+3a+1 \\ \times\ 3a^2-2a+4 \\ \hline 8a^2+12a+4 \\ -4a^3-6a^2-2a \\ +6a^4+9a^3+3a^2 \\ \hline 6a^4+5a^3+5a^2+10a+4 \end{array}$$

55. $(x+6)(x^3+5x^2-4x-4)$
$=x(x^3+5x^2-4x-4)$
$+6(x^3+5x^2-4x-4)$
$=x^4+5x^3-4x^2-4x+6x^3+30x^2-24x-24$
$=x^4+11x^3+26x^2-28x-24$

57. $(3n+1)(3n-1)=9n^2-3n+3n-1$
$=9n^2-1$

59. $(r^2-r+3)(r^2-4r-5)$
$=r^2(r^2-4r-5)-r(r^2-4r-5)+3(r^2-4r-5)$
$=r^4-4r^3-5r^2-r^3+4r^2+5r+3r^2-12r-15$
$=r^4-5r^3+2r^2-7r-15$

61. $(5t-1)^2=(5t-1)(5t-1)=25t^2-5t-5t+1$
$=25t^2-10t+1$

63. $3x(x-2)=3x\cdot x-3x\cdot 2=3x^2-6x$

APPLICATIONS

65. $A=lw=(x+2)(x-2)$
$=x^2-2x+2x-4=(x^2-4)ft^2$

67. $A=lw=(3x-1)(2x+1)$
$=6x^2+3x-2x-1=(6x^2+x-1)cm^2$

69. $A=lw=(7x+3)(5x+4)$
$=35x^2+28x+15x+12$
$=(35x^2+43x+12)in^2$

71. $\frac{1}{2}\cdot h(h-4)=\frac{1}{2}hh-\frac{1}{2}h\cdot 4=\left(\frac{1}{2}h^2-2h\right)in^2$

WRITING

73. To multiply two binomials, multiply each term of one binomial by both terms of the other binomial, and then combine like terms.

75. They are not equal because, since $(x+1)^2=(x+1)(x+1)$, the "OI" part of the FOIL process is missed.

REVIEW

77. four and ninety-one thousandths

79.
$$
\begin{array}{r}
0.109375 \\
64\overline{)7.000000} \\
\underline{-64} \\
60 \\
\underline{-0} \\
600 \\
\underline{-576} \\
240 \\
\underline{-192} \\
480 \\
\underline{-448} \\
320 \\
\underline{-320} \\
0
\end{array}
$$

81.
$$
\begin{array}{r}
^{1\,1\;\;\,1} \\
56.090 \\
78.000 \\
+\ \ 0.567 \\
\hline
134.657
\end{array}
$$

83. $\sqrt{16}+\sqrt{36}=4+6=10$

Chapter 10 Review

1. a. base n, exponent 12

b. base $2x$, exponent 6

c. base r, exponent 4

d. base $(y-7)$, exponent 3

2. a. $m\cdot m\cdot m\cdot m\cdot m=m^5$

b. $-3\cdot x\cdot x\cdot x\cdot x=-3x^4$

c. $a\cdot a\cdot b\cdot b\cdot b\cdot b=a^2b^4$

d. $(pq)(pq)(pq)=(pq)^3=p^3q^3$

3. a. $x^2 \cdot x^2 = x^{2+2} = x^4$

b. $x^2 + x^2 = 2x^2$

c. $x \cdot x^2 = x^{1+2} = x^3$

d. $x + x^2$ doesn't simplify

4. a. The base remains the same - $3^2 \cdot 3^4 = 3^6$.

b. Multiply, rather than add the exponents.

$(3^2)^4 = 3^8$

5. $7^4 \cdot 7^8 = 7^{4+8} = 7^{12}$

6. $mmnn^2 = m^{1+1}n^{1+2} = m^2n^3$

7. $(y^7)^3 = y^{7\cdot 3} = y^{21}$

8. $(3x)^4 = 3^4x^4 = 81x^4$

9. $(6^3)^{12} = 6^{3\cdot 12} = 6^{36}$

10. $-b^3b^4b^5 = -b^{3+4+5} = -b^{12}$

11. $(-16s^3)^2 s^4 = (-16)^2 (s^3)^2 s^4 = 256s^6s^4 = 256s^{10}$

12. $(2.1x^2y)^2 = (2.1)^2(x^2)^2 y^2 = 4.41x^4y^2$

13. $\left[(-9)^3\right]^5 = (-9)^{15}$

14. $(a^5)^3(a^2)^4 = a^{15} \cdot a^8 = a^{23}$

15. $(2x^2x^3)^3 = (2x^5)^3 = 2^3(x^5)^3 = 8x^{15}$

16. $(m^2m^3)^2(n^2n^4)^3 = (m^5)^2(n^6)^3 = m^{10}n^{18}$

17. $(3a^4)^2(2a^3)^3 = 3^2(a^4)^2 2^3(a^3)^3$

$= 9a^8 \cdot 8a^9 = 72a^{17}$

18. $x^{100} \cdot x^{100} = x^{200}$

19. $(4m^3)^3(2m^2)^2 = 4^3(m^3)^3 2^2(m^2)^2$

$= 64m^9 \cdot 4m^4 = 256m^{13}$

20. $(3t^4)^3(2t^5)^2 = 3^3(t^4)^3 2^2(t^5)^2$

$= 27t^{12} \cdot 4t^{10} = 108t^{22}$

21. trinomial

22. monomial

23. binomial

24. trinomial

25. 3

26. 4

27. 5

28. 5

29. $2t^2 + t - 2 = 2(-3)^2 + (-3) - 2$

$= 2 \cdot 9 - 3 - 2 = 18 - 3 - 2 = 13$

30. $h = -16t^2 + 12t + 20 = -16 \cdot 1^2 + 12 \cdot 1 + 20$

$= -16 + 12 + 20 = 16\,ft$

31. $3x^3 + 2x^3 = 5x^3$

32. $\frac{1}{2}p^2 + \frac{5}{2}p^2 + \frac{7}{2}p^2 = \frac{1+5+7}{p^2}p^2 = \frac{13}{2}p^2$

33. $(3x-1) + (6x+5) = 3x - 1 + 6x + 5$

$= 9x + 4$

34. $(3x^2 - 2x + 4) + (-x^2 - 1)$

$= 3x^2 - 2x + 4 - x^2 - 1 = 2x^2 - 2x + 3$

35.
$$\begin{array}{r} 5x-2 \\ +\underline{3x+5} \\ 8x+3 \end{array}$$

36.
$$\begin{array}{r} 3x^2-2x+7 \\ +\ \underline{-5x^2+3x-5} \\ -2x^2+\ x+2 \end{array}$$

37. $16p^3-9p^3=7p^3$

38. $4y^2-9y^2=-5y^2$

39. $(2.5x+4)-(1.4x+12)$
$=2.5x+4-1.4x-12=1.1x-8$

40. $(3z^2-z+4)-(2z^2+3z-2)$
$=3z^2-z+4-2z^2-3z+2$
$=z^2-4z+6$

41.
$$\begin{array}{r} 5x-2 \\ \underline{-(3x+5)} \\ 2x-7 \end{array}$$

42.
$$\begin{array}{r} 3x^2-2x+7 \\ \underline{-(-5x^2+3x-5)} \\ 8x^2-5x+12 \end{array}$$

43. $3x^2\cdot 5x^3=3\cdot 5\cdot x^2\cdot x^3=15x^5$

44. $(3z^2)(-2z^2)=3(-2)z^2z^2=-6z^4$

45. $2x^2(3x+2)=2x^2\cdot 3x+2x^2\cdot 2$
$=6x^3+4x^2$

46. $-5t^3(7t^2-6t-2)$
$=-5t^3\cdot 7t^2-(-5t^3)\cdot 6t-(-5t^3)\cdot 2$
$=-35t^5+30t^4+10t^3$

47. $(2x-1)(3x+2)=6x^2+4x-3x-2$
$=6x^2+x-2$

48. $(5t+4)(7t-6)=35t^2-30t+28t-24$
$=35t^2-2t-24$

49. $(3x+2)(2x^2-x+1)$
$=3x(2x^2-x+1)+2(2x^2-x+1)$
$=6x^3-3x^2+3x+4x^2-2x+2$
$=6x^3+x^2+x+2$

50. $(2r-3)(3r^2+2r-3)$
$=2r(3r^2+2r-3)-3(3r^2+2r-3)$
$=6r^3+4r^2-6r-9r^2-6r+9$
$=6r^3-5r^2-12r+9$

51.
$$\begin{array}{r} 5x^2-2x+3 \\ \times\qquad\quad \underline{3x+5} \\ 25x^2-10x+15 \\ \underline{+15x^3-\ 6x^2+\ 9x\qquad} \\ 15x^3+19x^2-\quad x+15 \end{array}$$

52.
$$\begin{array}{r} 3x^2-2x-1 \\ \underline{\times\ 5x-2} \\ -6x^2+4x+2 \\ \underline{+15x^3-10x^2-5x\qquad} \\ 15x^3-16x^2-\ x+2 \end{array}$$

53. $(x+2)^2=(x+2)(x+2)$
$=x^2+2x+2x+4=x^2+4x+4$

54. $(8a-3)^2=(8a-3)(8a-3)$
$=64a^2-24a-24a+9$
$=64a^2-48a+9$

Chapter 10 Test

1. In the expression 7^5, 7 is the base and 5 is the exponent.

2. Expressions such as x^4, 10^3, and $(5t)^2$ are called exponential expressions.

3. A polynomial is a term or a sum of terms in which all variables have whole-number exponents and no variable appears in a denominator.

4. A monomial is a polynomial with exactly one term. A binomial is a polynomial with exactly two terms. A trinomial is a polynomial with exactly three terms.

5. The degree of the term $3x^7$ is 7 because x appears as a factor 7 times: $3 \cdot x \cdot x \cdot x \cdot x \cdot x \cdot x \cdot x$.

6. To evaluate the polynomial $x^2 - 2x + 1$ for $x = 6$, we substitute 6 for x and follow the order of operations rule.

7. $(b^3 - b^2 - 9b + 1) + (b^3 - b^2 - 9b + 1)$ is the sum of two polynomials.

8. $(b^2 - 9b + 11) - (4b^2 - 14b)$ is the difference of a trinomial and a binomial.

9. In the acronym FOIL, F stands for first terms, O for outer terms, I for inner terms, and L for last terms.

10. $(2a - 4)(3a^2 + 5a - 1)$ is the product of a binomial and a trinomial.

11. $(2x + 3)^2$ is the square of a binomial.

12. To multiply two polynomials, multiply each term of one polynomial by each term of the other polynomial, and then combine like terms.

13. a. base 6, exponent 5 b. base b, exponent 4

14. a. $a^2 + a^2 = 2a^2$

b. $a^2 \cdot a^2 = a^{2+2} = a^4$

c. $a^2 + a$ doesn't simplify

d. $a^2 \cdot a = a^{2+1} = a^3$

15. $h^2 h^4 = h^{2+4} = h^6$

16. $(m^{10})^2 = m^{10 \cdot 2} = m^{20}$

17. $b^2 \cdot b \cdot b^5 = b^{2+1+5} = b^8$

18. $(x^3)^4 (x^2)^3 = x^{12} \cdot x^6 = x^{18}$

19. $(a^2 b^3)(a^4 b^7) = a^6 b^{10}$

20. $(12a^9 b)^2 = 12^2 (a^9)^2 b^2 = 144a^{18} b^2$

21. $(2x^2)^3 (3x^3)^3 = 2^3 (x^2)^3 3^3 (x^3)^3$
$= 8x^6 \cdot 27x^9 = 216x^{15}$

22. $(t^2 t^3)^3 = (t^5)^3 = t^{15}$

23. binomial

24. monomial

25. trinomial

26. binomial

27. 6

28. 7

29. $3x^2 - 2x + 4 = 3 \cdot 3^2 - 2 \cdot 3 + 4$
$= 3 \cdot 9 - 6 + 4 = 27 - 6 + 4 = 25$

30. $-2r^2 - r + 3 = -2(-1)^2 - (-1) + 3$
$= -2 \cdot 1 + 1 + 3 = -2 + 1 + 3 = 2$

31. $(2.1p^2 - 2p - 2) - (3.3p^2 - 5p - 2)$
$= 2.1p^2 - 2p - 2 - 3.3p^2 + 5p + 2$
$= -1.2p^2 + 3p$

32. $(-2x^3)(4x^2) = -2 \cdot 4 \cdot x^3 \cdot x^2 = -8x^5$

33. $(3x^2 + 2x) + (2x^2 - 5x + 4)$
$= 3x^2 + 2x + 2x^2 - 5x + 4$
$= 5x^2 - 3x + 4$

34. $(2x - 5)(3x + 4) = 6x^2 + 8x - 15x - 20$
$= 6x^2 - 7x - 20$

35.
$$\begin{array}{r} 3d^2 - 3d + 7.2 \\ -(-5d^2 + 6d - 5.3) \\ \hline 8d^2 - 9d + 12.5 \end{array}$$

36. $3y^2(y^2-2y+3)=3y^2\cdot y^2-3y^2\cdot 2y+3y^2\cdot 3$
$=3y^4-6y^3+9y^2$

37.
$$\begin{array}{r} 4x^2-5x+5 \\ +\underline{3x^2+7x-7} \\ 7x^2+2x-2 \end{array}$$

38. $(2x-3)(x^2-2x+4)$
$=2x(x^2-2x+4)-3(x^2-2x+4)$
$=2x^3-4x^2+8x-3x^2+6x-12$
$=2x^3-7x^2+14x-12$

39. $P=4s=4(x+4)=(4x+16)in$

40. Keep the common base 5 and add the exponents. Do not multiply the common bases to get 25.

Chapters 1 – 10 Cumulative Review

1. The largest offer was \$8,995.

2. 2,110,000

3.
$$\begin{array}{r} {}^{1\,1\;\;2\,3} \\ 23,419 \\ 8,099 \\ 458 \\ +\quad 58 \\ \hline 32,034 \end{array}$$

4.
$$\begin{array}{r} 3\not{5},\not{0}\,\not{2}\,\not{1} \quad (4;\ 10\to 9;\ 12\to 11;\ 11) \\ -23,9\,9\,9 \\ \hline 11,0\,2\,2 \end{array}$$

5. a. $P=204+97+204+97=602\,ft$

b. $A=lw=204\cdot 97=19,788\,ft^2$

6.
$$\begin{array}{r} 33 \\ 41\overline{)1363} \\ -\underline{123} \\ 133 \\ -\underline{123} \\ 10 \end{array}$$
$33R10$

7. a. $220=22\cdot 10=2\cdot 11\cdot 2\cdot 5=2^2\cdot 5\cdot 11$

b. 1, 2, 3, 4, 6, 12

8. $16=2\cdot 2\cdot 2\cdot 2$
$24=2\cdot 2\cdot 2\cdot 3$
$LCM(16,24)=2\cdot 2\cdot 2\cdot 2\cdot 3=48$
$GCF(16,24)=2\cdot 2\cdot 2=8$

9. $\dfrac{(3+5)^2+2}{2(8-5)}=\dfrac{8^2+2}{2(3)}=\dfrac{64+2}{6}=\dfrac{66}{6}=11$

10. a. $y-81=243$
$y-81+81=243+81$
$y=324$
Check: $y-81=243$
$324-81\stackrel{?}{=}243$
$243=243$

b. $81y=243$
$\dfrac{81y}{81}=\dfrac{243}{81}$
$y=3$
Check: $81y=243$
$81\cdot 3\stackrel{?}{=}243$
$243=243$

11. a. $\{\dots,-2,-1,0,1,2,\dots\}$

b. $-(-3)=3$

12. a. $-16+4=-12$

b. $16-(-4)=16+4=20$

c. $-16(4)=-64$

d. $\frac{-16}{-4}=\frac{16}{4}=4$

e. $-4^2=-4\cdot 4=-16$

f. $(-4)^2=(-4)(-4)=16$

13. $30-55-75-2\cdot 20$

$=-25-75-40=-\$140$

The current balance is -\$140.

14. $10-4\left|6-(-3)^2\right|=10-4|6-9|$

$=10-4|-3|=10-4\cdot 3=10-12=-2$

15.

$-x+2=13$

$-x+2-2=13-2$

$-x=11$

$\frac{-x}{-1}=\frac{11}{-1}$

$x=-11$

Check: $-x+2=13$

$-(-11)+2\overset{?}{=}13$

$11+2\overset{?}{=}13$

$13=13$

16.

$4+\frac{x}{-5}-6=-1$

$\frac{x}{-5}-2=-1$

$\frac{x}{-5}-2+2=-1+2$

$\frac{x}{-5}=1$

$-5\left(\frac{x}{-5}\right)=-5\cdot 1$

$x=-5$

Check: $4+\frac{x}{-5}-6=-1$

$4+\frac{-5}{-5}-6\overset{?}{=}-1$

$4+1-6\overset{?}{=}-1$

$5-6\overset{?}{=}-1$

$-1=-1$

17. x = the profit in 2008

$-63+x=-17$

$-63+x+63=-17+63$

$x=46$

The 2008 profit was \$46 million.

18. a. $k+1$ b. $m-1$

19. $x-5$

20. $d=r\cdot t=475\cdot 3=1{,}425mi$

21. a. $-10(-10a)=[-10(-10)]a=100a$

b. $-(-4y+6)=4y-6$

c. $-3(-6y-8)-12-4(5-y)$

$=18y+24-12-20+4y=22y-8$

22. $50x$ ¢

23.
$$5r-24=r+5r+2r$$
$$5r-24=8r$$
$$5r-24-5r=8r-5r$$
$$-24=3r$$
$$\frac{-24}{3}=\frac{3r}{3}$$
$$-8=r$$
Check: $5r-24=r+5r+2r$
$$5(-8)-24\overset{?}{=}(-8)+5(-8)+2(-8)$$
$$-40-24\overset{?}{=}-8-40-16$$
$$-64=-64$$

24.
$$2(4a+8)=-3(2-3a)-3a$$
$$8a+16=-6+9a-3a$$
$$8a+16=-6+6a$$
$$8a+16-6a=-6+6a-6a$$
$$2a+16=-6$$
$$2a+16-16=-6-16$$
$$2a=-22$$
$$\frac{2a}{2}=\frac{-22}{2}$$
$$a=-11$$
Check: $2(4a+8)=-3(2-3a)-3a$
$$2[4(-11)+8]\overset{?}{=}-3[2-3(-11)]-3(-11)$$
$$2(-44+8)\overset{?}{=}-3(2+33)+33$$
$$2(-36)\overset{?}{=}-3(35)+33$$
$$-72\overset{?}{=}-105+33$$
$$-72=-72$$

25. $d=$ the number of dimes

$25-d$ = the number of quarters
$$5(10)+10d+25(25-d)=600$$
$$50+10d+625-25d=600$$
$$675-15d=600$$
$$675-15d-675=600-675$$
$$-15d=-75$$
$$\frac{-15d}{-15}=\frac{-75}{-15}$$
$$d=5$$
There were 5 dimes and 25 – 5 = 20 quarters.

26.
$$\frac{35a^2}{28a}=\frac{5\cdot \cancel{7}\cdot \cancel{a}\cdot a}{4\cdot \cancel{7}\cdot \cancel{a}}=\frac{5a}{4}$$

27.

a. $\frac{3}{8}=\frac{3}{8}\cdot\frac{6}{6}=\frac{18}{48}$

b. $\frac{8}{9}$

c. $7\frac{1}{2}=\frac{2\cdot 7+1}{2}=\frac{15}{2}$

28.
$$\frac{1}{6}(54)=9oz$$

29.
$$-\frac{5}{77}\left(\frac{33}{50}\right)=-\frac{5\cdot 3\cdot 11}{7\cdot 11\cdot 5\cdot 10}$$
$$=-\frac{\cancel{5}\cdot 3\cdot \cancel{11}}{7\cdot \cancel{11}\cdot \cancel{5}\cdot 10}=-\frac{3}{70}$$

30.
$$\frac{15b^3}{16c^5}\div\frac{45b}{8c}=\frac{15b^3}{16c^5}\cdot\frac{8c}{45b}$$
$$=\frac{\cancel{15}\cdot \cancel{b}\cdot b^2\cdot \cancel{8}\cdot \cancel{c}}{2\cdot \cancel{8}\cdot \cancel{c}\cdot c^4\cdot 3\cdot \cancel{15}\cdot \cancel{b}}=\frac{b^2}{6c^4}$$

31.
$$\frac{3}{4}-\frac{3}{5}=\frac{3}{4}\cdot\frac{5}{5}-\frac{3}{5}\cdot\frac{4}{4}=\frac{15}{20}-\frac{12}{20}=\frac{3}{20}$$

32.
$$\frac{m}{9}-\frac{4}{5}=\frac{m}{9}\cdot\frac{5}{5}-\frac{4}{5}\cdot\frac{9}{9}$$
$$=\frac{5m}{45}-\frac{36}{45}=\frac{5m-36}{45}$$

33.
$$-\frac{6}{25}\left(2\frac{7}{24}\right)=-\frac{6}{25}\cdot\frac{55}{24}$$
$$=-\frac{\not{6}\cdot\not{5}\cdot 11}{\not{5}\cdot 5\cdot 4\cdot\not{6}}=-\frac{11}{20}$$

34.
$$45\frac{2}{3}+96\frac{4}{5}=45\frac{10}{15}+96\frac{12}{15}$$
$$=141\frac{22}{15}=141+1\frac{7}{15}=142\frac{7}{15}$$

35.
$$\frac{7-\frac{2}{3}}{4\frac{5}{6}}=\frac{\frac{21}{3}-\frac{2}{3}}{\frac{29}{6}}=\frac{\frac{19}{3}}{\frac{29}{6}}$$
$$=\frac{19}{3}\cdot\frac{6}{29}=\frac{19\cdot 2}{29}=\frac{38}{29}=1\frac{9}{29}$$

36.
$$\frac{1}{8}\left(\frac{3}{4}\right)=\frac{3}{32}\ fl\ oz$$

37.
$$17\frac{1}{2}-3\frac{3}{4}=\frac{35}{2}-\frac{15}{4}=\frac{70}{4}-\frac{15}{4}$$
$$=\frac{55}{4}=13\frac{3}{4}\text{ cups}$$

38.
$$a+b^2c=\frac{3}{4}+\left(-\frac{1}{3}\right)^2\left(\frac{5}{4}\right)$$
$$=\frac{3}{4}+\frac{1}{9}\cdot\frac{5}{4}=\frac{3}{4}\cdot\frac{9}{9}+\frac{1}{9}\cdot\frac{5}{4}$$
$$=\frac{27}{36}+\frac{5}{36}=\frac{32}{36}=\frac{8}{9}$$

39.
$$\frac{x}{12}+\frac{2}{3}=\frac{x}{6}$$
$$12\cdot\frac{x}{12}+12\cdot\frac{2}{3}=12\cdot\frac{x}{6}$$
$$x+8=2x$$
$$x+8-x=2x-x$$
$$8=x$$

Check: $\frac{x}{12}+\frac{2}{3}=\frac{x}{6}$
$$\frac{8}{12}+\frac{2}{3}\stackrel{?}{=}\frac{8}{6}$$
$$\frac{2}{3}+\frac{2}{3}\stackrel{?}{=}\frac{4}{3}$$
$$\frac{4}{3}=\frac{4}{3}$$

40.
$$\frac{2}{3}q-1=-6$$
$$\frac{2}{3}q-1+1=-6+1$$
$$\frac{2}{3}q=-5$$
$$\frac{3}{2}\cdot\frac{2}{3}q=\frac{3}{2}(-5)$$
$$q=\frac{-15}{2}$$

Check: $\frac{2}{3}q-1=-6$
$$\frac{2}{3}\left(-\frac{15}{2}\right)-1\stackrel{?}{=}-6$$
$$-\frac{15}{3}-1\stackrel{?}{=}-6$$
$$-5-1\stackrel{?}{=}-6$$
$$-6=-6$$

41.
$$\frac{3}{5}(275)=3(55)=165\text{ students}$$

42.

a. $\pi\approx 3.1416$

b. $154.34>143.339999$

43. a. six million, five hundred ten thousand, three hundred forty-five and seven hundred ninety-eight thousandths

b. $7{,}000+400+90+8+\frac{6}{10}+\frac{4}{100}+\frac{6}{1{,}000}+\frac{1}{10{,}000}$

44.

$$\begin{array}{r} \scriptstyle 1\,1 \\ 106.780 \\ 35.000 \\ 3.400 \\ +\ 0.008 \\ \hline 145.188 \end{array}$$

45.

$$\begin{array}{r} 5{,}091.50 \\ -\ 1{,}287.89 \\ \hline 3{,}803.61 \end{array}$$

46. $-8.8+(-7.3-9.5)$

$=-8.8+(-16.8)=-25.6$

47. A negative times a negative is a positive.

$5.5\cdot 3.1=17.05$ so $-5.5(-3.1)=17.05$

48. $\frac{0.0742}{1.4}=\frac{742}{14000}=\frac{7\cdot 106}{7\cdot 2000}$

$=\frac{106}{2000}=\frac{2\cdot 53}{2\cdot 1000}=\frac{53}{1000}=0.053$

49. $\frac{7}{8}(9.7+15.8)=\frac{7}{8}(25.5)$

$=0.875(25.5)=22.3125$

50. $2.17(1{,}250)=\$2{,}712.50$

51. a. $899{,}708$

b. 0.899708

52. Roughly $18{,}000\div 9=2{,}000$

53. $\frac{(-1.3)^2+6.7}{-0.9}=\frac{1.69+6.7}{-0.9}$

$=\frac{8.39}{-0.9}=-9.3222...\approx -9.32$

54.

$$\begin{array}{r} 0.133 \\ 15\overline{)2.000} \\ -15 \\ \hline 50 \\ -45 \\ \hline 50 \\ -45 \\ \hline \end{array}$$

$\frac{2}{15}=0.1\overline{3}$

55. a. $2\sqrt{121}-3\sqrt{64}=2\cdot 11-3\cdot 8$

$=22-24=-2$

b. $\sqrt{\frac{49}{81}}=\sqrt{\left(\frac{7}{9}\right)^2}=\frac{7}{9}$

56.

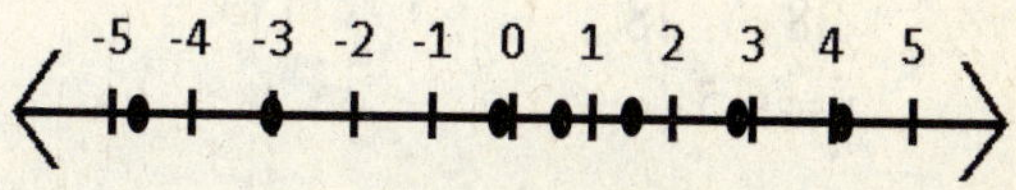

57. $1.7y+1.24=-1.4y-0.62$

$1.7y+1.24+1.4y=-1.4y-0.62+1.4y$

$3.1y+1.24=-0.62$

$3.1y+1.24-1.24=-0.62-1.24$

$3.1y=-1.86$

$\frac{3.1y}{3.1}=\frac{-1.86}{3.1}$

$y=-0.6$

Check: $1.7y+1.24=-1.4y-0.62$

$1.7(-0.6)+1.24\stackrel{?}{=}-1.4(-0.6)-0.62$

$-1.02+1.24\stackrel{?}{=}0.84-0.62$

$0.22=0.22$

58. b = the number of balloons she can buy

$15.15+0.05b=20$

$15.15+0.05b-15.15=20-15.15$

$0.05b=4.85$

$\frac{0.05b}{0.05}=\frac{4.85}{0.05}$

$b=97$

She can by 97 balloons.

59.

a. $\frac{3cm}{7cm} = \frac{3\cancel{cm}}{7\cancel{cm}} = \frac{3}{7}$

b. $\frac{13wk}{1yr} = \frac{13wk}{52wk} = \frac{\cancel{13wk}}{4 \cdot \cancel{13wk}} = \frac{1}{4}$

60.

Smaller board: $\frac{24}{400} = \$0.06 / in^2$

Larger board: $\frac{\$42}{600in^2} = \$0.07 / in^2$

The smaller board is a better deal.

61.

$\frac{x}{14} = \frac{13}{28}$

$28 \cdot x = 13 \cdot 14$

$\frac{28x}{28} = \frac{182}{28}$

$x = 6.5$

62.

$\frac{3}{1,000} = \frac{375}{x}$

$3 \cdot x = 375 \cdot 1,000$

$\frac{3x}{3} = \frac{375,000}{3}$

$x = 125,000$

63.

$\frac{\frac{1}{4}}{3} = \frac{6\frac{1}{4}}{x}$

$\frac{1}{4} \cdot x = \frac{25}{4} \cdot 3$

$\frac{x}{4} = \frac{75}{4}$

$x = 75\,ft$

64.

a. $\frac{168in}{1} \cdot \frac{1ft}{12in} = \frac{168}{12} ft = 14ft$

b. $\frac{212oz}{1} \cdot \frac{1lb}{16oz} = \frac{212}{16} lb = 13.25lb$

c. $\frac{30gal}{1} \cdot \frac{4qt}{1gal} = 30 \cdot 4qt = 120qt$

d. $\frac{12.5h}{1} \cdot \frac{60\min}{1h} = 12.5(60)\min = 750\min$

65. a. move 3 right: $1.538kg = 1,538g$

b. move 3 left: $500mL = 0.5L$

c. move 5 left: $0.3cm = 0.000003km$

66. move 3 left: $240,000m = 240km$

67. $\frac{10lb}{1} \cdot \frac{1kg}{2.2lb} = \frac{10}{2.2} kg \approx 4.5kg$

68.

$F = \frac{9}{5}C + 32 = \frac{9}{5}(75) + 32$

$= 135 + 32 = 167°F$

69.

Percent	Decimal	Fraction
57%	0.57	$\frac{57}{100}$
0.1%	0.001	$\frac{1}{1,000}$
$33\frac{1}{3}\%$	$0.\overline{3}$	$\frac{1}{3}$

70. a. 93% b. 100% - 93% = 7%

71. $0.15(450) = 67.5$

72.

$24.6 = 0.205x$

$\frac{24.6}{0.205} = \frac{0.205x}{0.205}$

$120 = x$

73. $51 = x \cdot 60$

$$\frac{51}{60} = \frac{60x}{60}$$

$0.85 = x$

51 is 85% of 60.

74. $\text{Discount} = 25\%(820) = 0.25(820) = \$205$

$\text{Sale price} = \$820 - 205 = \$615$

75. $0.0625(18,550) = \$1,159.38$

76. There was an increase of $\$750 - \$125 = \$625$.

$625 = 125 \cdot x$

$$\frac{625}{125} = \frac{125x}{125}$$

$5 = x$

The value increased by 500%.

77. Estimate 10% of 140 is $14.

Half of that is $7, so 15% is roughly $14 + $7 = $21.

78. $I = P \cdot r \cdot t = 1,400(0.09)(0.5) = \$63$

$\$1,500 + \$63 = \$1,563$ to be repaid.

79. a. about $7.5(50,000) = 375,000$ vehicles

b. about $5(50,000) = 250,000$ vehicles

c. $6.5(50,000) - 6(50,000)$

$= 325,000 - 300,000 = 25,000$ vehicles

80. a. $100\% - 40\% - 42\% = 18\%$

b. $0.40(7,300,000) = 2,920,000$

81. a. Food: about $20.5 billion

b. about $2.2 billion

c. about $\$13.5 - \$4 = \$9.5$ billion

82. $$\text{mean} = \frac{0.85 + 0.87 + \cdots + 0.85}{8} = \frac{6.88}{8} = 0.86oz$$

$$\text{median} = \frac{0.85 + 0.86}{2} = 0.855oz$$

$0.84, 0.85, 0.85, 0.85, 0.86, 0.87, 0.88, 0.88$

mode: $0.85oz$

83.

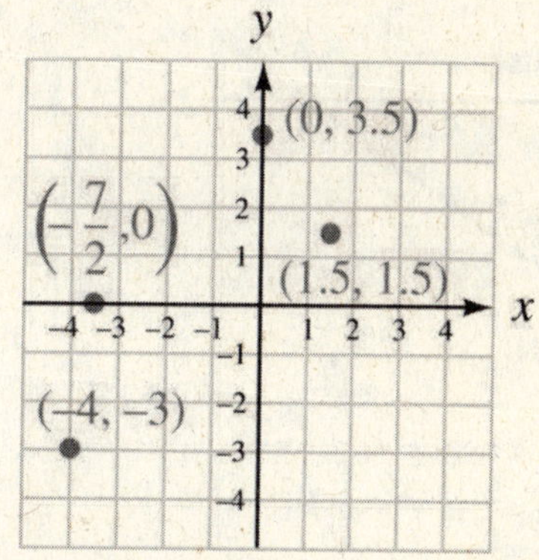

84. $3x - y = -8$

$3(-2) - (1) \stackrel{?}{=} -8$

$-6 - 1 \stackrel{?}{=} -8$

$-7 = -8$

$(-2, 1)$ is not a solution.

85.

$x = 0$:	$x = 2$
$y = -2(0)$	$y = -2(2)$
$y = 0$	$y = -4$
$(0, 0)$	$(2, -4)$

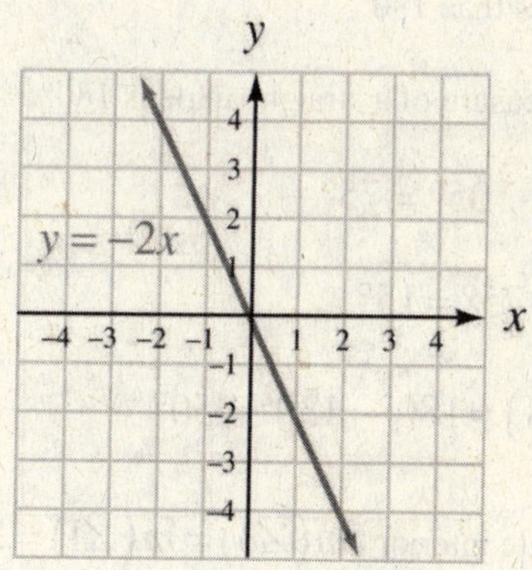

86.

$x=0$	$y=0$
$2(0)-3y=12$	$2x-3(0)=12$
$-3y=12$	$2x=12$
$y=-4$	$x=6$
$(0,-4)$	$(6,0)$

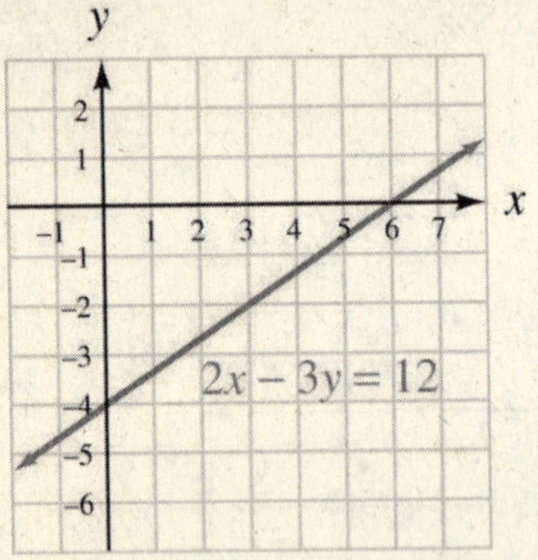

87.

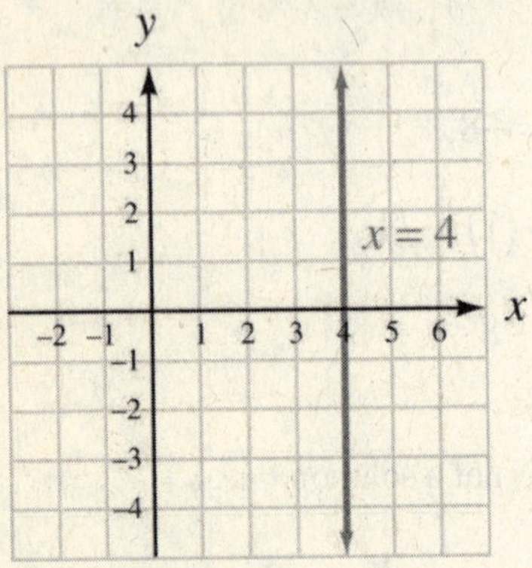

88. a. The measure of an <u>acute</u> angle is less than 90°.

b. The measure of a <u>right</u> angle is 90°.

c. The measure of an <u>obtuse</u> angle is greater than 90° but less than 180°.

d. The measure of a straight angle is <u>180°</u>.

89. a. $180°-105°=75°$

b. $90°-75°=15°$

90. a. $m(\angle 1)=180°-130°=50°$

b. alternate interior: $m(\angle 3)=m(\angle 1)=50°$

c. alternate interior: $m(\angle 2)=130°$

d. vertical angles: $m(\angle 4)=m(\angle 3)=50°$

91. a. similar triangles: $m(\angle 1)=75°$

b. isosceles: $2\cdot 75°+m(\angle C)=180°$

$150°+m(\angle C)=180°$

$m(\angle C)=30°$

c. supplemental: $m(\angle 2)=180°-75°=105°$

d. supplemental: $m(\angle 3)=180°-75°=105°$

92. $x°=180°-90°-44°=46°$

$y°=180°-x°=180°-46°=134°$

93. $34°+2x=180°$

$34°+2x-34°=180°-34°$

$2x=146°$

$\frac{2x}{2}=\frac{146°}{2}$

$x=73°$

94. $a^2+b^2=c^2$

$10^2+24^2=c^2$

$100+576=c^2$

$676=c^2$

$\sqrt{676}=c$

$26m=c$

95. $a^2+b^2\stackrel{?}{=}c^2$

$16^2+63^2\stackrel{?}{=}65^2$

$256+3,969\stackrel{?}{=}4,225$

$4,225=4,225$

It is a right triangle.

96. $\frac{t}{35}=\frac{6}{5}$

$t\cdot 5=6\cdot 35$

$\frac{5t}{5}=\frac{210}{5}$

$t=42\,ft$

97. $S = 180°(n-2) = 180°(5-2)$
$= 180° \cdot 3 = 540°$

98. $P = 4s = 4 \cdot 12 = 48m$
$A = s^2 = 12^2 = 144m^2$

99. $A = \frac{1}{2}bh = \frac{1}{2} \cdot 14 \cdot 18 = 7 \cdot 18 = 126 ft^2$

100. $A = \frac{1}{2}h(b_1 + b_2) = \frac{1}{2} \cdot 7(12+14)$
$= \frac{7}{2}(26) = 7 \cdot 13 = 91 in^2$

101. $(1ft)^2 = (1ft)(1ft) = (12in)(12in) = 144in^2$

102. $C = \pi d = 14\pi\, cm \approx 43.98cm$
$A = \pi r^2 = \pi \cdot 7^2 = 49\pi\, cm^2 \approx 153.94cm^2$

103. Rectangle area: $A = lw = 20.2(19.2) = 387.84 yd^2$

Circle area: $A = \pi r^2 = \pi \cdot 9.6^2 \approx 289.53 yd^2$

Shaded area: $387.84 - 289.53 = 98.31 yd^2$

104. $V = lwh = 15 \cdot 24 \cdot 18 = 6,480 in^3$

105. $V = \frac{4}{3}\pi r^3 = \frac{4}{3} \cdot \pi \cdot 9^3 = \frac{4}{3} \cdot \pi \cdot 729$
$= 972\pi in^3 \approx 3,053.63 in^3$

106. $V = \frac{1}{3}\pi r^2 h = \frac{1}{3}\pi \cdot 4^2 \cdot 9 = 16 \cdot \pi \cdot 3$
$= 48\pi\, m^3 \approx 150.80m^3$

107. $V = \pi r^2 h = \pi \cdot 1^2 \cdot 20$
$= 20\pi\, ft^3 \approx 62.83 ft^3$

108. $(1ft)^3 = (12in)^3 = 12^3 in^3 = 1,728 in^3$

109. $s^4 \cdot s^5 = s^{4+5} = s^9$

110. $(a^5)^7 = a^{5 \cdot 7} = a^{35}$

111. $(y^5)^2 (y^4)^3 = y^{10} \cdot y^{12} = y^{22}$

112. $(2b^3c^6)^3 = 2^3 (b^3)^3 (c^6)^3 = 8b^9c^{18}$

113. trinomial, degree 2

114. $0.5t^3 - t^2 - 4t = 0.5(4)^3 - (4)^2 - 4(4)$
$= 0.5 \cdot 64 - 16 - 16 = 32 - 16 - 16 = 0$

115. $(5x^2 - 2x + 4) - (3x^2 - 5)$
$= 5x^2 - 2x + 4 - 3x^2 + 5$
$= 2x^2 - 2x + 9$

116. $(6.2a^3 + 7.1a^2 - 4.1a) + (3.8a^3 - 4.1a)$
$= 6.2a^3 + 7.1a^2 - 4.1a + 3.8a^3 - 4.1a$
$= 10a^3 + 7.1a^2 - 8.2a$

117. $-3h^9(-5h) = -3(-5) \cdot h^9 \cdot h = 15h^{10}$

118. $-3p(2p^2 + 3p - 4)$
$= -3p(2p^2) + (-3p)(3p) - (-3p)(4)$
$= -6p^3 + (-9p^2) - (-12p) = -6p^3 - 9p^2 + 12p$

119. $(3x+5)(2x-1) = 6x^2 - 3x + 10x - 5$
$= 6x^2 + 7x - 5$

120. $(2y-7)^2 = (2y-7)(2y-7)$
$= 4y^2 - 14y - 14y + 49 = 4y^2 - 28y + 49$

Appendix 1

Section A1.1: Probability

VOCABULARY

1. An experiment is any process for which the result is uncertain.

2. The results of an experiment are called outcomes.

3. For any experiment, the set of all possible outcomes is called a sample space.

4. An event is one or more outcomes of an experiment.

5. Favorable outcomes of an experiment satisfy the requirements of that particular event.

6. The probability of an event is a measure of the likelihood (or chance) of its occurring.

CONCEPTS

7. If a sample space of an experiment has n distinct and equally likely outcomes, and E is an event that occurs in s favorable ways, the probability of event E is $P(E)=\frac{s}{n}$.

8. The probability of an event is always a number from 0 to 1, including 0 and 1.

9. If an event cannot happen, its probability is 0.

10. If an event is certain to happen, its probability is 1.

11. a. There are 52 cards in a standard deck.
    b. There are 26 red cards in a standard deck.
    c. There are 13 cards in a standard deck.
    d. There are 3 face cards in each suit.

12. There are six outcomes possible when rolling a die. They are $\{1, 2, 3, 4, 5, 6\}$.

NOTATION

13. When writing a sample space, we list all possible outcomes within braces { }.

14. We read $P(E)$ as "the probability of event E."

GUIDED PRACTICE

15. {(1, H), (2, H), (3, H), (4, H), (5, H), (6, H), (1, T), (2, T), (3, T), (4, T), (5, T), (6, T)}

16. {(H, H, H), (H, H, T), (H, T, H), (T, H, H), (H, T, T), (T, H, T), (T, T, H), (T, T, T)}

17. {a, e, i, o, u}

18. {0, 1, 2, 3, 4, 5, 6, 7, 8, 9}

19. $E = \{2, 4, 6\}$

20. E = {king of diamonds, queen of diamonds, jack of diamonds, king of hearts, queen of hearts, jack of hearts}

21. E = {(H, H), (H, T), (T, H)}

22. E = {(T, 1), (T, 3), (T, 5)}

23. $n=52$
    $E=\{\text{black card}\}\rightarrow s=26$
    $P(\text{black card})=\frac{26}{52}=\frac{1}{2}$

24. $n=52$
    $E=\{\text{diamond}\}\rightarrow s=13$
    $P(\text{diamond})=\frac{13}{52}=\frac{1}{4}$

25. $n=52$
    $E=\{\text{face card}\}\rightarrow s=12$
    $P(\text{face card})=\frac{12}{52}=\frac{3}{13}$

26. $n=52$
    $E=\{\text{Jack or King}\}\rightarrow s=8$
    $P(\text{Jack or King})=\frac{8}{52}=\frac{2}{13}$

27. $n = 36$

$E = \{\text{sum of } 6\} = \{(1, 5), (5, 1), (4, 2), (2, 4), (3,3)\}$

$\rightarrow s = 5$

$P(\text{sum of } 6) = \dfrac{5}{36}$

28. $n = 36$

$E = \{\text{sum of } 10\} = \{(4, 6), (6, 4), (5, 5)\} \rightarrow s = 3$

$P(\text{sum of } 10) = \dfrac{3}{36} = \dfrac{1}{12}$

29. $n = 36$

$E = \{\text{sum of } 13\} = \varnothing \rightarrow s = 0$

$P(\text{sum of } 13) = \dfrac{0}{36} = 0$

30. $n = 36$

$E = \{\text{sum less than } 13\} = \{(1, 1), (1, 2), ...\}$

$\rightarrow s = 36$

$P(\text{sum less than } 13) = \dfrac{36}{36} = 1$

31. $n = 6$

$E = \{\text{roll a } 2\} = \{2\} \rightarrow s = 1$

$P(\text{roll a } 2) = \dfrac{1}{6}$

32. $n = 6$

$E = \{\text{roll a number} > 4\} = \{5, 6\} \rightarrow s = 2$

$P(\text{roll a number} > 4) = \dfrac{2}{6} = \dfrac{1}{3}$

33. $n = 6$

$E = \{\text{roll a number} > 1 \text{ but} < 6\} = \{2, 3, 4, 5\} \rightarrow s = 4$

$P(\text{roll a number} > 1 \text{ but} < 6) = \dfrac{4}{6} = \dfrac{2}{3}$

34. $n = 6$

$E = \{\text{roll an odd number}\} = \{1, 3, 5\} \rightarrow s = 3$

$P(\text{roll an odd number}) = \dfrac{3}{6} = \dfrac{1}{2}$

35. $n = 42$

$E = \{\text{number} < 20\} = \{1, 2, 3, ..., 19\} \rightarrow s = 19$

$P(\text{number} < 20) = \dfrac{19}{42}$

36. $n = 42$

$E = \{\text{number} < 50\} = \{1, 2, 3, ..., 42\} \rightarrow s = 42$

$P(\text{number} < 50) = \dfrac{42}{42} = 1$

37. $n = 42$

$E = \{\text{number is prime}\}$

$= \{2, 3, 5, 7, 11, 13, 17, 19, 23, 29, 31, 37, 41\}$

$\rightarrow s = 13$

$P(\text{number is prime}) = \dfrac{13}{42}$

38. $n = 42$

$E = \{\text{number} < 10 \text{ or} > 40\}$

$= \{1, 2, 3, ..., 9, 41, 42\} \rightarrow s = 11$

$P(\text{number} < 10 \text{ or} > 40) = \dfrac{11}{42}$

39. $n = 12$

$E = \{\text{red egg}\} \rightarrow s = 5$

$P(\text{red egg}) = \dfrac{5}{12}$

40. $n = 6$

$E = \{\text{orange cube}\} \rightarrow s = 5$

$P(\text{orange cube}) = \dfrac{5}{6}$

41. $n = 12$

$E = \{\text{tails, even number}\} = \{(T, 2), (T, 4), (T, 6)\}$

$\rightarrow s = 3$

$P(\text{tails, even number}) = \dfrac{3}{12} = \dfrac{1}{4}$

42. $n = 12$

$E = \{\text{heads, number} > 2\}$

$= \{(\text{H}, 3), (\text{H}, 4), (\text{H}, 5)\ (\text{H}, 6)\} \rightarrow s = 4$

$P(\text{heads, number} > 2) = \frac{4}{12} = \frac{1}{3}$

43. $n = 90$

$E = \{\text{chip is not red}\} \rightarrow s = 50$

$P(\text{chip is not red}) = \frac{50}{90} = \frac{5}{9}$

44. $n = 90$

$E = \{\text{chip is neither red nor black}\}$

$= \{\text{chip is blue}\} \rightarrow s = 20$

$P(\text{chip is blue}) = \frac{20}{90} = \frac{2}{9}$

45. $n = 90$

$E = \{\text{chip is not blue}\} \rightarrow s = 70$

$P(\text{chip is not blue}) = \frac{70}{90} = \frac{7}{9}$

46. $n = 90$

$E = \{\text{chip is red or blue}\} \rightarrow s = 60$

$P(\text{chip is red or blue}) = \frac{60}{90} = \frac{2}{3}$

47. $n = 8$

$E = \{\text{spinner stops on red}\} \rightarrow s = 3$

$P(\text{spinner stops on red}) = \frac{3}{8}$

48. $n = 8$

$E = \{\text{spinner stops on green}\} \rightarrow s = 2$

$P(\text{spinner stops on green}) = \frac{2}{8} = \frac{1}{4}$

49. $n = 8$

$E = \{\text{spinner stops on brown}\} \rightarrow s = 0$

$P(\text{spinner stops on brown}) = \frac{0}{8} = 0$

50. $n = 8$

$E = \{\text{spinner stops on yellow}\} \rightarrow s = 1$

$P(\text{spinner stops on yellow}) = \frac{1}{8}$

51. a. {(S, S, S, S), (S, S, S, F), (S, S, F, S), (S, F, S, S), (F, S, S, S), (S, S, F, F), (S, F, F, S), (F, F, S, S), (F, S, F, S), (S, F, S, F), (F, S, S, F), (S, F, F, F), (F, S, F, F), (F, F, S, F), (F, F, F, S), (F, F, F, F)}

b. $n = 16$

$E = \{\text{all survive}\} = \{(\text{S, S, S, S})\} \rightarrow s = 1$

$P(\text{all survive}) = \frac{1}{16}$

c. $n = 16$

$E = \{\text{exactly one survives}\}$

$= \{(\text{S, F, F, F}), (\text{F, S, F, F}), (\text{F, F, S, F}), (\text{F, F, F, S})\}$

$\rightarrow s = 4$

$P(\text{exactly one survives}) = \frac{1}{4}$

d. $n = 16$

$E = \{\text{exactly two survive}\}$

$= \{(\text{S, S, F, F}), (\text{S, F, F, S}), (\text{F, F, S, S}),$

$(\text{F, S, F, S}), (\text{S, F, S, F}), (\text{F, S, S, F})\}$

$\rightarrow s = 6$

$P(\text{exactly two survive}) = \frac{6}{16} = \frac{3}{8}$

e. $n = 16$

$E = \{\text{exactly three survive}\}$

$= \{(\text{S, S, S, F}), (\text{S, F, S, S}), (\text{F, S, S, S}), (\text{S, S, F, S})\}$

$\rightarrow s = 4$

$P(\text{exactly three survive}) = \frac{4}{16} = \frac{1}{4}$

f. $n = 16$

$E = \{\text{none survive}\} = \{(\text{F, F, F, F})\}$

$\rightarrow s = 1$

$P(\text{none survive}) = \frac{1}{16}$

52. $\frac{1}{16} + \frac{1}{4} + \frac{3}{8} + \frac{1}{4} + \frac{1}{16}$

$= \frac{1}{16} + \frac{4}{16} + \frac{6}{16} + \frac{4}{16} + \frac{1}{16} = \frac{16}{16} = 1$

53. $P(\text{person favors legislation}) = \dfrac{176}{282} = \dfrac{88}{141}$

54. $P(\text{doctor opposes legislation}) = \dfrac{32}{119}$

55. $P(\text{person who opposes is a lawyer}) = \dfrac{15}{71}$

56. $P(\text{person has no opinion}) = \dfrac{35}{282}$

57. Because $0 \le s \le n$, it follows that $0 \le \dfrac{s}{n} \le 1$.

This implies that all probabilities have values from 0 to 1, including 0 and 1. An event that cannot happen has a probability of 0 and an event that is certain to happen has a probability of 1.

58. Probability is a measure of the likelihood that a particular event will occur.

Appendix 2

Section A2.1: Inductive and Deductive Reasoning

VOCABULARY

1. Inductive reasoning draws general conclusions from specific observations.

2. Deductive reasoning moves from the general case to the specific.

CONCEPTS

3. circular

4. decreasing

5. alternating

6. increasing

7. alternating

8. circular

9. 10 A.M.

10. a. 11 students

    b. 18 students

    c. 10 + 11 = 21 students

GUIDED PRACTICE

11. add 4 each time: 13 + 4 = 17

12. add 9 each time: 38 + 9 = 47

13. add 4, add 5, add 6, add 7… : 20 + 7 = 27

14. add 2, add 4, add 6, add 8… : 18 + 8 = 26

15. subtract 3 each time: 6 – 3 = 3

16. subtract 4 each time: 69 – 4 = 65

17. subtract 2, subtract 3, subtract 4, subtract 5… : -12 – 5 = -17

18. subtract 9, subtract 8, subtract 9, subtract 8, subtract 9… : -33 – 9 = -42

19. the pattern: add 4 letters, subtract 1: N + OPQ + **R**

20. the pattern: add 5 letters, subtract 4: F + GHIJ + **K**

21. the pattern: subtract 1 letter, add 2: f – **e**

22. the pattern: subtract 3 letters, add 2: w – vu – **t**

23. 

24. 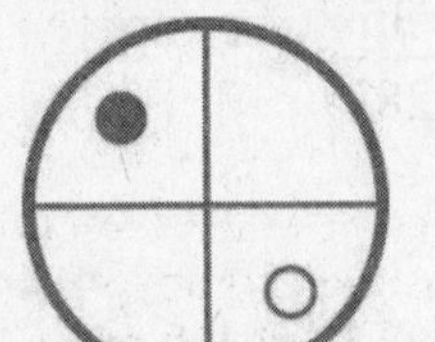

25. 

26. 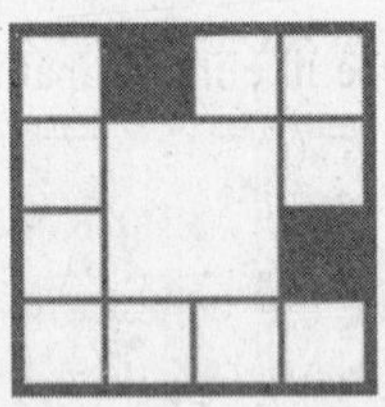

27. 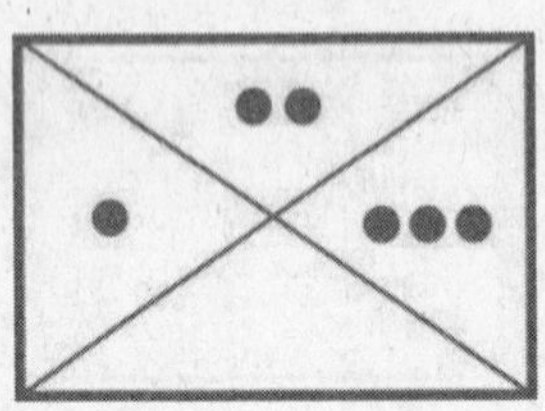

28. 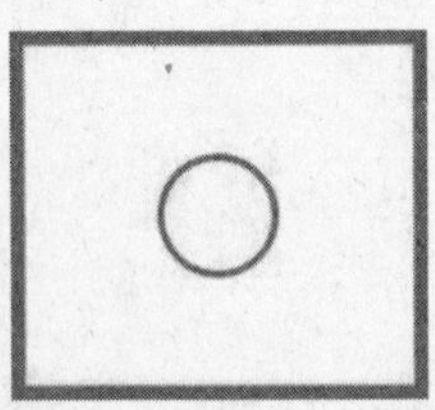

29. 

30.

31. The remaining classes are chemistry and zoology. The remaining professors are C and D. Since D doesn't want to teach zoology, D must be teaching chemistry.

32. The remaining displays are 2 and 4. The remaining companies are B and C. Since C doesn't want display 2, C gets display 4.

33. Since John and Paula are already married, the teacher and the baker must be Luis and Maria. Since Luis is the baker, Maria must be the teacher.

34. Think of the slots as numbered 1 – 2 – 3 – 4 from left to right. Since the Buick is on the left end, it is not the car on the right end. Since the Ford is between the Mercedes and the Dodge the other three slots must be M – F – D or D – F – M. Since the Mercedes is not next to the Buick it must be the second situation, so the Mercedes is parked in the far right slot.

35. The 63 and 47 both include the 16 managers with experience in both. 63 + 47 – 16 = 94 managers with experience in at least one category. There are 100 – 94 = 6 managers with experience in neither category.

36. The 23 and 35 both include the 6 students who used both purchasing methods. 23 + 35 – 6 = 52 students who purchased their books on-line and/or at the bookstore. There are 60 – 52 = 8 students who did neither.

37. The 11 and the 15 both include the 8 students with a brother and a sister. There are 11 + 15 – 8 = 18 children with at least one sibling, leaving 27 – 18 = 9 children with no siblings.

38. The 21 and 11 both include the 5 students with a cat and a dog. There are 21 + 11 – 5 = 27 students with at least one pet, leaving 35 – 27 = 8 students that have neither a dog nor a cat.

TRY IT YOURSELF

39. Add two letters and change the case: g – h - **I**

40. The next letter is repeated one more time: TTT – **UUUU**

41. Alternating letters from the end and beginning of the alphabet: C – **W**

42. Every even term is an N: D – **N**

43.

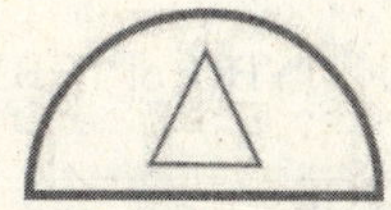

44.

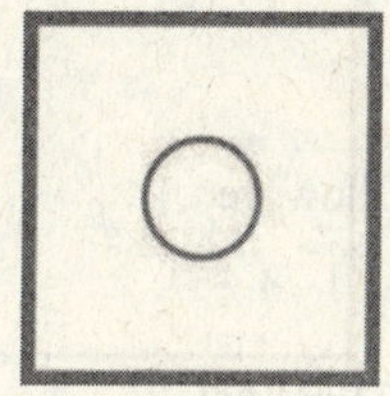

45. Subtract a letter, then add 4: L - **K**

46. Add 4 letters, the subtract 1: n – **m**

47. +16, -15, +14, -13, +12, -11, +10: -4 + 10 = 6

48. +3 , -2 , +3 , -2 … 5 + 3 = 8

49. -4 , +2 , -4 , +2 … 1 + 2 = 3

50. +0.3 , -0.2 , +0.3 , -0.2 … 1.8 – 0.2 = 1.6

51. -1 , -2 , -1 , -2 … -9 – 2 = -11

52. -3 , -4 , -5 , … -17 – 8 = -25

53. +2 , +1 , -2, +2 , +1 , -2 … 11 – 2 =9

54. -2 , -1 , +4 , -2 , -1 , +4 … 9 + 4 = 13

55. From point 3: 1,2,3,T

From point 2, since one end is taken: M,2,3,T

Since the lion can't be next to the tiger: M,L,3,T

The zebra is in cage 3.

56. From point 1: C,2,3,4

From point 2: C,H,3,4

From point 3: C,H,P,S

The sheep is in the last stall.

57. From point 1: B is not last.

From point 2: 1,D,A,C or 1,C,A,D
(Since B is not last, he must be first.)

From point 3: 1,D,A,C

The placement was B , D , A , C

58. From point 1: GBY4, YBG4, 1YBG, or 1GBY

From point 2: GBYR or RYGB

From point 3: GBYR

The order is green, blue, yellow, red.

APPLICATIONS

59. The 35 is included twice in the total.

$997+103-35=1{,}065$ have served before.

$20{,}000-1{,}065=18{,}935$ have never served.

60. The 19 is included twice in the total.

$124+27-19=132$ people clicked either choice.

$178-132=46$ people clicked 'Neither'.

61. The 2 is included twice in the total.

$4+7-2=9$ planets are rocky or have moons.

$9-9=0$ planets have neither.

62. From 4, Andres is not the gardener. From 6, Barry is not the gardener, so Carl is the gardener. From 5, neither Carl nor Barry is the painter, so Andres is the painter.

From 1, Andres can't be the jeweler.

From 3, Since Andres is the painter, he can't be the chauffeur.

From 4, Andres can't be the musician, so he must be the barber.

From 4, the musician and the gardener are different people, so Barry is the musician.

From 2, the musician is not the chauffeur, so Carl is the chauffeur.

That leaves only jeweler for Barry.

In summary: Andres: painter and barber; Barry: musician and jeweler, Carl: gardener and chauffeur.

WRITING

63. Inductive reasoning draws general conclusions from specific observations. Deductive reasoning moves from the general case to the specific.

64. From knowing how some batters hit, I can buy tickets to a section where I am likely to catch a ball hit by an opposing batter visiting for an away game.

65. From having watched drivers in the rain, I conclude people don't drive as well when it is raining.

66. 60 people were surveyed about their e-mail usage. 30 people replied that they used it only for personal matters. 10 people use e-mail for personal and work issues. How many people used e-mail only for work?